MW00843964

Image Correlation for Shape, Motion and Deformation Measurements

Michael A. Sutton · Jean-José Orteu
Hubert W. Schreier

Image Correlation for Shape, Motion and Deformation Measurements

Basic Concepts, Theory and Applications

Michael A. Sutton
University of South Carolina
Department of Mechanical Engineering
Columbia, SC 29208
USA
sutton@sc.edu

Jean-José Orteu
Ecole des Mines d'Albi
Campus Jarlard
Albi 81013
France
jean-jose.orteu@enstimac.fr

Hubert W. Schreier
Correlated Solutions, Inc.
120 Kaminer Way Parkway Suite A
Columbia, SC 29210
USA
schreier@correlatedsolutions.com

ISBN: 978-0-387-78746-6 e-ISBN: 978-0-387-78747-3
DOI: 10.1007/978-0-387-78747-3

Library of Congress Control Number: 2008938227

Printed on acid-free paper

springer.com

To Shameka, Isabel and Fiona Schreier; Elizabeth Ann Sutton, Michelle MK Sutton Spigner and Elizabeth Sutton; Catherine, Benoît and Julie Orteu, the authors wish to express their gratitude to all family members for having persevered through the 19 months of effort necessary to complete this manuscript. Finally, to our parents, through the publication of this book we hope to thank them and honor them for all their support over the years, including those that are no longer with us; Frances Rosalie Kaffenberger Sutton (1916–2002) and Percy Albert Sutton (1902–1979).

Financial support for the research described in this book was provided by a number of sources. We gratefully acknowledge the National Science Foundation, NASA Langley and NASA Headquarters, Army Research Office, Air Force Office of Scientific Research, Wright Patterson Air Force Research Lab, General Motors, Sonoco Products, University of Maryland, Sandia National Laboratory, University of South Carolina and Ecole des Mines d'Albi. Access to the facilities of the

Cornell Nanofabrication Facility and Center for Electron Microscopy at the University of South Carolina is also gratefully acknowledged.

Michael A. Sutton University of South Carolina
Jean-José Orteu Ecole des Mines d'Albi
Hubert W. Schreier Correlated Solutions, Incorporated

Preface

The impetus for the enclosed book is the nexus of accelerating interest and use of computer-vision-based measurement methods and the accumulation of sufficient theoretical, computational and experimental findings for development of a comprehensive treatment of the subject. With over five decades of active R&D in the field of computer vision for experimental measurements, either as originators, developers or practitioners of the method(s), the authors have both the commitment and the background necessary to provide the readership with a balanced, comprehensive treatment of the subject and its application to real-world problems.

As envisioned by the authors, the book is a reference book for engineers, scientists and educators seeking to employ advanced, image-based, non-contacting measurement methods to measure the shape and deformation of a material undergoing thermal, mechanical or variable environmental conditions. Since the methods have been shown to be broadly applicable in fields such as civil engineering, mechanical engineering, material science, electronic packaging, biomedical, manufacturing, joining, photogrammetry and others, the text should serve as a reference for all faculty, research scientists and students employing these methods in their investigations.

Encompassing both basic theoretical formulations in the areas of image correlation and computer vision and also recent developments and practical applications in two-dimensional, three-dimensional and volumetric image correlation methods in the fields of non-contacting measurements and experimental mechanics, the level of expertise is envisioned as an advanced supplement for an upper-level undergraduate class or as a companion text for a graduate-level class in measurements, experimental mechanics or non-contacting, vision-based image analysis methods with special emphasis in solids. Though the material contains a complete summary of concepts and also background material needed to develop a strong foundation in vision-based methods, the text does not include problem sets.

The scope of the work includes aspects of the broad area of non-contacting measurement of shape and deformation using images of an object, with special emphasis on computer vision and volumetric imaging. Specifics addressed in the various chapters include (a) ray optics, (b) single camera computer vision and calibration concepts including distortion correction, (c) multi-camera, stereo-vision principles and calibration methods, (d) digital image correlation for image matching, including error assessment and (e) experimental details for single-camera, multi-camera and volumetric imaging applications.

Theoretical developments presented in the book are based on general concepts obtained from a wide range of sources, some of which are original to the authors. Basic optics concepts were abstracted from a combination of textbooks. Single camera and multi-camera calibration concepts were abstracted from archival research papers, research books and textbooks. Digital image correlation principles for image matching were developed using research papers, references, and text books, or are original with the authors and presented herein for the first time. Volumetric image correlation principles were primarily extracted from research articles. Error assessment developments use basic probabilistic concepts, but are otherwise original to the authors. Most experiments and associated developments for both two-dimensional and three-dimensional computer vision experiments are based on work performed by the authors and their students. All appendices were abstracted from a combination of text books, research articles and reference books.

Chapter 1 provides an introduction and in-depth literature survey. Chapter 2 covers basic optical developments applicable for modeling vision systems. Chapters 3 and 4 provide the theoretical foundation for calibration and measurements using single camera and stereo camera vision systems. Chapter 5 outlines the essential concepts underlying digital image correlation for motion measurements. Specific items discussed include (a) image matching methods, (b) subset shape functions, (c) intensity pattern metrics, (d) intensity pattern interpolation for discretely sampled patterns and (e) quantitative error estimates in 2D motion. In the latter section, the authors highlight original contributions that provide quantitative metrics for the errors that are expected during image correlation during simple translation experiments. Developments include probabilistic estimates for the mean and variance in the measurements due to the effects of intensity interpolation, intensity pattern noise and intensity pattern contrast, as well as appropriate interpretations of the results using signal-processing concepts. Chapter 6 presents details regarding the application of a single camera imaging system for in-plane measurement of deformations, including the potentially deleterious effects of out-of-plane motion. Applications include (a) successful measurement of the stress–strain response of an aluminum alloy using the surface strain field measured on a planar specimen undergoing uniaxial tension by 2D-DIC, (b) details of extensive experimental studies that clearly highlight the effects of out-of-plane motion on single-camera motion measurements, (c) use of a far-field microscope for micro-scale measurements of crack tip deformations with nanometer accuracy, (d) determination of material properties through an inverse methodology enabled by full-field 2D-DIC deformation measurements and (e) use of a scanning electron microscope for deformation measurements in a 20×20 μm field of view, including the development and application of nanoscale patterns. Chapter 7 describes in detail the application of stereo-vision camera imaging systems for general three-dimensional measurement of surface deformations. Applications include (a) a synchronized four-camera stereovision system for the measurement of large, out-of-plane deformations in polymer beam specimens, (b) dynamic deformation measurements using high speed stereo cameras and (c) development and use of 3D stereomicroscopy systems for microscale deformation and shape measurements. Chapter 8 presents both theoretical foundations and experimental results from a series of volumetric image correlation measurements in a polymeric foam undergoing compression loading. Chapter 9 provides a brief discussion of methods to estimate errors during stereovision measurements using the results obtained in Chapter 5 for image based motion bias and variability. Chapter 10 provides a summary of practical considerations regarding both 2D and 3D image correlation measurement methods. Included in this lengthy chapter are (a) an engineering approach for estimating the appropriate speckle size in a pattern along with examples using the approach, (b) methods for patterning a material and issues related to pattern adherence, (c) a simple approach to estimate the speckle size in an image, (d) an engineering approach for determining the depth of field and field of view along with an example, (e) a simple formula to estimate the effect of out of plane motion on image measurements, (f) estimation of errors in 2D image-plane matching and (g) measurement issues at both high and low magnification. Appendices A through I provide the background information that the authors believe is important to provide the foundation necessary for both essential concepts in computer vision and the practical application of the method for shape and deformation measurements. Areas covered include continuum mechanics; linear algebra; surface strain estimation; non-linear optimization; basic concepts in statistics and probability; introduction to projective geometry; rotation tensor formulations; spline functions and triangulation.

In closing, the authors have believed for a long time in the usefulness and broad applicability of vision-based methods. Hence, though the effort required has been significant, the authors are delighted that the book a reality. We have enjoyed the opportunity to make a contribution that should further expand opportunities for those with measurement needs that can be addressed by digital image correlation methods.

Columbia
South Carolina and Albi, France
January 1, 2009

Michael A. Sutton
Jean José Orteu
Hubert W. Schreier

Acknowledgements

The authors are indebted to (a) Professors Brian K. Bay and Michel Bornet for the technical material provided regarding their research in volumetric image correlation and (b) Professor Fabrice Pierron and Dr. B. Guo for the parameter identification and virtual fields method material that formed the basis for Section 6.5. In addition, special recognition is noted for Professor Bay and Professor Pierron and Dr. Guo for providing unpublished experimental data from CT imaging and 2D surface imaging, respectively, related to their independent polymeric foam material studies.

The work would not have been completed without the efforts of so many students and colleagues. In the Department of Theoretical and Applied Mechanics at the University of Illinois, special thanks are due to fellow PhD students, Dr. Frederick T. Mendenhall III and Dr. Michael A. Tafralian, for their willingness to befriend and educate their colleague regarding recent advances in non-contacting measurements. At the University of South Carolina, student contributors include Michael Boone, Byron Amstutz, Jinseng Jang, Majid Babai, Vikrant Tiwari, Stephen R. McNeill, Jeffrey Helm, Nicolas Cornille, Jin Liu, Gang Han, Junhui Yan, Xiaodan Ke, Wenzhong Zhao, Peng Cheng and Philip T.C. Chu. At Ecole des Mines d'Albi, student contributors include Dr. Nicolas Cornille and Dr. Dorian Garcia.

Academic colleagues at the University of South Carolina that have meant so much in these endeavors include Yuh J. Chao, William F. Ranson, Walter H. Peters III, Stephen R. McNeill, as well as John L. Turner, Ning Li, Bangchang Yang, Jed S. Lyons and Anthony P. Reynolds. Academic colleagues in France that have provided support and guidance throughout this effort include Prof. Michel Devy at LAAS-CNRS in Toulouse and Dr. Laurent Robert at Ecole des Mines d'Albi.

Colleagues in the Society for Experimental Mechanics providing advice and person suggestions are especially noted, including Michael and William Fourney, James Dally, C.W. Smith, Albert Kobayashi, Michael Y.Y. Hung and William Sharpe. Special recognition is given to Prof. Charles E. Taylor and his lovely wife, Nikki, for their unwavering support over three decades.

Overview

As envisioned by the authors, the book is a reference book encompassing the basic theoretical formulations in the areas of image correlation and computer vision, recent developments and practical applications of two-dimensional, three-dimensional and volumetric image correlation methods in the fields of non-contacting measurements and experimental mechanics. Since the methods have been shown to be broadly applicable in fields such as civil, mechanical, material science, electronic packaging, biomedical, manufacturing, joining, photogrammetry and others, the text should serve as a reference for all faculty, research scientists and students employing these methods in their investigations. Though the material contains a complete summary of concepts and also background material needed to develop a strong foundation in vision-based methods, the current version provides limited examples and does not contain problem sets. Therefore, the text should serve as a reference for the researcher, or provide the material needed for a first course in the area. Chapter 1 provides an introduction and in-depth literature survey. Chapter 2 covers basic optical developments applicable for modeling vision systems. Chapters 3 and 4 provide the theoretical foundation for calibration and measurements using single camera and stereo camera vision systems. Chapter 5 outlines the concepts underlying digital image correlation for motion measurements. Specific items discussed include (a) image matching methods, (b) subset shape functions, (c) intensity pattern metrics, (d) intensity pattern interpolation for discretely sampled patterns and (e) error estimates in 2D motion. In the latter section, the authors highlight original contributions that provide quantitative metrics for the errors that are expected during image correlation during simple translation experiments. Developments include probabilistic estimates for the mean and variance in the measurements due to the effects of intensity interpolation, intensity pattern noise and intensity pattern contrast, as well as appropriate interpretations of the results using signal-processing concepts. Chapter 6 presents details regarding the application of a single camera imaging system for in-plane measurement of deformations, including the potentially deleterious effects of out-of-plane motion. Applications include (a) successful measurement of the stress-strain response of an aluminum alloy using the surface strain field measured on a planar specimen undergoing uniaxial tension by 2D-DIC, (b) details of extensive experimental studies that clearly highlight the effects of out-of-plane motion on single-camera motion measurements, (c) use of a far-field microscope for microscale measurements of crack tip deformations with nanometer accuracy, (d) determination of material properties through an inverse methodology enabled by full-field 2D-DIC deformation measurements and (e) use of a scanning electron microscope for deformation measurements in a $20 \times 20\,\mu\text{m}$ field of view, including the development and application of nanoscale patterns. Chapter 7 describes in detail the application of stereo-vision camera imaging systems for general three-dimensional measurement of surface deformations. Applications include (a) a synchronized four-camera stereovision system for the measurement of large, out-of-plane deformations in polymer beam specimens, (b) dynamic deformation measurements using high speed stereo cameras and (c) development and use of 3D stereomicroscopy systems for microscale deformation and shape measurements. Chapter 8 presents both theoretical foundations and experimental results from a series of volumetric image correlation measurements in a polymeric foam undergoing compression loading. Chapter 9 provides a brief discussion of methods to estimate errors that are expected during stereovision measurements. Developments include approaches to obtain probabilistic estimates for the mean and variance in the measurements due to the effects of intensity interpolation, intensity pattern noise and intensity pattern contrast on image-plane matching locations for corresponding points. Chapter 10 provides a summary of practical considerations regarding both 2D and 3D image correlation measurement methods. Included in this lengthy chapter are (a) an engineering approach for estimating the appropriate speckle size in a pattern along with examples using the approach, (b) methods for patterning a material and issues related to pattern adherence, (c) a simple approach to estimate the speckle size in an image, (d) an engineering approach for determining the depth of field and field of view along with an example, (e) a simple formula to estimate the effect of out of plane motion on image measurements, (f) estimation of errors in 2D image-plane matching and (g) measurement issues at both high and low magnification. Appendices A through I provide the background information that the authors believe is important to provide the foundation necessary for both essential concepts in computer vision and the practical application of the method for shape and deformation measurements. Areas covered include continuum mechanics; linear algebra; surface strain estimation; non-linear optimization; basic concepts in statistics and probability; introduction to projective geometry; rotation tensor formulations; optimal estimates for intensity scale and offset, spline functions and triangulation.

Contents

1 Introduction 1
- 1.1 Literature Survey 1
  - 1.1.1 Early History 1
  - 1.1.2 Photogrammetry, 1850–Present 1
  - 1.1.3 Digital Image Correlation – Background and Related Activities 2
  - 1.1.4 Digital Image Correlation 3
- 1.2 Discussion 10

2 Elements of Geometrical Optics 13
- 2.1 Optics of a Camera 13
  - 2.1.1 Thin Lens 13
  - 2.1.2 Thick Lens 22
  - 2.1.3 Interpretation 24
  - 2.1.4 Front Image Plane or Symmetry Interpretation 25

3 Single Camera Models and Calibration Procedures in Computer Vision 27
- 3.1 Camera Models 27
  - 3.1.1 Pinhole Camera 27
  - 3.1.2 Distortion 33
- 3.2 Calibration – Full Camera Model Parameter Estimation 44
  - 3.2.1 Camera Parameter Estimation without Distortion 44
  - 3.2.2 Bundle Adjustment for Distortion and Camera Parameter Estimation 48
  - 3.2.3 Equations and Procedures for Optimization Structure of Matrices 51
  - 3.2.4 Solution Details 55
  - 3.2.5 Discussion 60

4 Two-Dimensional and Three-Dimensional Computer Vision 65
- 4.1 Two-Dimensional Computer Vision 65

4.1.1 Model . . . 65
4.1.2 Bundle Adjustment for Plane-to-Plane Imaging . . . 67
4.1.3 Solution Details . . . 67
4.1.4 Discussion . . . 69
4.2 Three-Dimensional Computer Vision . . . 70
4.2.1 Geometry of a Stereovision System . . . 71
4.2.2 Epipolar Constraint . . . 72
4.2.3 Calibration in Stereovision . . . 73
4.2.4 Discussion . . . 79

5 Digital Image Correlation (DIC) . . . 81
5.1 Introduction to Image Matching . . . 81
5.1.1 The Aperture Problem . . . 81
5.1.2 The Correspondence Problem . . . 82
5.1.3 Speckle Pattern . . . 83
5.2 Image Matching Methods . . . 84
5.2.1 Differential Methods . . . 84
5.2.2 Template Matching . . . 87
5.3 Subset Shape Functions . . . 88
5.3.1 Polynomial Shape Functions . . . 90
5.3.2 Shape Functions for Stereo Matching . . . 92
5.3.3 Summary . . . 95
5.4 Optimization Criteria for Pattern Matching . . . 95
5.4.1 Offset in Lighting . . . 96
5.4.2 Scale in Lighting . . . 97
5.4.3 Offset and Scale in Lighting . . . 97
5.4.4 Concluding Remarks . . . 98
5.5 Efficient Solution Methods . . . 99
5.5.1 Efficient Update Rules for Planar Motion . . . 99
5.5.2 Extension to General Shape Functions . . . 101
5.6 Matching Bias . . . 103
5.6.1 Interpolation Bias . . . 103
5.6.2 Bias Due to Noise . . . 110
5.7 Statistical Error Analysis . . . 113
5.7.1 Derivation for the One-Dimensional Case . . . 113
5.7.2 Confidence Margins from the Covariance Matrix . . . 116

6 In-Plane Measurements . . . 119
6.1 Constraints and Applicability . . . 119
6.1.1 Object Planarity Constraints . . . 119
6.1.2 Object Deformation Constraints . . . 119
6.2 Uniaxial Tension of Planar Specimen . . . 120
6.2.1 Experimental Considerations . . . 120
6.2.2 Imaging Considerations . . . 120
6.2.3 Experimental Results . . . 122

6.2.4 Discussion 124
6.3 Out-of-Plane Motion 127
6.3.1 Standard Lens Systems for Single Camera Measurements . . 127
6.3.2 Telecentric Lens System for Single Camera Measurements . 128
6.3.3 Out-of-Plane Translation Experiments 128
6.3.4 Stereo-vision Calibration 130
6.3.5 Experimental Results 133
6.3.6 Discussion 133
6.3.7 Remarks 137
6.4 Development and Application of Far-Field Microscope for Microscale Displacement and Strain Measurements 138
6.4.1 Problem Description: Measurement of Crack Closure Load During Fatigue Crack Growth 138
6.4.2 Fatigue Specimen Geometry, Material and Surface Preparation 139
6.4.3 Validation Specimen and Preparation 140
6.4.4 Pattern Application 140
6.4.5 Optical Setup for Imaging 142
6.4.6 Validation Experiment 143
6.4.7 Fatigue Experiment 144
6.4.8 Post-processing to Determine COD 145
6.4.9 Experimental Results 146
6.4.10 Discussion 148
6.5 Inverse Methods: Material Property Measurements Using Full-Field Deformations and the Virtual Fields Method 150
6.5.1 Virtual Fields Method 150
6.5.2 Derivation of VFM Equations for Cantilever Beam Undergoing Planar Deformations 151
6.5.3 Virtual Fields for Cantilever Beam Specimen 152
6.5.4 Experimental Studies 154
6.5.5 Smoothing by 2D Finite Element Methods 156
6.5.6 Material Property Results 157
6.5.7 Discussion 157
6.6 Accurate 2D Deformation Measurements in a Scanning Electron Microscope: Basic Concepts and Application 159
6.6.1 Imaging in an SEM 159
6.6.2 Pattern Development and Application 160
6.6.3 Digital Image Correlation to Quantify SEM Imaging System Errors 162
6.6.4 Digital Image Correlation for Elastic Deformation Measurements 167
6.6.5 Tensile Experiment 169
6.6.6 Experimental Results 170
6.6.7 Discussion 171
6.6.8 Remarks 172

7 Stereo-vision System Applications ... 175
7.1 Stereovision System Design Considerations ... 175
7.2 Quasi-Static Experiment: Four-Camera Stereo-Vision System for Large Deformation Measurements in Specimen Subjected to Out-of-Plane Bending ... 176
7.2.1 Problem Description: Compression and Out-of-Plane Bending of Flawed Polymer Sheets ... 176
7.2.2 Geometry of Specimen ... 177
7.2.3 Experimental Considerations and Arrangement of Two Stereovision Systems ... 177
7.2.4 Calibration of the Camera Systems ... 179
7.2.5 Post-processing and Special Considerations ... 181
7.2.6 Experiments ... 186
7.2.7 Experimental Results ... 186
7.2.8 Discussion and Practical Considerations ... 188
7.3 Dynamic Experiment: High Speed Stereovision System for Large Deformation Measurements of Specimen Subjected to Combined Tension–Torsion Impact Loading ... 191
7.3.1 Problem Description: Impact Loading of Single Edge Cracked Specimen Subjected to Mixed Mode I/III Impact Loading ... 192
7.3.2 Specimen Geometry and Preparation ... 192
7.3.3 Experimental Considerations and Arrangement of Two Stereovision Systems ... 192
7.3.4 Calibration of the Stereovision System ... 194
7.3.5 Post-processing ... 195
7.3.6 Experiments ... 195
7.3.7 Experimental Results ... 195
7.3.8 Discussion and Practical Considerations ... 199
7.4 Development and Application of 3D Digital Image Correlation Principles in Stereomicroscopy for Microscale Shape and Deformation Measurements ... 199
7.4.1 Problem Description: Shape and Deformation Measurements for Small, Soft Specimens Subjected to Mechanical Loading ... 200
7.4.2 Specimen Geometry and Preparation ... 200
7.4.3 Optical Components in Stereo-Microscope and Experimental Considerations ... 202
7.4.4 Distortion Correction and Stereo Calibration ... 202
7.4.5 Post Processing ... 204
7.4.6 Validation Experiment ... 204
7.4.7 Mouse Carotid Experiments ... 205
7.4.8 Results ... 206
7.4.9 Discussion ... 206

8 Volumetric Digital Image Correlation (VDIC) ... 209
8.1 Imaging and Discrete Pattern Recording ... 209
8.2 System Calibration for VDIC ... 211
8.3 Volumetric Shape and Deformation Measurements ... 213
8.4 Volumetric Shape Functions ... 214
8.5 Volumetric Image Reconstruction ... 218
8.6 Case Study: Quantifying Micro-damage in Polymeric Foam Undergoing Uniaxial Compression ... 218
8.6.1 Background ... 218
8.6.2 Specimen and Experimental Considerations ... 219
8.6.3 CT Imaging and VDIC Parameters ... 222
8.6.4 Experimental Results ... 222
8.6.5 CT-Imaging Discussion ... 224

9 Error Estimation in Stereo-vision ... 225
9.1 Sub-optimal Position Estimator ... 225
9.2 Optimal Position Estimator ... 226
9.3 Variance in 3D Position ... 228
9.4 Discussion ... 228

10 Practical Considerations for Accurate Measurements with DIC ... 229
10.1 General Imaging Considerations ... 229
10.1.1 Depth of Field and Field of View ... 229
10.1.2 Image Artifacts ... 231
10.1.3 Subset Patterning ... 233
10.1.4 Exposure Time ... 243
10.2 Practical Considerations for 2D-DIC Measurements ... 244
10.2.1 Out-of-Plane Motion ... 244
10.2.2 Depth of Field at High Magnification ... 245
10.2.3 Measurement Accuracy and Intensity Noise ... 245
10.2.4 Thermal Effects at Moderate Temperatures ... 246
10.2.5 Other Factors ... 246
10.3 Practical Considerations for 3D-DIC Measurements ... 247
10.3.1 Camera and Lens Selection ... 247
10.3.2 Stereo-vision System Configuration ... 248
10.3.3 Individual Camera Calibration ... 250
10.3.4 Single Lens Stereo-vision System ... 250
10.4 Stereo-vision Measurement Process ... 252
10.4.1 Subset Selection and Matching ... 252

A Continuum Mechanics Formulation for Deformations ... 255
A.1 Strain Tensors ... 257
A.2 Strain Rate Tensors ... 258
A.3 Lagrangian ... 260
A.4 Eulerian ... 260

B Elements of Linear Algebra ... 261
B.1 Orthonormal Bases for Column Space, C_A ... 263
B.2 Orthonormal Bases for Row Space, R_A ... 263
B.3 Matrices Constructed from Rotation and Warping Components ... 264
B.4 Singular Value Decomposition of a Matrix ... 266

C Method for Local Surface Strain Estimation ... 269

D Methods of Non-linear Optimization and Parameter Estimation ... 273
D.1 Levenberg–Marquardt and Non-linear Optimization ... 273
D.1.1 Mathematical Background ... 273
D.1.2 Hessian Matrix Approximation and Least Squares ... 274
D.1.3 Gradient Search Process ... 275
D.1.4 Local Quadratic Functional Form ... 275
D.1.5 Combined Steepest Descent and Newton–Raphson ... 276
D.1.6 Least Squares with 2D Image Positions ... 277
D.2 Least Squares for Optimal Parameter Estimation ... 279

E Terminology in Statistics and Probability ... 281
E.1 Expectation ... 281
E.2 Mean Value ... 281
E.3 Variance ... 282
E.4 Approximate Expectation and Variance Expressions ... 283

F Basics of Projective Geometry ... 285
F.1 Homogeneous Coordinates ... 285
F.2 Why are Homogeneous Coordinates Interesting in Computer Vision: An Example ... 285
F.3 Lines in $\mathcal{P}^2$ ... 286

G Rotation Tensor Formulations ... 287

H Spline Functions ... 289
H.1 Two-Dimensional Case: Spline Curve ... 289
H.1.1 Spline Definition ... 289
H.1.2 B-Spline Definition ... 290
H.2 Three-Dimensional Case: Spline Surface ... 292
H.3 Spline Derivatives ... 293

I Triangulation – Location of 3D Position with Skew Rays ... 295

References ... 299

Index ... 317

Chapter 1
Introduction

1.1 Literature Survey

As used in this article, the term "digital image correlation" refers to the class of non-contacting methods that acquire images of an object, store images in digital form and perform image analysis to extract full-field shape, deformation and/or motion measurements. Digital image registration (i.e. matching) has been performed with many types of object-based patterns, including lines, grids, dots and random arrays. One of the most commonly used approaches employs random patterns and compares sub-regions throughout the image to obtain a full-field of measurements. The patterns may occur on solid surfaces or may be a collection of particles in a fluid medium.

1.1.1 Early History

The early history of image-based measurements appears to reside in the field of photogrammetry, for which there is a wealth of literature. As noted by Doyle [65] and Gruner [96], the discussions of perspective and imagery date back to the writings of Leonardo da Vinci in 1480 and his related studies in 1492. Key developments over the next 3 centuries, including the work of Heinrich Lambert who developed the basic mathematics relating perspective and imaging (The Free Perspective, 1759), had their greatest impact after photography was invented. Attributed to Niepce (1765–1833), the first practical photographs were made by Daguerre in 1837.

1.1.2 Photogrammetry, 1850–Present

With the invention and refinement of photographic methods, developments in the field of photogrammetry have often been separated into four relatively distinct

M.A. Sutton et al., *Image Correlation for Shape, Motion and Deformation Measurements: Basic Concepts, Theory and Applications,* DOI: 10.1007/978-0-387-78747-3_1,

phases (Konecny [144]). These phases are (1) plane photogrammetry (1850–1900), (2) analog photogrammetry (1900–1950), (3) analytical photogrammetry (1950–1985) and (4) digital photogrammetry (1985–present). From these four phases, the most enduring contributions are in the mathematical developments.

Specifically, the relationship between projective geometry and perspective imaging developed by Sturms and Haick (1883), the fundamental geometry of photogrammetry described by Sebastian Finsterwald (1899), the projective equations and their differentials for stereo-imaging which are fundamental to analytical photogrammetry developed by Otto von Gruber (1924), analytical solutions to the equations of photogrammetry in terms of direction cosines given by Earl Church (1945) and development of the principles of modern multi-station analytical photogrammetry using matrix notation by Dr. Hellmut Schmid (1953).

The work of Schmid is particularly relevant in that he not only developed the equations but also performed a "rigorously correct least squares solution" using any number of perspective views, along with a detailed study of error propagation. It is important to note that the early focus of photogrammetry was to extract 3D shape of objects through multi-view comparison of photographic records.

1.1.3 Digital Image Correlation – Background and Related Activities

Some of the first work in the area of image correlation was performed in the early 1950s by Gilbert Hobrough (1919–2002), who compared analog representations of photographs to register features from various views [17]. In 1961, Hobrough designed and built an instrument to "correlate high-resolution reconnaissance photography with high precision survey photography in order to enable more precise measurement of changeable ground conditions" [17], thereby being one of the first investigators to attempt a form of digital image correlation to extract positional information from the image correlation/matching process.

As digitized images became available in the 1960s and 1970s, researchers in artificial intelligence and robotics began to develop vision-based algorithms and stereo-vision methodologies in parallel with photogrammetry applications for aerial photographs. Rosenfeld [243] provides an extensive bibliography and a summary of the developments from 1955–1979. As noted by Rosenfeld, the areas of primary emphasis in this research community during the early years of digital image processing were (a) character recognition, (b) microscopy, (c) medicine and radiology and (d) photogrammetry/aerial photography, with engineering applications for shape and deformation measurements either non-existent or rare.

While digital image analysis methods were undergoing explosive growth, much of the field of experimental solid mechanics was focused on applying recently developed laser technology. Holography [82, 100, 117, 154], laser speckle [57], laser speckle photography [18, 19, 76, 175, 319], laser speckle interferometry [180, 267], speckle shearing interferometry [125, 153], holographic interferometry [327], moiré interferometry [224] and ultra high density moiré interferometry [225] are typical examples of the type of measurement techniques developed for use with coherent light sources. A particularly interesting example using laser speckle was developed by Yamaguchi in 1986 [343], and employed a linear sensor to determine in-plane translation. In most cases, the measurement data (surface slopes, displacements, displacement gradient combinations) is embedded in the photographic medium, typically in the form of a fringe pattern. Since the recording process is generally nonlinear, resulting in difficulties in extracting partial fringe positions with high accuracy, the most common process employed by experimental mechanicians was a laborious determination of estimates for fringe center locations at a few points.

1.1.4 Digital Image Correlation

Given the difficulties encountered by experimental mechanicians during the post-processing of photographically recorded measurement data, and the burgeoning growth of image processing methods in the vision community, it was natural for researchers to employ the recent progress in digital imaging technology and develop (a) methods for digitally recording images containing measurement data, (b) algorithms to analyze the digital images and extract the measurement data and (c) approaches for automating the entire process. In many cases, the characteristic pattern used to compare subsets and extract full-field information was obtained by either coherent light illumination or through application of a high contrast pattern with incoherent illumination, resulting in a full-field, random pattern, or white light speckle pattern.

1.1.4.1 Two-Dimensional Measurements, 1982–1999

One of the earliest papers to propose the use of computer-based image acquisition and deformation measurements in material systems was written by Peters and Ranson in 1982 [215]. Interestingly, the original application envisioned developing and digitally recording a full-field pattern by subjecting an object to ultrasonic waves both before loading (the reference image) and during the loading process (the deformed image). By recording the resulting reflected wave pattern prior to and after applying load, the authors proposed a method for analyzing the resulting full-field recorded digital "images". The method suggested a comparison of the digital images for various small regions (known as subsets) throughout the images before and after deformation, locating the positions of each of these subsets after deformation through digital image analysis. As part of the envisioned approach, the authors suggested using fundamental continuum mechanics concepts governing the deformation of small areas as part of the "matching process".

Using this approach, in 1983 Sutton et al. [288] developed numerical algorithms and performed preliminary experiments using optically recorded images to show that the approach, known today as 2D Digital Image Correlation (2D-DIC), was feasible when using optically recorded images. Anderson et al. [13] performed rigid body motion measurements using the algorithms, demonstrating that both planar translations and rotations can be reliably estimated through 2D-DIC image matching. Chu et al. in 1985 [48] performed a series of experiments to demonstrate that the method can be applied to quantify rotations and deformations in solids. To improve the speed of the analysis process, in 1986 Sutton et al. [280] demonstrated the use of gradient search methods so that subset-matching could be performed with sub-pixel accuracy throughout the image to obtain a dense set of full-field, two-dimensional displacement measurements. Tian and Huhns [305] discussed the use of several search methods capable of sub-pixel accuracy, showing that both coarse-fine and gradient search approaches are viable. Over the next decade, these procedures were validated, modified, improved and numerical algorithms refined [37, 294]. In the late 1980s, Sutton et al. [287] performed one-dimensional numerical simulations to provide initial estimates for the accuracy of deformation measurements in image correlation. In an effort to quantify internal deformations in a composite sheet, in 1989 Russell et al. [245] performed X-ray radiography before and after deformation. After digitizing the radiographs, 2D-DIC was used to determine the average through-thickness strain fields.

In the decade of the 1990s, limited studies were performed to assess the accuracy of the method. For example, Sjödahl [262–264] discussed the accuracy of measurements in "electronic speckle photography". However, in most cases, investigators began applying the method to measure surface deformations in planar components. Research using 2D-DIC in fracture mechanics began in the early 1980s and continues to this day; summaries are provided in [291, 292]. Early work by McNeill et al. [189] demonstrated the use of 2D-DIC measurements for stress intensity factor estimation. Sutton et al. used local crack tip plastic zone measurements [295] to estimate the zone affected by "three-dimensional effects". Dawicke and Sutton [62] measured crack opening displacement with 2D-DIC. Using work performed by Lyons that extended 2D-DIC for strain measurements at high temperature [176], Liu et al. [161] made full-field creep measurements at 700°C in alloy 718 over several hundred hours. Amstutz et al. [11, 12] measured tensile and shear deformations in Arcan specimens undergoing mixed mode I/II loading. Han et al. developed a high magnification optical system to measure deformations around stationary [101] and growing [102] crack tips under nominally Mode I loading.

In addition to the work in fracture mechanics, investigators used 2D-DIC to understand material deformation behavior including metals [56, 307, 308], plastics [168, 314], wood [355], ceramics [49] and tensile loading of paper [40, 279]. In the late 1990s and 2000, investigators applied 2D-DIC to study damage in composites [85, 93, 99] and concrete [46, 47].

Prior to recent emphasis on nano-science and nanotechnology, efforts of investigators such as Davidson and Lankford [60, 61] in the 1980s extended 2D-DIC concepts to the microscale for large deformation measurements near fatigue flaws. Additional high magnification studies [241, 274, 278, 290] were performed using 2D-DIC. Of particular note is the effort to extend the work of Davidson and

Lankford, as described in the work of Sutton et al. [290]. The authors developed a fully automated image acquisition and analysis system using a Questar far-field microscope to image a 700×700 μm crack tip region. Riddell et al. [241] employed the system to measure crack tip deformations as part of a study focused on crack closure assessment.[1] Details regarding this micro-scale application are given in Section 6.4.

Though much of the development and application in 2D digital image correlation remained focused at the University of South Carolina through the mid-1990s, a unique extension of 2D-DIC for scanning tunneling microscopy (STM) was completed by Prof Knauss and his students. In 1998, Vendroux and Knauss [321–323] reported their STM work. By acquiring STM images and using a modified gradient search approach to perform digital image correlation, the authors evaluated sequential STM images to extract deformation measurements at the sub-micron scale. In the same time frame, Doumalin et al. [64] discussed the use of image correlation using images from a scanning electron microscope for deformation measurements at higher magnification.

While novel methods were being developed for quantitative measurements at structural and micro length scales in solid mechanics, in 1984 Peters et al. [216] applied 2D-DIC to measure the velocity field in a seeded, two-dimensional flow field, demonstrating that the approach can be used in fluid systems. In the field of fluid mechanics, investigators used a variety of illumination sources and various high speed imaging concepts such as rotating drum and rotating mirror (frame/streak) camera systems to extend the method into a wide range of areas. Today, particle image velocimetry (PIV) [6] and digital PIV are accepted methods for extracting 2D motion measurements from images of fluid particles, providing local velocity and motion measurements in fluid systems.

Most, if not all, of the studies described above employ direct image correlation principles for matching subsets and extracting full-field displacements. A parallel approach for performing the matching process that employed Fast Fourier Transforms (FFTs) was developed by Cheng et al. in 1993 [43]. Using well-known Fourier Transform properties relating frequency content to local displacement, the peak location in the frequency domain of an FFT was shown to be a viable alternative for estimating local displacement components for those applications where in-plane strains and rigid body rotations are small.

1.1.4.2 Two-Dimensional Measurements, 2000–Present

In the past several years, 2D-DIC has undergone explosive growth worldwide; the authors identified 400+ archival articles[2] using the method since 2000.

Schreier et al. [252] demonstrated the importance of the image reconstruction process when attempting to improve the accuracy of the matching process. His

---

[1] The system remains in active use in the laboratory of R.S. Piascik and S.W. Smith at NASA Langley Research Center, as of the publication of this book.

[2] Science Citation Index literature search identifying use of 2D-DIC as part of measurement process, 2000–2007.

work showed that higher order spline interpolation functions can be used effectively to reconstruct the image intensity pattern and minimize measurement bias during the matching process, resulting in image position accuracy of 0.01 pixels or better in both the x and y directions when distortions are removed from the images. In a companion work, extending work by Lu and Cary [167], Schreier and Sutton [253] confirmed that quadratic shape functions provide some advantages when performing the matching process, especially for non-uniform strain fields, without appreciable increase in computational time. A summary of many of the key concepts noted above is given in two recent publications [285,286]. Chapter 5 presents recent results for both bias and variability in 2D-DIC image displacement measurements.

Other researchers have presented modifications to various aspects of the 2D-DIC measurement process, including the search procedure [106, 352], correlation approach [123, 134, 210, 221, 304] and registration method [271, 273, 350]. With regard to registration methods, of note is the work of Cheng et al. [44] which proposed full-field deformation through pixel-by-pixel mapping using a B-spline functional form. Since that time, investigators have developed modified full-field DIC methods that combine analytical models with 2D-DIC measurements. For example, in a series of articles, Réthoré et al. [235–238] defined an "extended 2D-DIC" method using finite element concepts, with initial emphasis on crack specimen studies. Additional methods include a range of approaches [16, 135].

Inverse methods have been actively pursued in recent years for mechanical property estimation, including elastic properties [90, 91, 116, 150], properties in heterogeneous materials [163, 164], hyperelastic properties [92], micromechanics [113] and composites [179].

Even as additional studies have been performed to understand the theoretical fundamentals of the method, researchers and scientists have applied 2D-DIC in a breathtaking array of areas. For example, Chasiotis [41, 42], Jin [132], Li [159, 160, 341, 342] and others [270] have extended 2D-DIC for use in AFM systems, determining defomations and strains with spatial resolution on the order of 50 nm. As with the work of Davidson et al., the approach is most effective when deformations are large ($\varepsilon > 0.003$).

When relatively large deformations are expected, investigators have used a scanning electron microscope (SEM) with 2D-DIC [138, 146]. Recently, Sutton et al. [158, 282–284] have demonstrated that SEM images can be acquired with a high contrast random pattern [50, 255] under mechanical or thermal loading, corrected for spatial [130] and temporal distortions and used with 2D-DIC to extract elastic properties for fields of view on the order of $10 \times 10\,\mu$m. In this work, the variation is small so that local strains on the order of 0.0005 can be reliably determined. Details for correcting SEM images and peforming 2D-DIC for high magnification measurements with optimal accuracy are presented in Section 6.6.

In a recent study, Berfield et al. [28] performed experiments to acquire deformations on an internal plane containing a high contrast random pattern, extending 2D-DIC in a manner similar to seeding of fluids [6, 216].

Given the sheer volume of applications, it is only possible to present representative examples for how the method is being studied, modified and applied. Fracture studies in recent years have included measurements in functionally graded materials [1–4], composites and concrete [8, 269, 297, 330], composites [9, 10, 103, 192, 202], asphalt [32], metals [211], polymers [131], rubber [71], shape memory alloys [58], electronic components and joints [258, 259], brittle materials [77, 190, 191], rock [177], wood [312] and general fracture parameter estimation methods [234, 239, 344–346].

Material characterization studies have included 2D-DIC measurements in thin materials and films [127, 128, 230, 246], polymers [141, 169, 182, 213, 223], metals [184, 220], heterogeneous materials [261, 299, 309], wood [129, 198, 199, 313], bio-materials [23, 45, 120, 157, 240, 331, 332, 347, 348], shape memory alloys [38, 193, 196, 197, 248], composites [78, 79, 214, 226, 244, 303, 315], asphalt [256], ceramics [126], glass wool [29], mineral wool [115], rock [31], glass [121], foams [133, 152, 227, 333, 334, 353, 354], clay [142], sands/soils [72, 98, 162, 231, 232, 339], concrete [222], paint [275, 276] and electronic components and joints [55, 66, 139, 185, 212, 272, 340, 349]. An interesting application was performed by Louis et al. [166]. In a manner analogous to the work of Russell et al. [245] for composites, the authors used X-Ray radiography with 2D-DIC to estimate the average strains in sandstone [166].

1.1.4.3 Three-Dimensional Digital Image Correlation Measurements, Pre-2000

Since two-dimensional digital image correlation requires predominantly in-plane displacements and strains, relatively small out-of-plane motion will change the magnification and introduce errors in the measured in-plane displacement. The effect was clearly highlighted in publications [68, 168] that attempted to estimate the effect of curvature on the sensor plane projections, with relatively large errors evident in the estimated deformations. Section 6.3 quantitatively demonstrates the deleterious effect of out-of-plane motion on 2D image motion measurements, while also confirming that stereo-vision measurements account for such motion by measuring all components simultaneously [289].

As early as the 1960s, photogrammetry principles developed for shape and motion measurements were used to estimate plate deflections [337]. From 1970–1990, the concept of digital correlation for use in photogrammetry was highlighted [5, 34, 108, 137], and photogrammety used to estimate critical crack tip opening [143] and surface deformations through selective feature identification [329]. Morimoto and Fujigaki [194] discussed the use of multiple cameras with images of a deforming rectangular grid and FFT methods for image analysis and surface motion estimates. Combining stereo-vision principles with 2D-DIC concepts developed and used in single camera imaging, Chao et al. successfully developed, automated and applied a two-camera stereo vision system for the measurement of three-dimensional crack tip deformations [172, 173]. Cárdenas-Garcia et al. [39] considered parallel and converging stereo configurations with DIC to identify features.

To overcome some of the key limitations of the method used in these studies (square subsets remained square in both cameras, mismatch in the triangulation of corresponding points and manual motions in a calibration process that was laborious and time consuming), the stereo-vision method was modified [109] to include (a) the effects of perspective on subset shape, (b) use of a grid with a range of calibration motions and (c) appropriate constraints on the analysis to include the presence of epipolar lines. As part of the on-going US aging aircraft research program, the modified system was used to obtain deformation measurements on a range of center crack panel sizes.[3] Results from the combined computational and experimental effort were presented by Helm et al. [110–112].

In the late 1990s, Synnergren et al. [300–302] and Lacey et al. [145] employed stereovision to make deformation measurements. Andresen [15] used photogrammetry principles and a stereo system to record and analyze images of a grating undergoing large deformation. Of particular note is the set of experiments by Synnergren et al. where the investigators obtained flash X-ray images of a specimen from two directions before and during high-rate loading. By comparing features in both sets of X-rays, the investigators extracted 3D estimates for specimen deformations. Lockwood and Reynolds [165] used SEM imaging to obtain stereo images via in-situ specimen tilting to obtain surface images of the specimen from two orientations. Without patterning the surface, the authors used image correlation and clearly discernible surface features to extract the matching image locations. This data was used with a simplified stereo imaging model to extract approximate 3D fracture surface shape information.

As investigators in the US were focused on developing robust measurement systems for a range of applications, European investigators working in the field of computer vision were also making significant progress. For example, Faugeras and Devernay [68] used a Taylor's series expansion of the unknown disparity function between two stereoscopic images to define a "subset shape function". For example, the first and second order expansions lead to the affine and quadratic shape functions used in many applications. Devy et al. [63] developed an improved methodology for 3D calibration. In their work, the authors defined the classic sum of squared reprojection errors (differences between the model-based image plane locations and the measured image plane locations) and optimized over the set of intrinsic and extrinsic model parameters; the model also included a radial distortion correction parameter. Lavest et al. [148] introduced the bundle adjustment technique (known for decades in the photogrammetry community) to the computer vision field and proposed an accurate, flexible calibration method based on this approach.

Garcia et al. [84] proposed an improved method for the calibration of a stereovision sensor and were the first to demonstrate that seven constraints are needed to reduce the family of calibration parameter solutions to a unique set when using bundle adjustment. As part of their work, the authors also extended the single camera calibration procedure commonly used in computer vision to the calibration of a stereo-camera system. Known as a stereo-rig system, the approach used by the

---

[3] All experiments were performed in Bldg 1205 at NASA Langley Research Center in Hampton, VA.

authors included an efficient approach for calibration by bundle adjustment using a global minimization procedure, and demonstrated experimentally that the approach outperforms previous calibration approaches based on separate camera calibration.

1.1.4.4 Three-Dimensional Digital Image Correlation Measurements, 2000–Present

As a result of the developments which have occurred in recent years, three-dimensional digital image correlation now is being used for a wide range of applications on both large and small structures (e.g. [111, 187]), as well as in the area of fluid measurements, where similar developments have led investigators to modify 2D PIV setups and employ multiple camera systems (i.e., stereoscopic PIV systems) to resolve all three components of velocity of imaged fluid particles [118, 119]. Of particular note is the use of a dual distribution of speckle sizes using complementary colors within a single pattern [187]. In this work, the authors used green or red filters to extract either the large or small pattern, depending upon whether full field information or local information is to be emphasized. Method development has included stereo optical microscopy with a novel distortion correction approach [130, 254]; the method has been used with existing surface features [147]. Section 7.4 describes the use of stereomicroscopy for deformation and shape measurements on specimens ranging from a few millimeters to 300 μm. In recent years, high speed imaging systems have been developed for use in a variety of experimental studies. Basic studies of typical high speed imaging systems were performed by Tiwari et al. [306]. In this case, the authors have shown that results from their studies led the authors to conclude that ultra high speed cameras utilizing image intensifiers may introduce large errors in the image-based deformation measurements. Recently, Orteu et al. [206] have used stereovision images to obtain simultaneously both 3D motion measurements and temperature estimates through appropriate thermal calibration procedures.

Stereovision system applications have included measurements on flexible wings undergoing aerodynamic loading [7, 266], as well as measurement of shape and deformation on cylindrical surfaces [173, 174] and on balloon structures [328].

In addition to the early work noted previously, recent fracture studies have been performed under mixed mode I/II [174] and mixed mode I/III loading conditions [298], while other investigators measured side-necking in a fracture specimen [151]. Section 7.3 presents experimental details for application of high speed stereovision to the study of mixed mode I/III fracture in a ductile metallic specimen. For low ductility materials, such as concrete, Lecompte et al. considered the use of imaging for crack detection [149].

Material characterization has also been an active area including microscale studies in engineered materials [73, 74, 338] and bio-materials [201, 281, 311], foam [51, 97], ceramics [204, 242], composites [187, 195, 207, 208], polymers [136] and high rate events [250, 251, 260]. A recent area of focus has been the determination of residual stresses using surface deformation measurements [122, 186, 200].

Section 7.2 describes the use of four synchronized cameras in a multi-stereo-vision system application to quantify large deformations in an edge-cracked, highly ductile polymer undergoing out-of-plane bending.

1.1.4.5 Volumetric Digital Image Correlation (VDIC)

Volumetric imaging and image analysis has been an active area of research and development within the medical community for decades, resulting in mature imaging capabilities. Micro and Macro Computer Aided Tomography (CT), Magnetic Resonance Imaging (MRI), Confocal Imaging Microscopy (CIM) and Positron Emission Tomography (PET) are examples of technology that are currently available for full-volume imaging of bio-material systems.

In addition to imaging technology, there is a wide range of software available to view volumetric images and analyze the images to extract information of interest to investigators. As a recent example, the work of Weiss, Veress, Phatak et al. [219, 324, 325] is noteworthy. In their studies, the authors developed methods for comparing images that have minimal feature content to extract full-field estimates for the deformations required to "warp" the image into the required shape.

Even as medical investigators developed software to analyze images for their applications, Bay et al. [26, 27, 188, 265] proposed that the 2D-DIC concepts used successfully for 2D and 3D computer vision be extended and used to match small sub-volumes before and after undergoing loading to obtain a full volumetric field of three-dimensional motions. The authors applied these principles to make quantitative deformation measurements in bone, demonstrating the efficacy of the proposed "volumetric digital image correlation" approach.

Applications of the approach include measurements on sandwich structures [247], bio-materials [326], steel powders [318], metallic alloys [171] and rock [156]. An interesting approach was developed by Germaneau et al. [86–89], using internally scattered light to obtain a volumetric pattern for imaging and analysis. The approach is qualitatively similar to the CIM approach used by Franck et al. [80], where volumetric images of a bio-material are acquired through point-by-point imaging of laser illuminated positions. Chapter 8 presents both theoretical background and experimental details for application of Volumetric DIC to study the behavior of foam under compression.

1.2 Discussion

Chapter 2 presents the essential elements of geometric optics that are employed in the theoretical developments. The section includes (a) thin and thick lens approximations, (b) depth of field and field of view formulae and (c) the general form of the pinhole model arising from geometrical optics approximations.

In Chapter 3, the theoretical foundation for single camera perspective image models is detailed, along with the relationship of various coordinate transformations to the camera calibration process. Additional topics introduced in this section include (a) distortion and various parametric and non-parametric distortion models and (b) modern camera calibration approaches using bundle adjustment, distortion and optimization algorithms.

Chapter 4 presents two-dimensional and three-dimensional computer vision models with special emphasis on calibration procedures that are commonly employed.

Chapter 5 outlines fundamental concepts underlying digital image correlation for motion measurements. Topics include (a) image correspondence during matching, (b) matching methods including differential and template approaches, (c) optimization criteria with intensity pattern scaling and offset accommodation, (d) subset shape functions, (e) solution methods, (f) interpolation methods and (g) error estimates, including bias and variability estimation approaches.

Chapter 6 presents details regarding the application of a single camera imaging system for in-plane measurement of deformations, including the potentially deleterious effects of out-of-plane motion. Applications include (a) successful measurement of the stress–strain response of an aluminum alloy using the surface strain field measured on a planar specimen undergoing uniaxial tension by 2D-DIC, (b) details of extensive experimental studies that clearly highlight the effects of out-of-plane motion on single-camera motion measurements, (c) use of a far-field microscope for microscale measurements of crack tip deformations with nanometer accuracy, (d) determination of material properties through an inverse methodology enabled by full-field 2D-DIC deformation measurements and (e) use of a scanning electron microscope for deformation measurements in a 20×20 μm field of view, including the development and application of nanoscale patterns.

Chapter 7 describes in detail the application of stereo-vision camera imaging systems for general three-dimensional measurement of surface deformations. Applications include (a) a synchronized four-camera stereovision system for the measurement of large, out-of-plane deformations in polymer beam specimens, (b) dynamic deformation measurements using high speed stereo cameras and (c) development and use of 3D stereomicroscopy systems for microscale deformation and shape measurements.

Chapter 8 presents both theoretical foundations and experimental results from a series of volumetric image correlation measurements in a polymeric foam undergoing compression loading.

Chapter 9 provides a brief discussion of methods to estimate errors that are expected during stereovision measurements. Developments include approaches to obtain probabilistic estimates for the mean and variance in the measurements due to the effects of intensity interpolation, intensity pattern noise and intensity pattern contrast on image-plane matching locations for corresponding points.

Chapter 10 presents practical considerations for accurate measurements using digital image correlation methods. Topics include (a) general imaging issues including camera and lens selection, depth of field, field of view and image magnification,

(b) image artifacts, including image sensor defects, reflections and lens contamination, (c) speckle pattern application, (d) optimal speckle pattern structure, (e) optimal speckle size and engineering estimates with examples, (f) exposure time and image blurring, (g) out of plane motion and estimates for the effect on 2D measurements and (h) image matching errors and approaches for estimating errors in both 2D and stereo imaging.

Appendices A through I provide the background information that the authors believe is important to provide the foundation necessary for both essential concepts in computer vision and the practical application of the method for shape and deformation measurements. Areas covered include continuum mechanics; linear algebra; surface strain estimation; non-linear optimization; basic concepts in statistics and probability; introduction to projective geometry; rotation tensor formulations; optimal estimates for intensity scale and offset; spline functions and triangulation.

Chapter 2
Elements of Geometrical Optics

2.1 Optics of a Camera

Figure 2.1 shows a simple optical system that consists of a single, ideal lens, i.e. a thin lens. Such lenses are good approximations when (a) the angles and diameters of the focused, light rays are sufficiently small so that the Gauss approximation is appropriate and (b) geometrical aberrations and other optical defects can be neglected [107]. Next, we consider an optical system consisting of a single thick lens. Using ray optics to extract the basic formulae relating object positions to sensor locations, we will see that by neglecting the effect of blur due to defocus, these optical models can be combined into a single simple geometric model of perspective projection: the so-called pinhole model.[1]

2.1.1 Thin Lens

A ray incident on a surface is described by its angle of incidence relative to the surface normal. In this context, a thin lens, or an association of lenses that are thin, should have a thickness (distance along the optical axis between the two outer surfaces of the lens) that is negligible compared to its focal length or to any of its dimensions. Lenses whose thickness is not negligible are called thick lenses.

The Gauss, or paraxial, approximation assumes that (a) the incident light rays make a small angle, θ, relative to the optical axis of the lens and (b) the ray lies close to the optical axis as it traverses the imaging system. Under this condition, the following approximation is assumed to be valid: $\sin\theta \simeq \tan\theta \simeq \theta$. Using this approximation, ray tracing is relatively simple since (a) any ray going through the centre of the lens is not deviated, (b) all parallel rays pass through the focal plane of the lens and (c) rays passing through a thin lens are deviated at the lens' mid-plane.

[1] In-depth developments in optics and applied optics are available in books such as Klein [140] and Born and Wolf [35].

M.A. Sutton et al., *Image Correlation for Shape, Motion and Deformation Measurements: Basic Concepts, Theory and Applications*, DOI: 10.1007/978-0-387-78747-3_2,

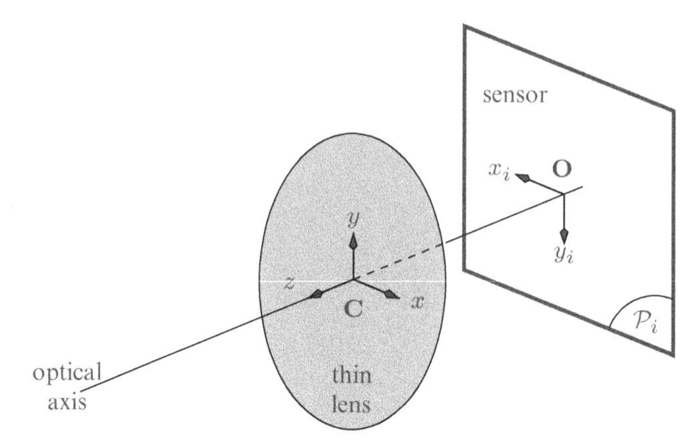

Fig. 2.1 Thin lens with image/sensor plane and associated coordinate frames

As shown in Fig. 2.1, a coordinate system, (x,y,z), has an origin at **C**, with the $+z$-direction along the optical axis and directed towards the object. This system is oftentimes referred to as the ***pinhole system***. Since the imaging process results in inversion, the sensor plane typically would employ an inverted coordinate system (x_s, y_s). The center of the sensor plane system, O, is the ***principal point*** and is located at the intersection with the optical axis.

Assuming symmetry in the imaging process about the optic axis, then without loss of generality the ray optics imaging process can be viewed by performing a transformation between object point and image point within a single plane. This is shown schematically in Fig. 2.2. Elements in Fig. 2.2 that require additional discussion are as follows:

- **Object point, M**, with coordinates (x,y,z) in the lens system
- **Lens center, C**, located at the intersection with the optical axis
- **Front or object focal point, F**, located a distance $z = +\bar{f}$ from **C**
- **Focal length,** $\bar{f}$, for the thin lens used in imaging
- **Rear or image focal point, F'**, located a distance $z = -\bar{f}$ from **C**
- **Ideal image point, M'**, located at optimal focus of rays
- **Image point in sensor plane (metric units), M''**, for object point **M**
- **Distance between the sensor plane and lens center**, γ.

The imaging process shown in Fig. 2.2 is typically associated with a pinhole camera model. Thus, using the coordinate system located at the lens center, **C**, and an object point **M** located at (x,y,z), the imaging process transforms **M** into an ideal image point, **M'** at (x',y',z'), using the following equations

$$\frac{y}{|\mathbf{CM}|} = \frac{-y'}{|\mathbf{CM}'|}$$
$$\frac{1}{|\mathbf{CM}|} + \frac{1}{|\mathbf{CM}'|} = \frac{1}{\bar{f}} \tag{2.1}$$

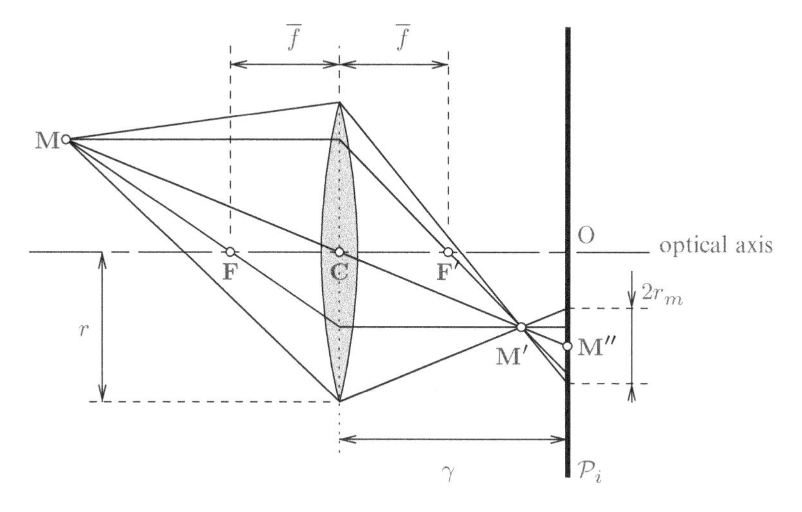

Fig. 2.2 A thin lens (side view)

The former describes the relationship between image and object coordinates for a simple lens system, where y in the image plane is positive when inverted. The latter is a version of the ***lensmaker's equation*** for ideal imaging of an object point. Combining these equations for the y-z components, and extending the approach to include the x-distance for **M** using imaging isotropy, we can write the following

$$x' = \frac{-\overline{f}x}{z-\overline{f}}$$
$$y' = \frac{-\overline{f}y}{z-\overline{f}} \tag{2.2}$$
$$z' = \frac{-\overline{f}z}{z-\overline{f}}$$

where z is the coordinate of **M** along the optic axis. In order that the image point be well-focused on the sensor plane, the sensor plane must be located where the rays intersect. This means that the distance γ between the sensor plane and the lens center must equal the z-component of the ideal image point **M'**, i.e., it is necessary that $\gamma = -z'$. This leads to the following equation to define the ***focus plane***

$$z = \frac{\gamma\overline{f}}{\gamma-\overline{f}} \tag{2.3}$$

Using an inverted coordinate system as shown in Fig. 2.1, the image point location, (x'_i, y'_i), in the sensor plane are positive and can be written as

$$
\begin{aligned}
x'_i &= x[\frac{\gamma - \overline{f}}{\overline{f}}] = \frac{x\gamma}{z} \\
y'_i &= y[\frac{\gamma - \overline{f}}{f}] = \frac{y\gamma}{z}
\end{aligned}
\tag{2.4}
$$

As shown in Fig. 2.2, the ideal image location for a general point **M** may not correspond with the sensor plane location. In practice, the image of **M** on the sensor plane will be slightly blurred, enlarged and shifted. The blurred region around point **M** is oftentimes known as the circle of confusion. Figure 2.3 shows a schematic of the blurring process.

To estimate the diameter of the defocus zone, consider rays emanating from point **M** with coordinates (x, y, z). Assuming that the lens has a radius r, then all of the points passing through the outer edge of the lens can be written in the form $(r\cos\beta, r\sin\beta, 0)$ with $0 \leq \beta \leq 2\pi$. These points are then imaged onto the sensor plane, where $z = -\gamma$, to form a circle. Using simple trigonometric formulae, and defining the defocused sensor plane coordinates by $(x", y", -\gamma)$, we have the following for the $y"$-component

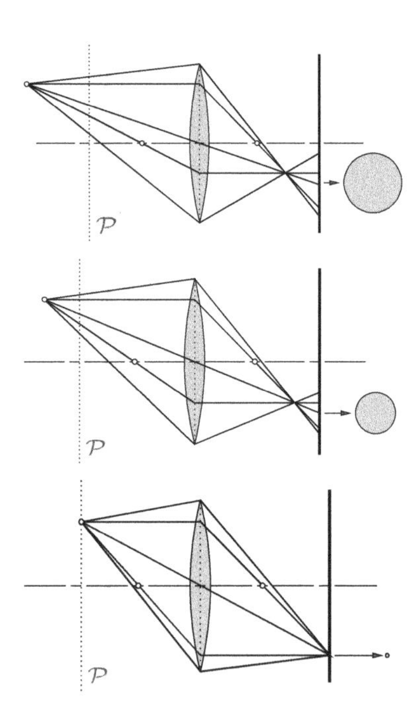

Fig. 2.3 Effect of defocus on image blur

$$\frac{-r\sin\beta+y'}{z'}=\frac{y'-y''}{\gamma+z'} \tag{2.5}$$

Using Eq. (2.2) to replace y' and z' and assuming isotropy in the imaging process, we can write the following equation for the points on the imaged circle

$$\mathbf{M}''=\begin{Bmatrix} x'' \\ y'' \\ z'' \end{Bmatrix}=\begin{Bmatrix} r\cos\beta+r\cos\beta\left(1+\frac{\gamma(f-z)}{fz}\right) \\ r\sin\beta+r\sin\beta\left(1+\frac{\gamma(f-z)}{fz}\right) \\ 0 \end{Bmatrix}+\begin{Bmatrix} -\frac{\gamma x}{z} \\ -\frac{\gamma y}{z} \\ -\gamma \end{Bmatrix} \tag{2.6}$$

The radius of the blurred image of point **M** on the sensor plane can be obtained using the vector in Eq. (2.6) as follows;

$$\begin{aligned} 2r_{M''} &= y''(\beta=90^\circ)-y''(\beta=270^\circ)=\left(r\sin(90^\circ)+r\sin(90^\circ)\left(1+\frac{\gamma(f-z)}{fz}\right)\right) \\ &\quad -\left(r\sin(270^\circ)+r\sin(270^\circ)\left(1+\frac{\gamma(f-z)}{fz}\right)\right) \\ \Rightarrow r_{M''} &= r_{\text{lens}}\left(1+\frac{\gamma(f-z)}{fz}\right) \end{aligned} \tag{2.7}$$

The image center for **M**, which is also the best estimate for the image point, **M"**, can be written as

$$\mathbf{M}''=\begin{Bmatrix} -\frac{\gamma x}{z} \\ -\frac{\gamma y}{z} \\ -\gamma \end{Bmatrix} \tag{2.8}$$

Equation (2.7) shows the effect of blurring. When an object point, **M**, does not lie on the focus plane of the lens, or when the sensor plane is shifted relative to the ideal focus position, the image in the sensor plane is a circular spot with radius $r_{M''}$.

The image spot radius can be reduced by using an ***aperture diaphragm***. The aperture diaphragm can be placed either in front of the lens or behind the lens. Figure 2.4 shows the decrease in blurring when placing an aperture diaphragm behind a lens.

2.1.1.1 Depth of Field and Field of View

Depth of Field (DOF)

The **depth of field** (DOF) is the distance in front of and beyond the object that appears to be in focus. As shown in Fig. 2.2, a simple lens system has a single focal surface, with ideally focused image locations given in Eq. (2.4). At any other distance, a point object is defocused, and will produce a blur spot shaped like the aperture, which for the purpose of analysis is usually assumed to be circular. When this circular spot is sufficiently small, it is indistinguishable from a point. In such

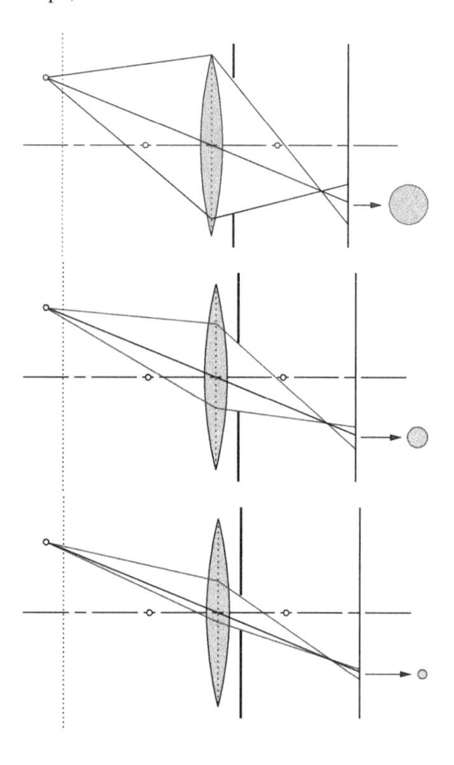

Fig. 2.4 Effect of rear aperture diaphragm on image blur. Decreasing aperture size will reduce image blur

cases, the point appears to be in focus and is considered to be an "acceptably sharp" image. The diameter of the circle increases with distance from the ideal focus (see Fig. 2.3); the largest circle that is indistinguishable from a point is known as the acceptable circle of confusion, or informally, simply as the **circle of confusion**.

As shown in Figs. 2.2–2.4, the DOF is determined by the object distance, the lens focal length, and the lens f-number (relative aperture). Except at close-up distances, DOF is approximately determined by the subject magnification and the lens f-number, $\mathcal{N} = \overline{f}/D_P$, where D_P is the diameter of the entrance pupil in the imaging system. DOF can be decreased by:

- Increasing image magnification by moving closer to the subject (e.g., microscope) for a fixed $\mathcal{N}$
- Increasing image magnification by increasing focal length for fixed object distance, s, for a fixed $\mathcal{N}$
- Decreasing $\mathcal{N}$ by increasing the aperture diameter, D_P

Conversely, at a fixed magnification the DOF can be increased by:

- Decreasing image magnification by moving the object further from the lens for fixed $\mathcal{N}$
- Decreasing image magnification by decreasing focal length for fixed object distance, s, for fixed $\mathcal{N}$
- Increasing $\mathcal{N}$ by decreasing D_P

Defining the hyperfocal distance, H, as the location of the object nearest to the camera lens such that the DOF extends to infinity, when the ideal focal surface is set to the hyperfocal distance, then the DOF extends from $H/2$ in front of the object to infinity. This configuration provides the largest DOF for a given $\mathcal{N}$.

Using the terminology given in Fig. 2.2, where $\overline{f}$ is the lens focal length and $c = 2\,r_m$ is the diameter of the circle of confusion for a given image format, we can estimate the hyperfocal distance H by:

$$H \simeq \frac{\overline{f}^2}{\mathcal{N}\,c} \tag{2.9}$$

One can directly relate H to the object distance, s, where the camera is ideally focused. Specifically, when s is large in comparison with the lens focal length, the distance D_N from the camera to the near limit of DOF and the distance D_F from the camera to the far limit of DOF are:

$$D_N \simeq \frac{H\,s}{H+s} \tag{2.10}$$

$$D_F \simeq \frac{H\,s}{H-s} \qquad \text{for } s < H \tag{2.11}$$

The depth of field $D_F - D_N$ can be written:

$$\text{DOF} = \frac{2\,H\,s^2}{H^2 - s^2} \qquad \text{for } s < H \tag{2.12}$$

Substituting for H and rearranging, DOF can be expressed as:

$$\text{DOF} = \frac{2\,\mathcal{N}\,c\,\overline{f}^2\,s^2}{\overline{f}^4 - \mathcal{N}^2\,c^2\,s^2} \tag{2.13}$$

Thus, for a given image format, depth of field is determined by three factors: the focal length of the lens, the f-number of the lens opening (the aperture), and the camera-to-subject distance.

Field of View (FOV)

The **field of view** (FOV) is the angular extent of a given scene that is imaged by a camera (see Fig. 2.5). For lenses projecting rectilinear (non-spatially-distorted) images of distant objects, the **focal length** and the **image format** dimensions completely define the **angle of view**.

As shown in Fig. 2.6, consider a camera-lens system that photographs an object at a distance S_1. The maximum FOV is obtained when the camera system forms an image that coincides with the dimension d of the frame (the image sensor). Treating

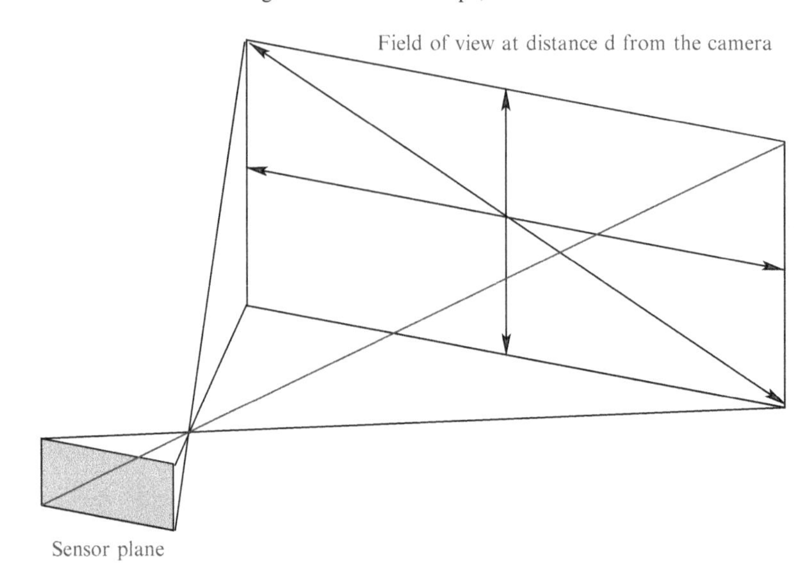

Fig. 2.5 Field of view

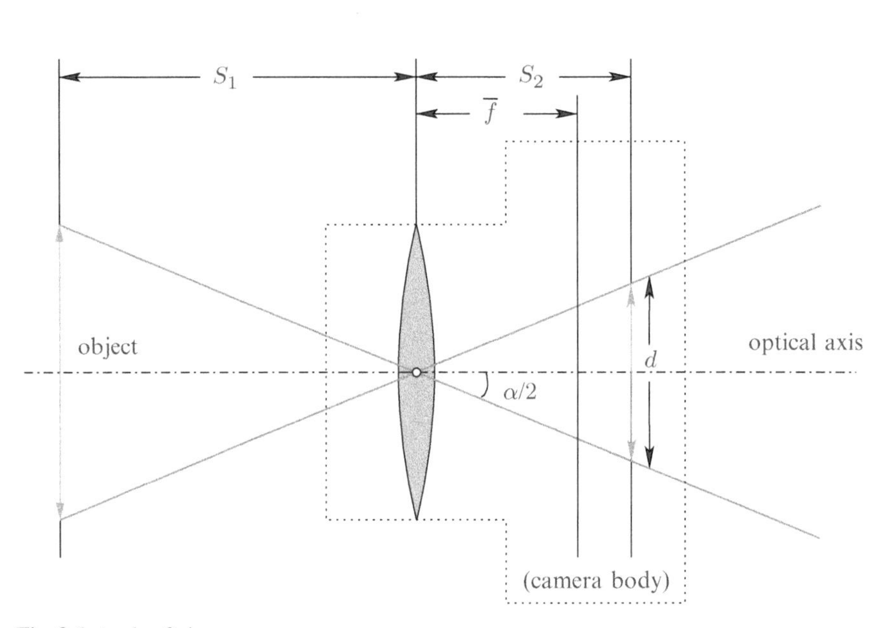

Fig. 2.6 Angle of view

the imaging system as if it were a pinhole camera, with pinhole located at distance S_2 from the image plane, one can define an angle, α, such that $\alpha/2$ is the angle between the optical axis of the lens and the ray joining its optical center to the edge of the image sensor, where α is the camera's **angle-of-view**, as it is the angle enclosing the largest object whose image can fit on the imaging (sensor) plane. The relationship between:

- The angle $\alpha/2$ (half of the angle-of-view)
- The "opposite" side of the right triangle, $d/2$ (half the image sensor dimension)
- The "adjacent" side, S_2 (distance from the lens to the image plane)

can be obtained using basic trigonometry resulting in

$$\tan(\alpha/2) = \frac{d/2}{S_2}, \tag{2.14}$$

which we can solve for α to obtain

$$\alpha = 2 \arctan \frac{d}{2S_2} \tag{2.15}$$

To project a sharp image of distant objects, S_2 needs to be equal to the focal length $\overline{f}$, which is attained by setting the lens for infinity focus. Then the angle of view is given by:

$$\alpha = 2 \arctan \frac{d}{2\overline{f}}. \tag{2.16}$$

2.1.1.2 Perspective Model for Transformation to Image Plane

Using Eqs. (2.2–2.4) with $z' = -\gamma$, imaging by a thin lens gives the following equations to transform object point $\mathbf{M}$ into the image plane

$$\mathbf{M}'' = \begin{Bmatrix} x'' \\ y'' \\ z'' \end{Bmatrix} = \begin{Bmatrix} -x\frac{\gamma}{z} \\ -y\frac{\gamma}{z} \\ -\gamma \end{Bmatrix} = \begin{bmatrix} -\frac{\gamma}{z} & 0 & 0 \\ 0 & -\frac{\gamma}{z} & 0 \\ 0 & 0 & -\gamma \end{bmatrix} \begin{Bmatrix} x \\ y \\ 1 \end{Bmatrix} \tag{2.17}$$

where all terms in Eq. (2.17) have metric units (e.g. mm). Since all images are inverted, it is natural to select an image plane coordinate system using inverted directions so that all image plane coordinates are positive in this system.

Equation (2.17) clearly demonstrates that the sensor positions are a non-linear function of the z-distance from the object to the lens system for this thin-lens imaging model. To remove the z-dependence from the matrix, Eq. (2.17) can be converted into a homogeneous form that employs a constant matrix manipulation, resulting in the following form for ***homogeneous image coordinates*** (see Appendix F).

$$\alpha \begin{Bmatrix} -x'' \\ -y'' \\ 1 \end{Bmatrix} = \begin{bmatrix} \gamma & 0 & 0 & 0 \\ 0 & \gamma & 0 & 0 \\ 0 & 0 & 1 & 0 \end{bmatrix} \begin{Bmatrix} x \\ y \\ z \\ 1 \end{Bmatrix} \tag{2.18}$$

where α is a scale factor. Equation (2.18) shows that the imaging process is a perspective model. Mathematically, the form in Eq. (2.18) is known as a *projective transformation* or *perspective projection* of the 3D position **M** into the sensor plane. A brief discussion of homogeneous coordinates is given in Appendix F.

2.1.2 Thick Lens

In many practical applications, the thickness of the lens is negligible compared to the focal length or other dimensions. However, when this approximation is no longer acceptable, the imaging equations for the thin lens model must be modified in order to take into account the effect of lens thickness on image plane coordinates.

Figure 2.7 shows the ray paths within a thick lens. A schematic of the imaging process is given in Fig. 2.8. Elements of the imaging process are as follows:

- **Front focal point, F, and Rear focal point, F'**
- **Front image principal plane,** P_0 and **Rear image principal plane,** P_1
- **Principal points, P and P'** locating P_0 and P_1 along the optical axis
- **Lens center, C,** located at **P** on the optical axis
- **Image distance from rear image plane,** γ
- **Front focal length,** $\overline{f}$, and **rear focal length,** $\overline{f}'$. Here, $\overline{f}' = \overline{f}$

Using this terminology, a ray emanating from **M** that is parallel to the optical axis is refracted at principal plane P_1 located at **P'** and imaged to **M'**. Similarly,

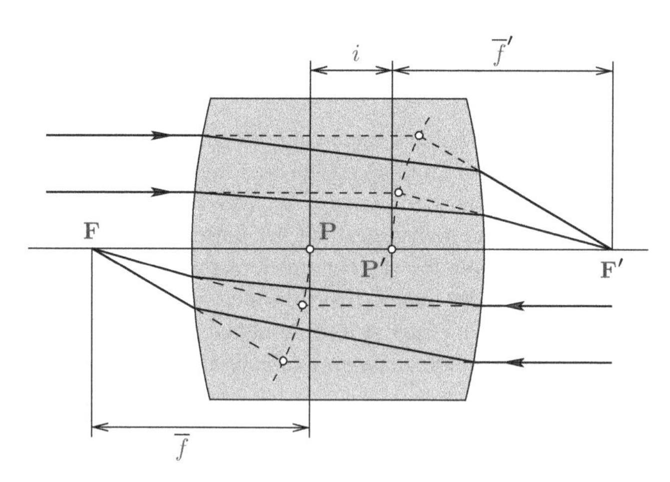

Fig. 2.7 Geometry of a thick lens

a ray emanating from **M** that passes through the **front focal point** is refracted at principal plane P_0 located at **P** and imaged to **M'**. Using these concepts, a form the lensmaker's imaging equation can be developed. Assuming that rays such as **MP** are parallel to the refracted rays like **P'M'**, we can write

$$\frac{1}{\|\mathbf{PM}\|}+\frac{1}{\|\mathbf{P'M'}\|}=\frac{1}{\overline{f}}$$
$$\frac{\mathbf{MN}}{\mathbf{NP}}=\frac{\mathbf{M'N'}}{\mathbf{N'P'}} \tag{2.19}$$

Following procedures similar to those given for a simple lens, the coordinates of the ideal image point **M'** are obtained as

$$x'=-\frac{\overline{f}x}{z-\overline{f}}$$
$$y'=-\frac{\overline{f}y}{z-\overline{f}} \tag{2.20}$$
$$z'=-i-\frac{\overline{f}z}{z-\overline{f}}$$

where (x',y',z') are ideal image coordinates. As shown in Fig. 2.8, the ideal image location for a general point **M** may not correspond with the sensor plane location. In practice, the image of **M** on the sensor plane will be slightly blurred, enlarged and shifted, in a similar manner to the thin lens equations. Using the same developments as for thin lenses, we can estimate the diameter of the defocus zone, as well as the center of the zone and obtain

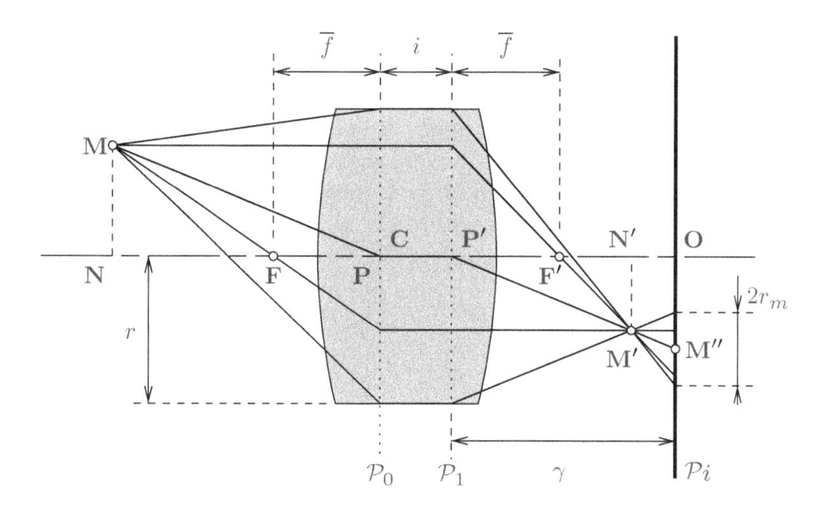

Fig. 2.8 Image process and ray-tracing for a thick lens

$$\mathbf{M}'' = \begin{Bmatrix} -\frac{\gamma x}{z} \\ -\frac{\gamma y}{z} \\ -\gamma - i \end{Bmatrix}; \qquad r_{M''} = r_{\text{lens}} \left(1 + \frac{\gamma(\bar{f} - z)}{\bar{f} z} \right) \tag{2.21}$$

In matrix form, the image location can be written as

$$\mathbf{M}'' = \begin{Bmatrix} x'' \\ y'' \\ z'' \end{Bmatrix} = \begin{bmatrix} -\frac{\gamma}{z} & 0 & 0 \\ 0 & -\frac{\gamma}{z} & 0 \\ 0 & 0 & -\gamma - i \end{bmatrix} \begin{Bmatrix} x \\ y \\ 1 \end{Bmatrix} \tag{2.22}$$

This can be further simplified by using inverted directions (see Fig. 2.1) and converting to homogeneous form to remove z-dependence from the matrix to obtain

$$\alpha \begin{Bmatrix} -x'' \\ -y'' \\ 1 \end{Bmatrix} = \begin{bmatrix} \gamma & 0 & 0 & 0 \\ 0 & \gamma & 0 & 0 \\ 0 & 0 & 1 & 0 \end{bmatrix} \begin{Bmatrix} x \\ y \\ z \\ 1 \end{Bmatrix} \tag{2.23}$$

2.1.3 Interpretation

For thin lens and thick lens imaging, Figs. 2.9-1 and 2.9-2 show an equivalent geometric model to describe the transformation from object coordinates **M** to image coordinates **M"** shown in Eqs. (2.17) and (2.22). Geometrically, the figures schematically represent the process of ***pinhole projection***, with all rays passing through the centerpoint of the lens. Known as a ***perspective projection model*** or ***pinhole projection model***, the parameter $\bar{f}$ in Figs. 2.2 and 2.8 has historically been referred to as the ***focal length of the pinhole system***, and is commonly designated by the variable f. To minimize confusion while also maintaining consistency with existing literature in the vision community, the authors have adopted (a) the symbol $\bar{f}$ to represent the focal length of a lens system and (b) the symbol f to represent the image distance, replacing γ throughout the remainder of the book.

Assuming isotropy in the imaging process, Fig. 2.9 indicates that the transformation of object point (x, y, z) into the sensor plane has the following form;

$$\mathbf{M}'' = \begin{Bmatrix} x'' \\ y'' \\ z'' \end{Bmatrix} = \begin{Bmatrix} -x\frac{f}{z} \\ -y\frac{f}{z} \\ -f \end{Bmatrix} = \begin{bmatrix} -\frac{f}{z} & 0 & 0 \\ 0 & -\frac{f}{z} & 0 \\ 0 & 0 & -f \end{bmatrix} \begin{Bmatrix} x \\ y \\ 1 \end{Bmatrix} \tag{2.24}$$

with a corresponding homogeneous form for the inverted coordinates shown in Fig. 2.9-2;

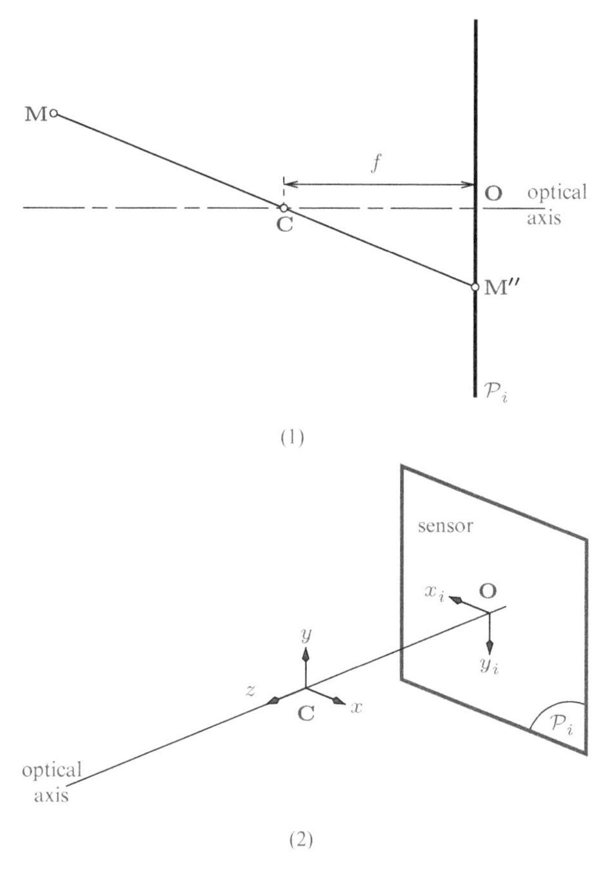

Fig. 2.9 Perspective projection imaging model and associated coordinate frames

$$\alpha \begin{Bmatrix} -x'' \\ -y'' \\ 1 \end{Bmatrix} = \begin{bmatrix} f & 0 & 0 & 0 \\ 0 & f & 0 & 0 \\ 0 & 0 & 1 & 0 \end{bmatrix} \begin{Bmatrix} x \\ y \\ z \\ 1 \end{Bmatrix} \tag{2.25}$$

2.1.4 Front Image Plane or Symmetry Interpretation

Consider the imaging process shown in Fig. 2.10-1, where the image plane is located at distance f behind the lens. The **front image plane model** is obtained by translating the image plane (also known as the ***retinal plane***) along the optic axis by a distance $+2f$. The final position of the front image plane, $\mathcal{P}_i$, is shown in Fig. 2.10-2. If it is now assumed that imaging occurs in the same manner as before, with the image point **M''** corresponding to the ray intersection with $\mathcal{P}_i$, the final equation is the same as shown in Eqs. (2.24, 2.25).

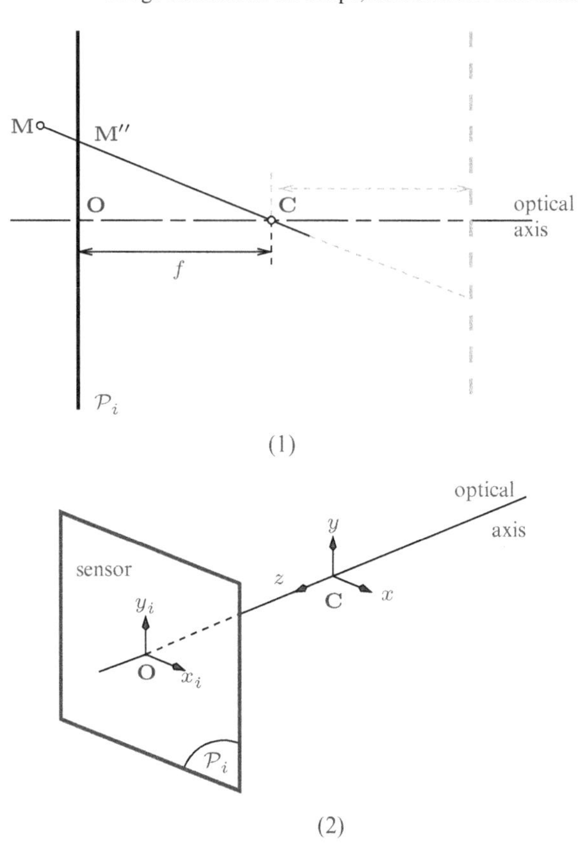

Fig. 2.10 Front image plane model and associated coordinate frames. Image symmetry with respect to *x*-*y* plane through the optical center, **C**, is assumed

This approach to develop the pinhole perspective imaging model is commonly used in the broad field of computer vision. An advantage of the approach is that the scene image is not inverted. In the remaining sections of the book, we will use the front image plane model in order to establish the camera model relating coordinates of a 3D point in space to the coordinates of its projection into the image plane.

Chapter 3
Single Camera Models and Calibration Procedures in Computer Vision

3.1 Camera Models

A camera is typically an opto-electronic device consisting of several sub-systems. First is the optics, consisting of elements such as a series of lenses, optical filters and shuttering elements that collect light from the object and focus the image onto the sensor plane. Second is the camera hardware. For example, when using a charge coupled device (CCD) camera to acquire digital images, the incident illumination is converted into an electrical signal. The third component (which may be an integrated element in the hardware) is the digitization and storage process. Using an A/D device, the CCD signal is converted into an array of discrete digital intensity data. In this section, we apply the pinhole projection models from Chapter 2 and develop a formulation that includes (a) rigid body transformations between several coordinate systems used to represent various elements in the imaging process, (b) the transformation between image plane coordinates and skewed sensor coordinates and (c) the effect of distortion on image positions.

3.1.1 Pinhole Camera

The classical geometric model for a camera is based on the pure perspective projection concepts presented in Chapter 2 and requires a total of three elementary transformations. Figure 3.1 presents a schematic of the geometry for a single camera system using a ***front image plane*** construction. A 3D point, **M**, has coordinates (X_W, Y_W, Z_W) in the world system, R_W. The camera or lens system, R_C, has its origin at the lens center, **C**, with coordinates given by (x, y, z). The image or retinal system, R_r, has coordinates (x_i, y_i); the image point corresponding to **M** is designated $\underline{\mathbf{m}}$. The sensor system, R_s, has coordinates (x_s, y_s) in units of pixels; sensor coordinates for $\underline{\mathbf{m}}$ are designated **m**.

M.A. Sutton et al., *Image Correlation for Shape, Motion and Deformation Measurements: Basic Concepts, Theory and Applications,* DOI: 10.1007/978-0-387-78747-3_3,

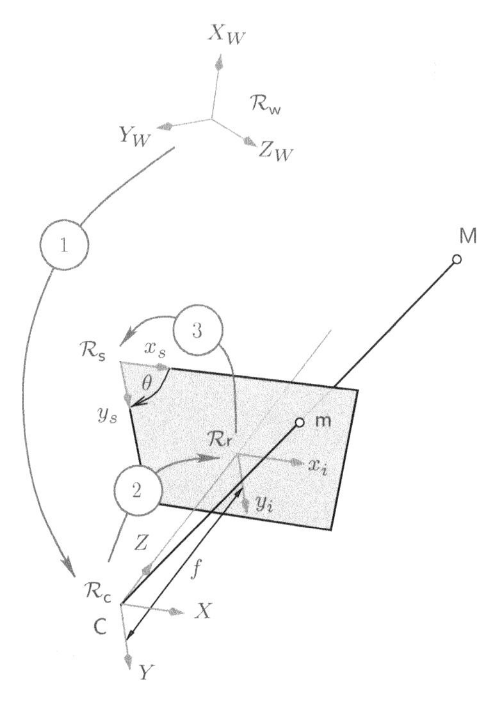

Fig. 3.1 The three elementary transformations of the pinhole camera model, and the associated coordinate systems. The first transformation relates the world coordinates of a scene point to coordinates in the camera system. The second one is the projection transformation of this point onto the retinal plane. The third one transform the point into the sensor coordinate system (pixel units)

3.1.1.1 Transformation Between World and Camera Systems

As shown in Fig. 3.1, ① represents a transformation between the coordinate systems R_W and R_C. This process requires both rotation and translation. Here, the rotation tensor is denoted $[\mathbf{R}]$ and the translation vector is $\mathbf{t}$. The transformation can be written as

$$
\begin{aligned}
\mathbf{M} &= \begin{Bmatrix} x \\ y \\ z \end{Bmatrix} = [\mathbf{R}] \bullet \begin{Bmatrix} X_W \\ Y_W \\ Z_W \end{Bmatrix} + \mathbf{t} \\
\tilde{\mathbf{M}} &\doteq \begin{Bmatrix} X \\ Y \\ Z \\ 1 \end{Bmatrix} = [\mathbf{T}] \bullet \begin{Bmatrix} X_W \\ Y_W \\ Z_W \\ 1 \end{Bmatrix} \\
[\mathbf{T}] &\doteq \begin{bmatrix} \mathbf{R} & \mathbf{t} \\ \mathbf{0} & \mathbf{1} \end{bmatrix}_{4\times 4}; \quad \mathbf{t} = \begin{Bmatrix} t_x \\ t_y \\ t_z \end{Bmatrix}; \quad [\mathbf{R}] = \begin{bmatrix} R_{11} & R_{12} & R_{13} \\ R_{21} & R_{22} & R_{23} \\ R_{31} & R_{32} & R_{33} \end{bmatrix}
\end{aligned}
\tag{3.1}
$$

where the symbol $\doteq$ is used to define a projective form for a specific variable.

3.1.1.2 Transformation Between Camera System and Image Plane

The second transformation, denoted by ② in Fig. 3.1, is between R_C and R_r. This is the imaging process that projects a 3-D point **M** onto the image plane to a point (x_i, y_i). The transformation is a pure perspective projection and can be written either in standard matrix form using a square matrix, [**P**], or in homogeneous form using results from Chapter 2 (e.g., Eq. (2.18)).

$$\underline{\tilde{\mathbf{m}}} \doteq \alpha \begin{Bmatrix} x_i \\ y_i \\ 1 \end{Bmatrix} = [P]_{3\times 4} \bullet \begin{Bmatrix} x \\ y \\ z \\ 1 \end{Bmatrix}$$

$$[P] \doteq \begin{bmatrix} f & 0 & 0 & 0 \\ 0 & f & 0 & 0 \\ 0 & 0 & 1 & 0 \end{bmatrix} \tag{3.2}$$

3.1.1.3 Transformation Between Image and Sensor Coordinates

The final transformation, designated ③ in Fig. 3.1, describes the sampling of the intensity field incident on the sensor array. As part of the transformation, metric positions are converted to pixel coordinates within the image.

Since the sensor array may be skewed, the transformation is somewhat more complex. As shown in Fig. 3.2, the sensor system is assumed to have non-orthogonal

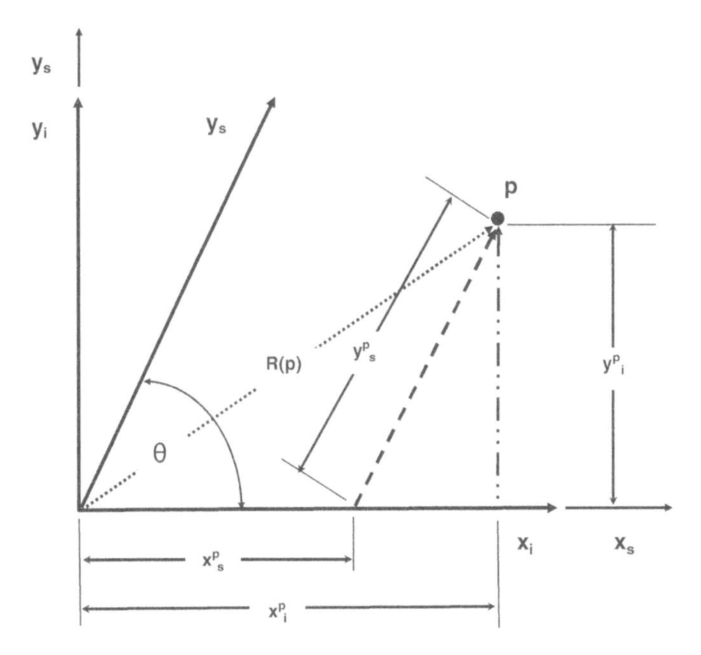

Fig. 3.2 Relationship between skewed sensor system and orthogonal image system

coordinate directions. The skew transformation between image and sensor coordinates can be written as

$$\begin{Bmatrix} \bar{\bar{x}} \\ \bar{\bar{y}} \end{Bmatrix} = \begin{bmatrix} 1 & -\cot\theta \\ 0 & \sin^{-1}\theta \end{bmatrix} \begin{Bmatrix} x_i \\ y_i \end{Bmatrix} \tag{3.3}$$

Assuming different scale factors in the sensor coordinate directions and designating them as (S_x, S_y) with units of pixels/unit metric length, the conversion of metric distances to sensor values (pixels) can be written in the form

$$\begin{Bmatrix} x_s \\ y_s \end{Bmatrix} = \begin{bmatrix} S_x & 0 \\ 0 & S_y \end{bmatrix} \begin{Bmatrix} \bar{\bar{x}} \\ \bar{\bar{y}} \end{Bmatrix} \tag{3.4}$$

Translating the skewed and scaled system to the local origin of the sensor system, while assuming that the translation $(\hat{c}_x, \hat{c}_y)$ is measured in the image system, the transformation is written as

$$\begin{Bmatrix} x_s \\ y_s \end{Bmatrix} = \begin{bmatrix} S_x & -S_x\cot\theta \\ 0 & \frac{S_y}{\sin\theta} \end{bmatrix} \begin{Bmatrix} x_i \\ y_i \end{Bmatrix} - \begin{Bmatrix} S_x\hat{c}_x - S_x\hat{c}_y\cot\theta \\ \frac{S_y\hat{c}_y}{\sin\theta} \end{Bmatrix} \tag{3.5}$$

Finally, the homogeneous form for the sensor coordinates is written as

$$\tilde{\mathbf{m}} \doteq \alpha \begin{Bmatrix} x_s \\ y_s \\ 1 \end{Bmatrix}$$

$$\begin{Bmatrix} x_s \\ y_s \\ 1 \end{Bmatrix} = \begin{bmatrix} S_x & -S_x\cot\theta & -S_x(\hat{c}_x - \hat{c}_y\cot\theta) \\ 0 & \frac{S_y}{\sin\theta} & -\frac{S_y\hat{c}_y}{\sin\theta} \\ 0 & 0 & 1 \end{bmatrix} \begin{Bmatrix} x_i \\ y_i \\ 1 \end{Bmatrix} = [\boldsymbol{A}] \begin{Bmatrix} x_i \\ y_i \\ 1 \end{Bmatrix} \tag{3.6}$$

The terms (x_s, y_s) and $(\hat{c}_x, \hat{c}_y)$ are measured in pixels and the skew angle, θ, is typically given in radians.

3.1.1.4 Overall Transformation from World to Sensor Coordinates

Equations (3.1) through (3.6) are used to transform **M** to **m**. Figure 3.3 presents a schematic of the process. Combining Eqs. (3.1) through (3.6), the final form of the combined transformation in matrix notation can be written as

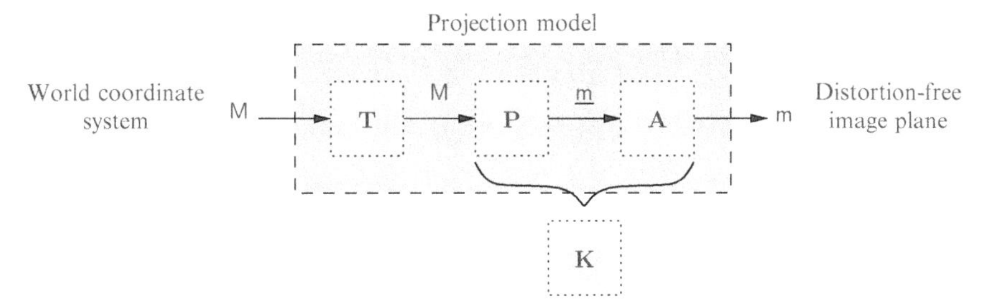

Fig. 3.3 Projection model diagram

$$\tilde{\mathbf{m}} = \alpha \begin{Bmatrix} x_s \\ y_s \\ 1 \end{Bmatrix} = [\Lambda] \bullet \begin{Bmatrix} X_W \\ Y_W \\ Z_W \\ 1 \end{Bmatrix}$$

$$[\boldsymbol{\Lambda}]_{3\times 4} = [\boldsymbol{K}]_{3\times 4} \bullet [\boldsymbol{T}]_{4\times 4}$$

$$[\boldsymbol{K}] = [\boldsymbol{A}]_{3\times 3} \bullet [\boldsymbol{P}]_{3\times 4} = \begin{bmatrix} fS_x & -fS_x \cot\theta & -S_x(\hat{c}_x - \hat{c}_y \cot\theta) & 0 \\ 0 & fS_y/\sin\theta & -S_y\hat{c}_y/\sin\theta & 0 \\ 0 & 0 & 1 & 0 \end{bmatrix} = \begin{bmatrix} f_x & f_s & c_x & 0 \\ 0 & f_y & c_y & 0 \\ 0 & 0 & 1 & 0 \end{bmatrix} \tag{3.7}$$

where :

$$f_x = fS_x$$

$$f_y = \frac{fS_y}{\sin\theta}$$

$$f_s = -fS_x \cot\theta$$

$$c_x = -S_x(\hat{c}_x - \hat{c}_y \cot\theta)$$

$$c_y = -\frac{S_y\hat{c}_y}{sin\theta}$$

The parameters (S_x, S_y) are horizontal and vertical scale factors (expressed in pixels/unit length), converting metric distance along the (x_s, y_s) directions into pixel units. As shown in Fig. 3.2, the angle θ models the non-orthogonality (skew) between the rows and columns in the image. If $\theta = \pi/2$, the expression is simplified to give the form most commonly presented in the literature, which includes $f_x = fS_x, f_y = fS_y$, and $f_s = 0$.

If the sensor positions are written in algebraic form in terms of the global three-dimensional position and the various parameters defined in Eq. (3.7), one arrives at the following expression for the combined transformation

$$\begin{Bmatrix} \alpha \\ \alpha x_s \\ \alpha y_s \end{Bmatrix} = \begin{bmatrix} \Lambda_{31}X_W + \Lambda_{32}Y_W + \Lambda_{33}Z_W + \Lambda_{34} \\ \Lambda_{11}X_W + \Lambda_{12}Y_W + \Lambda_{13}Z_W + \Lambda_{14} \\ \Lambda_{21}X_W + \Lambda_{22}Y_W + \Lambda_{23}Z_W + \Lambda_{24} \end{bmatrix} \quad (3.8)$$

where:

$$\Lambda_{11} = R_{11}f_x + R_{21}f_s + R_{31}c_x$$
$$\Lambda_{12} = R_{12}f_x + R_{22}f_s + R_{32}c_x$$
$$\Lambda_{13} = R_{13}f_x + R_{23}f_s + R_{33}c_x$$
$$\Lambda_{14} = f_xt_x + f_st_y + c_xt_z$$

$$\Lambda_{21} = R_{21}f_y + R_{31}c_y$$
$$\Lambda_{22} = R_{22}f_y + R_{32}c_y$$
$$\Lambda_{23} = R_{23}f_y + R_{33}c_y$$
$$\Lambda_{24} = f_yt_y + c_yt_z$$

$$\Lambda_{31} = R_{31}$$
$$\Lambda_{32} = R_{32}$$
$$\Lambda_{33} = R_{33}$$
$$\Lambda_{34} = t_z$$

$$\alpha = R_{31}X_w + R_{32}Y_w + R_{33}Z_w + t_z$$

3.1.1.5 Discussion

The matrix $[\Lambda]$ in Eq. (3.7) is a form of projective transformation that converts world coordinates into a scaled sensor position. With regard to this formulation, several points are noted. The components in $[\Lambda]$ are not a function of the global position of the point **M**. Rather, the components are only a function of camera parameters (e.g. position and orientation of world system, scale factors). Thus, the transformation described by Eq. (3.7) is *linear with respect to sensor and world coordinates.* The terms in $[\Lambda]$ are relatively complex combinations of the various camera parameters. Even when all terms in $[\Lambda]$ are determined, additional analysis is required to separate the parameters into manageable groups that can be quantified. As shown in Eq. (3.7) $[\Lambda]$ is composed of two sub-matrices; a 3×3 matrix that is nominally invertible and a 3×1 vector.

Inspection of Eq. (3.8) shows that the first equation relates the scale factor (α) to world coordinates. This equation can be used to eliminate α, resulting in two equations relating sensor image positions to metric world coordinates for the

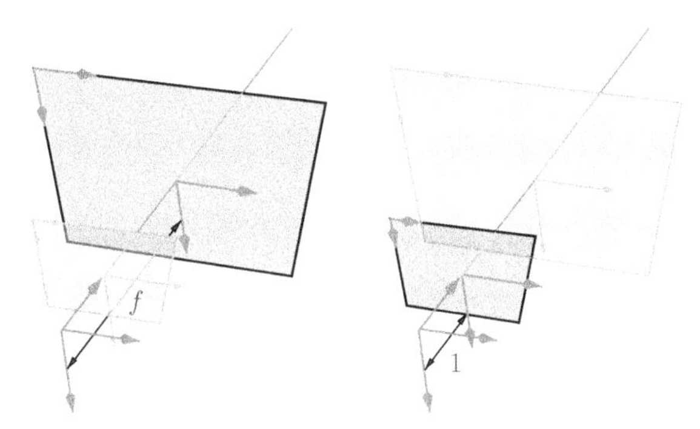

Fig. 3.4 Two camera models which produce the same images. *Left*: the pinhole focal length is equal to f and the sensor elements are $S_x^{-1} \times S_y^{-1}$ sized. *Right*: the focal length is equal to 1 and the sensor elements are $(fS_x)^{-1} \times (fS_y)^{-1}$ sized

corresponding 3D point. Equation (3.9) presents a modified version of the transformation that eliminates the scale factor.

$$\left\{ \begin{matrix} x_s \\ y_s \end{matrix} \right\} = \left\{ \begin{matrix} c_x + f_x \frac{R_{11}X_W + R_{12}Y_W + R_{13}Z_W + t_x}{R_{31}X_W + R_{32}Y_W + R_{33}Z_W + t_z} + f_s \frac{R_{21}X_W + R_{22}Y_W + R_{23}Z_W + t_y}{R_{31}X_W + R_{32}Y_W + R_{33}Z_W + t_z} \\ c_y + f_y \frac{R_{21}X_W + R_{22}Y_W + R_{23}Z_W + t_y}{R_{31}X_W + R_{32}Y_W + R_{33}Z_W + t_z} \end{matrix} \right\} \tag{3.9}$$

Inspection of Eqs. (3.7–3.9) shows that f is always coupled with other parameters within $[\boldsymbol{A}]$ and cannot be separately determined without additional information. The geometric interpretation is shown in Fig. 3.4. This camera model cannot differentiate between images produced by (a) a camera with f and sensors with size (S_x^{-1}, S_y^{-1}) and (b) a camera with $f = 1$ and sensors with size $(fS_x)^{-1}$ and $(fS_y)^{-1}$.

Finally, a basic result from projective geometry [320] is that a transformation that maps points to points and straight lines into straight lines is a real projective transformation. Since Eq. (3.7) demonstrates that a pinhole camera model is a projective transformation, three-dimensional points and straight lines are transformed into two-dimensional points and straight lines in the sensor plane in the absence of image distortions.

3.1.2 Distortion

The pinhole camera model described in the previous section is based on the Gaussian ray tracing approximation. Thus, it is an idealized representation for physical imaging systems, performing projective transformation between object points and image plane projections. In the "homogeneous" form shown in Eq. (3.8), the ideal imaging process is a linear transformation between object points and image locations.

In practical applications, it is very difficult or impossible to construct an optical system in which all lenses are perfectly parallel, their optical centers perfectly

Fig. 3.5 Distortion in image is visually observable when well defined vertical or horizontal features with high contrast are present, but much less noticeable when viewing regions with random contrast variations

aligned and their curvature meets the form required for ideal Gaussian estimation. As a result, a ray does not intersect the image plane at the point described by the Gaussian approximation. Deviations in the actual image positions may be several pixels relative to the ideal model.

Since natural scenes (e.g., trees, mountains, skies, landscapes) have shapes that vary significantly with time of day, weather or natural events such as landslides, slight perturbations in shape or orientation are generally indiscernible within the overall scene structure. Conversely, in some situations large distortions are clearly visible and may actually be added to an image for artistic purposes, providing unusual structure to an otherwise "standard view". Figure 3.5 shows a situation where the high contrast vertical features in the overall scene (e.g. railroad tracks) provide clear visual evidence of the presence of distortion. Conversely, Fig. 3.5 also shows that the same distortions are not readily recognized in those regions having minimal feature content and random variations in contrast.

Given the potential for distortions to create substantial measurement errors when using the pinhole model to predict image locations, in all applications where accuracy is at a premium it is essential that distortions be properly modeled, measured and adequately removed from image-based measurements.

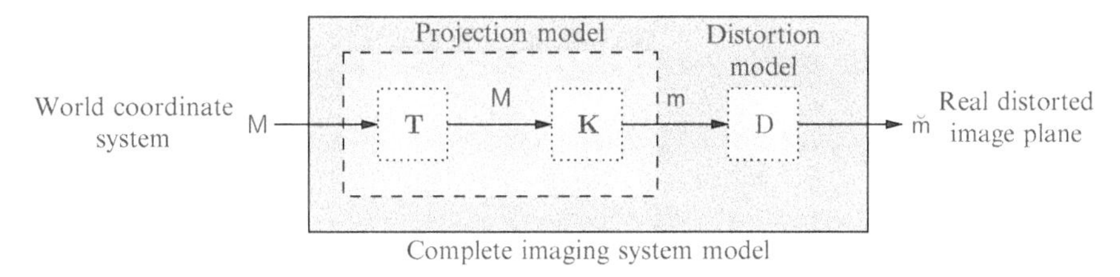

Fig. 3.6 Schematic complete imaging system model including projection and distortion transformations

Physically, distortions occur when the image location (x_i, y_i) does not match the pinhole prediction. Thus, one could introduce a distortion model prior to performing the pinhole model prediction using **[A]**. Since digitized sensor images are the primary data used to estimate distortions, the more convenient approach (and mathematically equivalent) is to perform the distortion correction after performing the pinhole estimation.

As shown in Fig. 3.6, image distortion can be expressed as a transformation of the pinhole projection estimation, mapping the ideal, distortion-free image point **m** into a real distorted location, $\breve{\mathbf{m}} = \overline{\mathbf{D}}(\mathbf{m})$. To date, two main approaches have been used to quantify image distortions. The first one, and by far the most widespread approach, models distortion by adding a distortion vector term to the pinhole model. This approach commonly leads to sophisticated parametric highly non-linear camera models. Parametric models are commonly used to model distortions in simple lens systems [53, 54, 109, 205, 257, 293, 296], as discussed in the following section.

3.1.2.1 Geometric Aberrations and Parametric Distortion Models

For a general three-dimensional point (X_W, Y_W, Z_W), let $\mathbf{x_s}$ be the ideal sensor position (i.e. location predicted by pinhole camera model such as Eq. (3.9)) and let **X** or $\mathbf{X_s}$ be the distorted sensor location. Then, using the terminology defined in Fig. 3.6, the following relationship can be defined;

$$\begin{aligned} \breve{\mathbf{m}} &= \overline{\mathbf{D}}(\mathbf{m}) = \mathbf{m} + \mathbf{D}(\mathbf{x_s}, \boldsymbol{k_i}), \\ &\Rightarrow \mathbf{X} = \mathbf{x_s}(\beta) + \mathbf{D}(\mathbf{x_s}, \boldsymbol{k_i}) \end{aligned} \tag{3.10}$$

where :

$\boldsymbol{k_i}$ = vector of distortion parameters for model, $i = 1, 2, \ldots, N$

$\mathbf{X} = (X_1, X_2), (X_{s1}, Y_{s1})$ = distorted 2D sensor position

$\mathbf{x_s} = (x_s, y_s)$ = ideal 2D sensor position defined by Eq. (3.9)

$\mathbf{D} = (D_x, D_y)$ = sensor plane 2D distortion displacement vector

β = parameter vector defined by terms in Eq. (3.9)

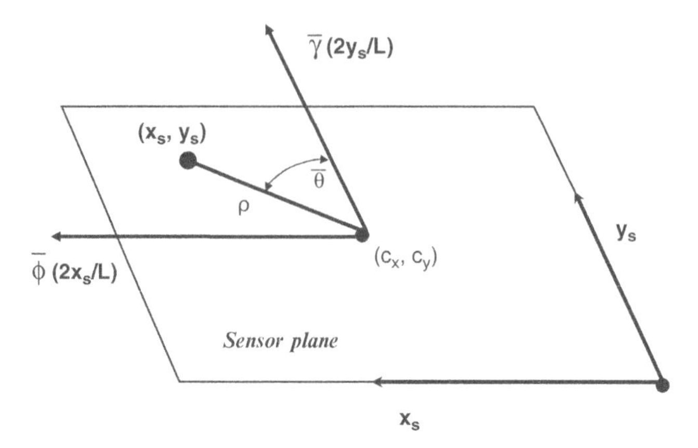

Fig. 3.7 Coordinates used to define spatial distribution for components of Seidel lens distortion

The most common approach in developing parametric distortion models is to include the effects of each individual type of distortion and additively combine these effects to obtain the total image distortion.

Figure 3.7 shows the coordinate system for the graphical presentation. Typical geometric distortions are based on Seidel lens distortions and include distortions such as (a) spherical, (b) coma, (c) astigmatism, (d) curvature of field and (e) linear. The equations for each type are as provided in the following paragraphs.

Spherical

Spherical distortion in a lens system is shown graphically in Fig. 3.8. This distortion occurs due to slight focusing errors along the optical axis direction, and is assumed to be (a) axi-symmetric relative to an axis that is perpendicular to the sensor plane and passes through the center of the sensor plane and (b) a quartic function of radial distance.

$$\mathbf{D_S}(x_s, y_s) = \kappa_1 \rho^4 \mathbf{e_r} \tag{3.11}$$

$\mathbf{e_r}$ = radial unit vector

κ_1 = amplitude of spherical distortion (pixels^{-3})

$\rho = [(x_s - c_x)^2 + (y_s - c_y)^2]^{1/2}$

Coma Distortion

Coma, shown graphically in Fig. 3.9, occurs when the lens (or the object) is tilted relative to the optic axis so that image points for locations offset from the center of

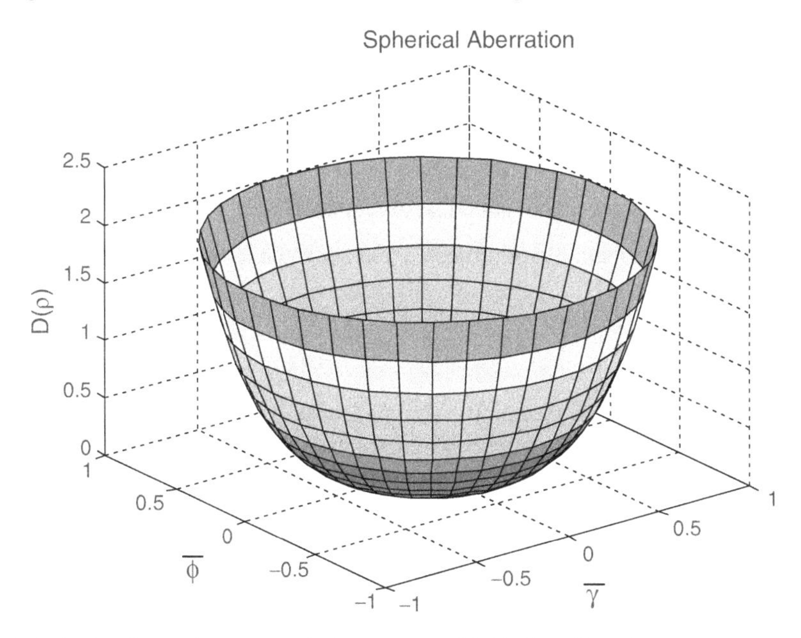

Fig. 3.8 Spatial distribution for spherical component of Seidel lens distortion

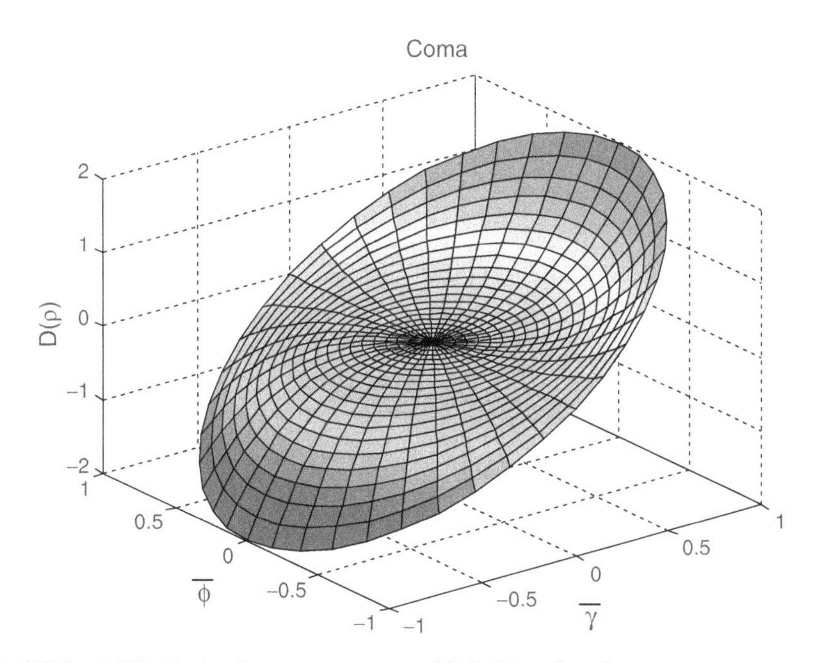

Fig. 3.9 Spatial distribution for coma component of Seidel lens distortion

the entrance pupil are focused at incorrect locations, with the error increasing with distance from the tilt axis.

$$\mathbf{D_C}(x_s, y_s) = \kappa_2 \rho^3 \cos(\bar{\theta} - \bar{\theta}_C)\mathbf{e_r} \tag{3.12}$$

κ_2 = amplitude of spherical distortion (pixels^{-2})

$\bar{\theta}_C$ = orientation of projected lens tilt angle in sensor plane

Astigmatism

Astigmatism is shown in Fig. 3.10. Astigmatism occurs when the lens (or the object) is tilted relative to the optic axis so that image points for locations offset from the center of the entrance pupil are focused at incorrect locations, with the error increasing with distance from the tilt axis.

$$\mathbf{D_A}(x_s, y_s) = \kappa_3 \rho^2 \cos(\bar{\theta} - \bar{\theta}_A)\,\mathbf{e_r} \tag{3.13}$$

κ_3 = amplitude of the distortion in the sensor plane (pixels^{-1})

$\bar{\theta}_A$ = orientation of projected astigmatic plane within the sensor plane

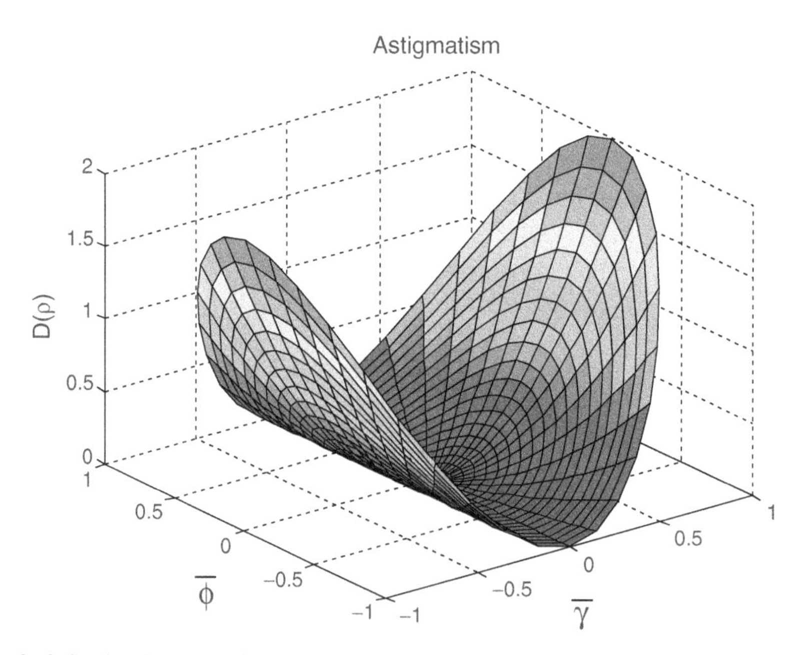

Fig. 3.10 Spatial distribution for astigmatism component of Seidel lens distortion

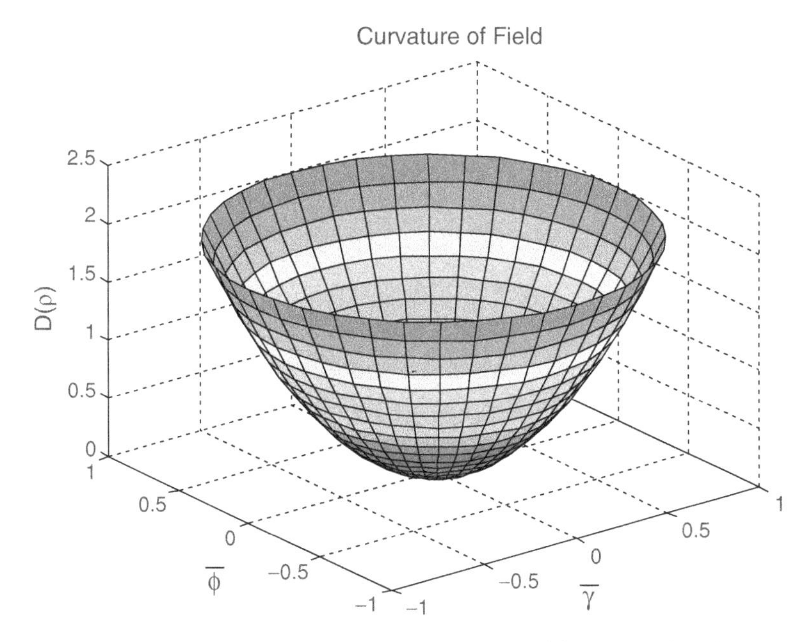

Fig. 3.11 Spatial distribution for curvature of field component of Seidel lens distortion

Curvature of Field

This component of Seidel distortion is shown in Fig. 3.11. Similar in form to spherical distortion, curvature of field is assumed to be symmetric relative to the optical axis and quadratic in radial position. Mathematically, the curvature of field distortion can be expressed as

$$\mathbf{D}_{\boldsymbol{COF}}(x_s, y_s) = \kappa_4 \rho^2 \mathbf{e_r} \tag{3.14}$$

κ_4 = amplitude of curvature of field distortion in the sensor plane (pixels^{-1})

Linear

Another component of Seidel distortion is assumed to be a linear function of radial position and non-symmetric with respect to angle in the sensor plane. Linear distortion is shown in Fig. 3.12. The linear component of Seidel distortion is expressed as

$$\mathbf{D_L}(x_s, y_s) = \kappa_5 \rho \cos(\bar{\theta} - \bar{\theta}_L)\, \mathbf{e_r} \tag{3.15}$$

κ_5 = amplitude of linear distortion in the sensor plane

$\bar{\theta}_L$ = angular orientation of linear distortion axis

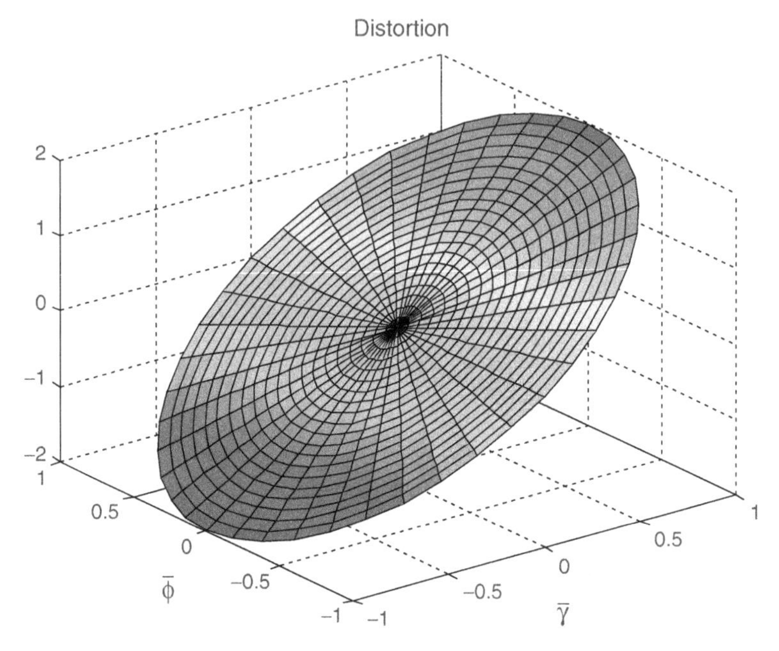

Fig. 3.12 Spatial distribution for linear component of Seidel lens distortion

Radial

Similar to spherical distortion in shape and symmetry, it is common to retain the first three odd powers of radial distance in this distortion model.

$$\mathbf{D_R}(x_s, y_s) = \kappa_6 \rho^3 e_r + \kappa_7 \rho^5 e_r + \kappa_8 \rho^7 \mathbf{e_r} \tag{3.16}$$

$\mathbf{e_r}$ = radial unit vector

$\kappa_6, \kappa_7, \kappa_8$ = coefficients for radial distortion with units (pixels^{-2}), (pixels^{-4}) and (pixels^{-6}), respectively

De-centering

Termed a minor distortion relative to either spherical or radial distortions, only the first order term is generally retained to describe this distortion model. The distortion is

$$\mathbf{D_d}(x_s, y_s) = \kappa_9 \rho^2 [3\sin(\bar{\theta} - \bar{\theta}_d)\mathbf{e_r} + \cos(\bar{\theta} - \bar{\theta}_d)\mathbf{e_t}] \tag{3.17}$$

$\mathbf{e_t}$ = tangential unit vector; $\mathbf{e_t} \bullet \mathbf{e_r} = 0$

κ_9 = decentering distortion coefficient with units (pixels^{-1})

$\bar{\theta}_d$ = orientation of axis for maximum tangential distortion

Since one cannot readily separate the distortion components described above when viewing a distorted image, the only practical approach is to define the total distortion and write

$$\mathbf{D}(\mathbf{x_s},\kappa_i) = D_S + D_R + D_C + D_A + D_{COF} + D_L + D_d \tag{3.18}$$

$\kappa_1 \rightarrow \kappa_{10}$ = distortion parameters for all individual components
$\bar{\theta}_C, \bar{\theta}_A, \bar{\theta}_L, \bar{\theta}_d$ = angle parameters for all components

Equation (3.18) can be written as follows:

$$\begin{aligned}\mathbf{D}(\mathbf{x_s},\kappa_i) = (&\rho[\kappa_5\cos(\bar{\theta}-\bar{\theta}_L)] + \rho^2[\kappa_3\cos(\bar{\theta}-\bar{\theta}_A) + \kappa_4 + 3\kappa_9\sin(\bar{\theta}-\bar{\theta}_d)] \\ &+\rho^3[\kappa_2\cos(\bar{\theta}-\bar{\theta}_C) + \kappa_6] + \rho^4[\kappa_1] + \rho^5[\kappa_7] + \rho^7[\kappa_8])\,\mathbf{e_r} + \rho^2[\kappa_9\cos(\bar{\theta}-\bar{\theta}_d)]\mathbf{e_t}\end{aligned} \tag{3.19}$$

Inspection of Eq. (3.19) shows that (a) six of the first seven powers of radial distance are present, (b) radial distance in the distortion models is measured from the distortion center, which is generally assumed to be the same location for all models and located at the projected lens center, (c_x, c_y) and (c) there are 13 unknown parameters to be obtained during the distortion estimation process.

3.1.2.2 Remarks

An obvious disadvantage of this approach for distortion modeling is the similarity in the various individual distortion components, making it difficult to accurately quantify individual distortion parameters. Another disadvantage is that Eq. (3.19) is highly non-linear with respect to the image positions. Such equations require accurate initial parameter estimates to reduce the potential for convergence to a local optimum, resulting in the potential for large residual errors if modest to high levels of distortion are present. Because of these difficulties, it is common practice to select one or two components of distortion for modeling purposes, thereby reducing the number of parameters that require accurate initial estimates. The terms typically selected include spherical, radial and linear distortion, though models including coma and other terms have been employed successfully.

A potentially more important limitation of this approach is that the pre-specified functional forms take into account classical distortions but are ineffective when used to estimate arbitrary lens aberrations or unknown (but deterministic) distortions in an imaging process. In particular, these methods have been shown to be inadequate for use in complex imaging systems such as a stereo optical microscope [254] or a scanning electron microscope [282–284].

3.1.2.3 A Priori Distortion Modeling

Given the difficulties associated with using multiple individual distortion terms to model a complex distortion field, additional approaches have been developed.

For example, Peuchot [217, 218] and Brand et al. [36] proposed a process for developing an overall distortion model to warp the recorded image and correct the embedded distortions so that the resulting image optimally obeys a distortion free pinhole projection model. Utilizing the straight line transformation properties of a pinhole cameral model, the authors imaged a planar target grid containing a series of straight line and grid intersection points (Fig. 3.17 shows a typical target grid and distorted image of the grid), with the grid intersection locations assumed to be known with high accuracy.

To perform the correction process, the investigators extracted the locations of each grid intersection in the distorted image with sub-pixel accuracy. The estimated image plane locations for all grid points are then transformed onto horizontal and vertical straight lines. The straight lines are then placed in correspondence with a square grid being of unit size (i.e., the four corner grid points noted as $\bar{G}_1 \longrightarrow \bar{G}_4$ are defined to be at $(0,0)$, $(1,0)$, $(1,1)$ and $(0,1)$). Through an interpolation process, all points in the deformed grid locations are mapped into the virtual, distortion-free coordinate system. The virtual system is a mapping of straight lines into straight lines so that the overall process is consistent with a pinhole camera model. Since the predictions are likely to be far from a pixel coordinate system, an additional transformation is performed to convert the unit shape into pixel coordinates that are consistent with the assumed target dimensions.

Though an attractive methodology in principle, to ensure that the method has optimal accuracy for distortion correction requires (a) a target having extremely straight lines and accurate and regular spacing (manufacturing such grids may be time-consuming, expensive and difficult to reproduce) and (b) accurate and unbiased estimation of grid intersections; typical accuracy is limited to 0.02 pixels. An additional limitation of the method is that single grid images with few grid intersections are typically used in this method. Such a procedure limits the ability to reconstruct the images more accurately by using a larger number of intersection points to reduce the effect of errors in individual intersection estimates.

More recently, Schreier at Correlated Solutions, Inc. [130] developed a completely generic approach to correct simple or complex image distortions in general image systems and improve the accuracy of the vision-based measurements. The procedure consists of warping the images such that the equivalent virtual camera will closely approximate a distortion-free imaging system model. The process of determining the magnitude of distortion at each image location through a set of simple translations and/or rotations is termed a priori distortion correction.

Unlike classic individual element distortion modeling described previously, this approach allows one to determine the pixel-by-pixel image distortions that are incurred due to the combined effects of all the elements in the optical imaging system. To extract full-field distortion corrections, a spline function representation for the pixel-by-pixel deviations from a pinhole camera prediction is employed; this approach can be used to represent any reasonably well behave distortion field [254]:

$$\mathbf{D}(\mathbf{m},\alpha) = \sum_{i=0}^{n_i}\sum_{j=0}^{n_j} \alpha_{i,j} N_i(x) N_j(y) \tag{3.20}$$

$$\alpha_{i,j} = \left(\alpha_{0,0}\cdots\alpha_{0,j}\cdots\alpha_{0,n_j};\alpha_{1,0}\cdots\alpha_{1,n_j};\cdots\cdots\cdots\alpha_{n_i,n_j}\right)$$

where $\alpha_{i,j}$ are the spline coefficients, N_i and N_j are the basis functions and n_i and n_j are equal to the degree of the spline plus the number of splines in the horizontal and vertical directions, respectively, with (x, y) representing coordinates in either the ideal or distorted sensor coordinate systems, as discussed in the following section. Appendix H provides additional information on splines.

3.1.2.4 Procedure for Correcting Distorted Images Using Distorted Image Positions

When a spline function is used to represent the full-field distortion, it would be more efficient if (a) the domain of definition for the distortion function does not change during the anticipated iterative distortion estimation process and (b) spline inversions are not required. Since the measured sensor coordinates are in the distorted state, (X_1, Y_1), an approach to meet these requirements would be to directly estimate the inverse $[\bar{\mathbf{D}}]^{-1}$ and use it to estimate the undistorted coordinates. Figure 3.13 shows a schematic of the modified procedure, where $[\mathbf{C}] = [\bar{\mathbf{D}}]^{-1}$.

Using this approach, we can write:

$$[\mathbf{K}]\,[\mathbf{T}]\,\mathbf{M} = [\mathbf{K}]\,[\mathbf{T}]\begin{Bmatrix} X_W \\ Y_W \\ Z_W \\ 1 \end{Bmatrix} \rightarrow \begin{Bmatrix} x_s \\ y_s \\ 1 \end{Bmatrix} \Leftrightarrow \begin{Bmatrix} x_s \\ y_s \\ 1 \end{Bmatrix} = [\mathbf{C}]\,\breve{\mathbf{m}} = [\mathbf{C}]\begin{Bmatrix} x_s^d \\ y_s^d \\ 1 \end{Bmatrix} \tag{3.21}$$

ideal pinhole projection into sensor plane — spline correction of distorted images

In this form, the basis functions $N_i(x_s)$ and $N_j(y_s)$ and, hence, **[C]** is defined on deformed image coordinates.

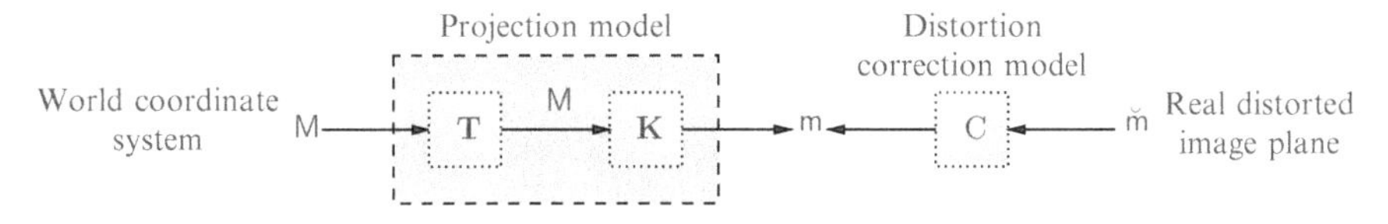

Fig. 3.13 Schematic imaging system model with correction distortion

3.1.2.5 Remarks

The a priori distortion method has advantages over the parametric approach. The most important of these is that the method is capable of more accurately quantifying distortions in both complex and simple optical systems. An additional advantage is accrued once the spline coefficients are determined. At that point, a simple pinhole camera model is combined with the corrected image coordinates to (a) significantly increase the accuracy of the vision system measurements and (b) decrease complexity of the imaging algorithms to a simple projective model by using the corrected image coordinates, making the process more robust and faster than non-linear parametric-based system models.

3.2 Calibration – Full Camera Model Parameter Estimation

The problem of camera calibration has found great interest in the computer vision community and a large number of calibration methods have been proposed over the years. Among the linear calibration methods, the algorithm proposed by Ravn et al. [229] is attractive since it is simple to implement and only requires a planar calibration target that is easily manufactured. Adapted to our pinhole model and assuming orthogonal sensor plane coordinates ($\theta = \pi/2$), the following development[1] provides a methodology for estimating all parameters in Eqs. (3.8) or (3.9). In this version, we consider that the 3D position is known and the corresponding image equations provide two equations that give the ideal sensor locations estimated by the pinhole camera model.

3.2.1 *Camera Parameter Estimation without Distortion*

Without loss of generality, we assume that the calibration target is planar and that the world coordinate system is aligned with the target plane; positions in the target plane are written $(X_W, Y_W, 0)$. Furthermore, it is assumed that distortions have been removed from the image positions so that a pinhole camera is appropriate for the image model. Simplifying Eq. (3.9) with $Z_W = 0$ and using a lowest common denominator, the two equations can be written in the form

$$\begin{aligned} x_s &= \frac{c_x(R_{31}X_W + R_{32}Y_W + t_z) + f_x(R_{11}X_W + R_{12}Y_W + t_x)}{R_{31}X_W + R_{32}Y_W + t_z} \\ y_s &= \frac{c_y(R_{31}X_W + R_{32}Y_W + t_z) + f_y(R_{21}X_W + R_{22}Y_W + t_y)}{R_{31}X_W + R_{32}Y_W + t_z} \end{aligned} \tag{3.22}$$

[1] The original development has been modified to correct problems that have been reported elsewhere in the literature.

Since these equations cannot directly be solved for the calibration parameters of interest using linear methods, we define a parameter vector $\{\eta\}$ containing combinations of the calibration parameters to obtain a set of linear equations in $\{\eta\}$

$$\begin{aligned} x_s &= \eta_1 + \eta_3 X_W + \eta_4 Y_W - \eta_7 x_s X_W - \eta_8 x_s Y_W \\ y_s &= \eta_2 + \eta_5 X_W + \eta_6 Y_W - \eta_7 y_s X_W - \eta_8 y_s Y_W \end{aligned} \tag{3.23}$$

where:

$$\begin{aligned} \eta_1 &= c_x + \tfrac{f_x t_x}{t_z} \\ \eta_2 &= c_y + \tfrac{f_y t_y}{t_z} \\ \eta_3 &= \tfrac{c_x R_{31} + f_x R_{11}}{t_z} \\ \eta_4 &= \tfrac{c_x R_{32} + f_x R_{12}}{t_z} \\ \eta_5 &= \tfrac{c_y R_{31} + f_y R_{21}}{t_z} \\ \eta_6 &= \tfrac{c_y R_{32} + f_y R_{22}}{t_z} \\ \eta_7 &= \tfrac{R_{31}}{t_z} \\ \eta_8 &= \tfrac{R_{32}}{t_z} \end{aligned}$$

Thus, the linear calibration problem becomes a two-step process. First, linear equations are solved to obtain parameter vectors $\{\eta\}_i$ for each view of the target. Second, the actual camera parameters such as the focal lengths and image center have to be extracted from the parameter vectors $\{\eta\}_i$.

3.2.1.1 Solution Structure Details

As shown in Eq. (3.23), there are eight unknown sets of camera parameters for each view of a calibration target. When more than four 3D points $(X_{W_i}, Y_{W_i}, 0)$ are imaged from a single view of a planar target onto the sensor plane, an over-determined set of equations based on Eq. (3.23) can be obtained and then solved as follows[2]:

$$\begin{aligned} [M]\{\eta\} &= \{b\} \\ [M^T M]\{\eta\} &= [M]^T\{b\} \\ \{\eta\} &= [M^T M]^{-1}[M]^T\{b\} \end{aligned}$$

[2] There are several classical methods that can be used to solve a set of linear, over-determined equations. These include (a) inverse of $A^T A$ shown in Eq. (3.24), (b) direct minimization using appropriate parameter metric and (c) iterative approaches.

$$[M] = \begin{bmatrix} 1 & 0 & X_{W_1} & Y_{W_1} & 0 & 0 & -x_{s1}X_{W_1} & -x_{s1}Y_{W_1} \\ 0 & 1 & 0 & 0 & X_{W_1} & Y_{W_1} & -y_{s1}X_{W_1} & -y_{s1}Y_{W_1} \\ 1 & 0 & X_{W_2} & Y_{W_2} & 0 & 0 & -x_{s2}X_{W_2} & -x_{s2}Y_{W_2} \\ 0 & 1 & 0 & 0 & X_{W_2} & Y_{W_2} & -y_{s2}X_{W_2} & -y_{s2}Y_{W_2} \\ 1 & 0 & X_{W_3} & Y_{W_3} & 0 & 0 & -x_{s3}X_{W_3} & -x_{s3}Y_{W_3} \\ 0 & 1 & 0 & 0 & X_{W_3} & Y_{W_3} & -y_{s3}X_{W_3} & -y_{s3}Y_{W_3} \\ \vdots & \vdots & \vdots & \vdots & \vdots & \vdots & \vdots & \vdots \\ 1 & 0 & X_{W_N} & Y_{W_N} & 0 & 0 & -x_{sN}X_{W_N} & -x_{sN}Y_{W_N} \\ 0 & 1 & 0 & 0 & X_{W_N} & Y_{W_N} & -y_{sN}X_{W_N} & -y_{sN}Y_{W_N} \end{bmatrix} ; \{b\} = \begin{Bmatrix} x_{s1} \\ y_{s1} \\ x_{s2} \\ y_{s2} \\ x_{s3} \\ y_{s3} \\ \vdots \\ x_{sN} \\ y_{sN} \end{Bmatrix} ; \{\eta\} = \begin{Bmatrix} \eta_1 \\ \eta_2 \\ \eta_3 \\ \eta_4 \\ \eta_5 \\ \eta_6 \\ \eta_7 \\ \eta_8 \end{Bmatrix}$$

$$[M^T M] = \begin{bmatrix} N & 0 & \sum_i X_{W_i} & \sum_i Y_{W_i} & 0 & 0 & -\sum_i (x_{si}X_{W_i}) & -\sum_i (x_{si}Y_{W_i}) \\ & N & 0 & 0 & \sum_i X_{W_i} & \sum_i Y_{W_i} & -\sum_i (y_{si}X_{W_i}) & -\sum_i (y_{si}Y_{W_i}) \\ & & \sum_i X_{W_i}^2 & \sum_i (X_{W_i}Y_{W_i}) & 0 & 0 & -\sum_i (x_{si}X_{W_i}^2) & -\sum_i (x_{si}X_{W_i}Y_{W_i}) \\ & & & \sum_i Y_{W_i}^2 & 0 & 0 & -\sum_i (x_{si}X_{W_i}Y_{W_i}) & -\sum_i (x_{si}Y_{W_i}^2) \\ & & & & \sum_i X_{W_i}^2 & \sum_i (X_{W_i}Y_{W_i}) & -\sum_i (y_{si}X_{W_i}^2) & -\sum_i (y_{si}X_{W_i}Y_{W_i}) \\ & & & & & \sum_i Y_{W_i}^2 & -\sum_i (y_{si}X_{W_i}Y_{W_i}) & -\sum_i (y_{si}Y_{W_i}^2) \\ & & \text{SYMMETRIC} & & & & \sum_i X_{W_i}^2 (x_{si}^2 + y_{si}^2) & \sum_i X_{W_i}Y_{W_i}(x_{si}^2 + y_{si}^2) \\ & & \text{MATRIX} & & & & & \sum_i Y_{W_i}^2 (x_{si}^2 + y_{si}^2) \end{bmatrix}$$

$$[M]^T \{b\} = \begin{Bmatrix} \sum_i X_{s_i} \\ \sum_i Y_{s_i} \\ \sum_i (x_{si}X_{W_i}) \\ \sum_i (x_{si}Y_{W_i}) \\ \sum_i (y_{si}X_{W_i}) \\ \sum_i (y_{si}Y_{W_i}) \\ -\sum_i X_{W_i}(x_{si}^2 + y_{si}^2) \\ -\sum_i Y_{W_i}(x_{si}^2 + y_{si}^2) \end{Bmatrix} \tag{3.24}$$

The procedure outlined above is fairly typical in the computer vision community. Parameter sets are defined so that a linear solution process can be completed. Once the set of eight grouped parameters are determined for each view of the calibration target, it is then necessary to further process the results so that the individual camera parameters can be determined. As shown in Eq. (3.24), there are four intrinsic parameters (c_x, c_y, f_x, f_y) and six extrinsic parameters (3D position (t_x, t_y, t_z) and three independent rotation angles) for each target view.

Assuming that $\{\eta\}_i$ is known for each target view, Ravn and his co-authors developed a second algorithmic approach for separating the parameters and determining their values. By combining these sets, assuming R_{ij} is an orthonormal matrix and ensuring that the right hand side vector is solely a function of intrinsic camera coordinates that are invariant with target motions, the authors obtained the following linear system:

$$\begin{bmatrix} (\eta_3\eta_8+\eta_4\eta_7) & \eta_5\eta_6 & (\eta_5\eta_8+\eta_6\eta_7) & \eta_7\eta_8 \\ 2(\eta_3\eta_7-\eta_4\eta_8) & \eta_5^2-\eta_6^2 & 2(\eta_5\eta_7-\eta_6\eta_8) & \eta_7^2-\eta_8^2 \end{bmatrix} \begin{Bmatrix} \bar{\xi}_1 \\ \bar{\xi}_2 \\ \bar{\xi}_3 \\ \bar{\xi}_4 \end{Bmatrix} = \begin{Bmatrix} -\eta_3\eta_4 \\ \eta_4^2-\eta_3^2 \end{Bmatrix} \tag{3.25}$$

where:

$$\bar{\xi}_1 = -c_x$$
$$\bar{\xi}_2 = (\tfrac{f_x}{f_y})^2$$
$$\bar{\xi}_3 = -c_y(\tfrac{f_x}{f_y})^2$$
$$\bar{\xi}_4 = c_y^2(\tfrac{f_x}{f_y})^2 + c_x^2 + f_x^2$$

Equation (3.25) shows that the values for $\{\eta\}$ are required for at least two views of the target plane to solve uniquely for the four components of $\{\bar{\xi}\}$. Once $\{\bar{\xi}\}$ is determined, it is straightforward to analytically solve for the individual camera parameters:

$$\begin{aligned} c_x &= -\bar{\xi}_1 \\ c_y &= -\tfrac{\bar{\xi}_3}{\bar{\xi}_2} \\ f_x &= \sqrt{\bar{\xi}_4 - \tfrac{\bar{\xi}_3^2}{\bar{\xi}_2} - \bar{\xi}_1^2} \\ f_y &= \sqrt{\tfrac{\bar{\xi}_4 - \frac{\bar{\xi}_3^2}{\bar{\xi}_2} - \bar{\xi}_1^2}{\bar{\xi}_2}} \end{aligned} \tag{3.26}$$

Finally, the 3D translation vector for each view is required. Values of $\{\eta\}$ for each view are used with results from Eq. (3.26), giving the following equations for the translation vector in each view:

$$\begin{Bmatrix} 0.5t_z^2 \\ t_x \\ t_y \end{Bmatrix} = \begin{Bmatrix} \left(\frac{(\eta_3-\eta_7c_x)^2+(\eta_4-\eta_8c_x)^2}{f_x^2} + \frac{(\eta_5-\eta_7c_y)^2+(\eta_6-\eta_8c_y)^2}{f_y^2} + \eta_7^2+\eta_8^2\right)^{-1} \\ (\eta_1-c_x)\frac{t_z}{f_x} \\ (\eta_2-c_y)\frac{t_z}{f_y} \end{Bmatrix} \tag{3.27}$$

The first two columns of the rigid body rotation tensor can be determined using these results by substituting directly into the definition for η in Eq. (3.24), with the final column determined from these using the cross product of these two columns to define a third orthogonal vector.

$$\begin{aligned}
R_{11} &= \frac{t_z(\eta_3-\eta_7 c_x)}{f_x} \quad ; \quad & R_{12} &= \frac{t_z(\eta_4-\eta_8 c_x)}{f_x} \\
R_{21} &= \frac{t_z(\eta_5-\eta_7 c_y)}{f_y} \quad ; \quad & R_{22} &= \frac{t_z(\eta_6-\eta_8 c_y)}{f_y} \\
R_{31} &= t_z\eta_7 \quad ; \quad & R_{32} &= t_z\eta_8 \\
R_{i3} &= \varepsilon_{ijk} R_{j1} R_{k2} & i,j,k &= 1,2,3
\end{aligned} \tag{3.28}$$

ε_{ijk} = permutation symbol, +1 (−1) if ijk is an even (odd) number of permutations of 123 and zero otherwise

Due to the fact that the solution process described above does not guarantee the orthonormality of the resulting rotation tensor $\boldsymbol{R}$, it has been shown that the best orthonormal approximation for the tensor is given by $\boldsymbol{R} = \boldsymbol{U}\boldsymbol{V}^T$, where the matrices $\boldsymbol{U}$ and $\boldsymbol{V}$ are obtained by performing singular value decomposition (SVD) of $\boldsymbol{R}$. See Appendix B for details regarding basic properties in linear algebra and the development of SVD.

3.2.2 Bundle Adjustment for Distortion and Camera Parameter Estimation

The procedure outlined in the previous section provides a direct approach for estimating camera parameters from a set of linear equations through a complex restructuring of unknowns into specific groupings. The procedure provides a theoretically sound approach for determining initial estimates for camera parameters in the absence of distortion and has been used for this purpose effectively in many practical applications. However, the linear calibration method does not permit the determination of the non-linear distortion parameters, and it requires that the calibration points on the target are known with a high degree of accuracy. These limitations can be overcome using the bundle adjustment technique developed for photogrammetry applications. The bundle adjustment method permits the calibration of non-linear camera models while simultaneously determining the shape of the calibration object up to a scale factor. This significantly reduces the requirements on the calibration target, since only the distance between two points has to be accurately known to resolve the scale factor.

Grouping the intrinsic parameters together, we define a vector as follows; $\mathbf{Int} = (f_x, f_y, c_x, c_y, f_s, \kappa_1, \kappa_2, \kappa_3)$. Grouping extrinsic parameters together for a given view, we define $\mathbf{Ext} = (n_x, n_y, n_z, t_x, t_y, t_z)$. Grouping the 3D points together for a given target, we define $\mathbf{p_T} = (p_1, p_2, \cdots, p_N) = [(X_{W_1}, Y_{W_1}, Z_{W_1}); (X_{W_2}, Y_{W_2}, Z_{W_2}); \cdots; (X_{W_N}, Y_{W_N}, Z_{W_N})$. When imaging any of the target points, say $\mathbf{p_1}$, in $\mathbf{p_T}$ then one can define a metric measure of the difference between model prediction of the projected sensor location using the pinhole camera model and the measured sensor location for an arbitrary three-dimensional point, $\mathbf{p_1}$.

In the forthcoming developments, the distorted sensor plane prediction, $\mathbf{X}$, is used interchangeable with (X_1, Y_1) and (X_s, Y_s). Furthermore, inspection of Eq. (3.9) shows that, embedded within X, are the pinhole model functions $(x_s(\boldsymbol{\beta}), y_s(\boldsymbol{\beta}))$ so that $\mathbf{X}$ is a function of both camera parameter and distortion parameters.

Since the pinhole camera functions are no longer explicitly evident, the authors opted to modify notation for the remainder of the book and define the quantities $\mathbf{x}, (x_1, x_2), (x_{s1}, y_{s1})$ and (x_s, y_s) as the "measured sensor position", replacing $\boldsymbol{x}_{\boldsymbol{s}}^{\boldsymbol{d}}$ in Eq. (3.21), with appropriate superscripts to denote the appropriate point $\mathbf{p_1}$ and camera view. This change improves consistency with previous literature with minimal confusion, since these quantities are not mathematical functions of camera parameters.

$$E(\mathbf{Int}, \mathbf{Ext}, \mathbf{p_1}) = \left\| [\mathrm{x_s} - \mathrm{X_s}(\mathbf{Int}, \mathbf{Ext}, \mathbf{p_1})]^2 + [\mathrm{y_s} - \mathrm{Y_s}(\mathbf{Int}, \mathbf{Ext}, \mathbf{p_1})]^2 \right\| \qquad (3.29)$$

where:

$\mathbf{X_s}(\mathbf{Int}, \mathbf{Ext}, \mathbf{p_1})$ = model for prediction of the measured x_s location on the sensor plane for point $\mathbf{p_1}$, as defined in Eq. (3.9)

$\mathbf{Y_s}(\mathbf{Int}, \mathbf{Ext}, \mathbf{p_1})$ = model for prediction of the measured y_s location on the sensor plane for point $\mathbf{p_1}$, as defined in Eq. (3.9)

$\mathbf{p_1}$ = 3D point with coordinates in the world and camera systems.

If there are $I = 1, 2, 3, \ldots, N$ points and there are $J = 1, 2, 3, \ldots, M$ views of an object, then the metric measure can be written in the following form [277]:

$$E(\mathbf{Int}, \mathbf{Ext}, \mathbf{p}) = \sum_{\mathrm{J}=1}^{\mathrm{M}} \sum_{\mathrm{I}=1}^{\mathrm{N}} \left\| [\mathrm{x_s^{IJ}} - \mathrm{X_s}(\mathbf{Int}, \mathbf{Ext}^{\mathrm{J}}, \mathbf{p_T}^{\mathrm{I}})]^2 + [\mathrm{y_s^{IJ}} - \mathrm{Y_s}(\mathbf{Int}, \mathbf{Ext}^{\mathrm{J}}, \mathbf{p_T}^{\mathrm{I}})]^2 \right\|^2 \qquad (3.30)$$

where:

$\mathbf{Ext}^{\mathrm{J}}$ = extrinsic camera parameters that change with each view, $J = 1, 2, \ldots, M$

$\mathbf{Int}$ = intrinsic camera parameters that do not change throughout calibration process

$\mathbf{p_T}^I$ = 3D world coordinates $(X_{W_I}, Y_{W_I}, Z_{W_I})$ for M views of N target points, $I = 1, 2, \ldots, N$

(x_s, y_s) = location of image point in the sensor plane for each 3D point and each view

In this form, the metric measure is defined for all image points. Note that, if all 3D points are known precisely, then the terms $\mathbf{p_T}^I$ are known and not variable.

3.2.2.1 Example Calibration Procedures

To obtain optimal estimates for the intrinsic and extrinsic camera parameters, we minimize Eq. (3.30) using an efficient, accurate iterative approach such as the Levenberg-Marquardt algorithm described in detail within Appendix D.

Table 3.1 shows results for three examples. In all examples, we assume a combined distortion model of the form:

$$\mathbf{D}(x_s, y_s) = \left(\kappa_1 \rho + \kappa_2 \rho^2 + \kappa_3 \rho^3\right) \mathbf{e_r} \tag{3.31}$$

where ρ is defined in Eq. (3.11). For this case, $\mathbf{Int} = (f_x, f_y, c_x, c_y, f_s, \kappa_1, \kappa_2, \kappa_3)$. Assuming that there are N target grid points to be re-estimated during the optimization process and M views, for this application there are $2MN$ equations relating predicted and measured 2D sensor positions to optimally estimate the following parameters:

- Eight intrinsic parameters
- $6M$ extrinsic parameters positioning the camera relative to the grid
- $3N$ coordinates of the target grid points.

Previous work has shown that seven constraints are required on the target points (e.g., the 3D coordinates of two points and the Z-coordinate of a third, non-collinear point to define rotation about the line defined by the two points are adequate) to ensure that a unique solution is obtained [83]. Thus, there are $2MN$ equations to solve for the $[3N + 6M + 1]$ unknowns in this case.

Solving such a large set of over-determined equations for an equally large number of unknowns can be exceedingly time intensive unless algorithms are developed and implemented in a manner that takes advantage of the unique structure of the matrices in the system of equations that result from the optimization process.

Table 3.1 Bundle adjustment examples with unknowns, equations and constraints

Parameters and equations	Example 1	Example 2	Example 3
Number of 3D points (N)	$8 \times 8 = 64$	$10 \times 10 = 100$	$6 \times 6 = 36$
Number of views (M)	10	7	5
Intrinsic parameters (8)	8	8	8
Extrinsic parameters ($6M$)	$10 \times 6 = 60$	$7 \times 6 = 42$	$5 \times 6 = 30$
3D components re-estimated ($3N$)	$64 \times 3 = 192$	$100 \times 3 = 300$	$36 \times 3 = 108$
Number of constraints	7	7	7
Total number of unknowns $((3N + 6M + 8) - 7)$	$260 - 7 = 253$	$350 - 7 = 343$	$146 - 7 = 139$
Equations ($2NM$)	$2 \times 64 \times 10 = 1{,}280$	$2 \times 100 \times 7 = 1{,}400$	$2 \times 36 \times 5 = 360$

3.2.3 Equations and Procedures for Optimization Structure of Matrices

As shown in Appendix D, the solution process requires (a) the gradient, $\boldsymbol{\nabla E}_i = \partial E/\partial\beta_i$ and (b) an approximation for the Hessian tensor, $(\boldsymbol{\nabla\nabla E})$. Next, the parameters are ordered so that

$$
\begin{aligned}
&(\beta_1 \rightarrow \beta_8) = (\mathrm{Int}_1 \rightarrow \mathrm{Int}_8)\\
&(\beta_9 \rightarrow \beta_{8+6M}) = (\mathrm{Ext}_1 \rightarrow \mathrm{Ext}_{6M})\\
&(\beta_{9+6M} \rightarrow \beta_{8+6M+3N}) = ((p_{x_1}, p_{y_1}, p_{z_1})\cdots(p_{x_{3N}}, p_{y_{3N}}, p_{z_{3N}}))
\end{aligned}
\tag{3.32}
$$

Using this ordering of the various parameters to be determined during the optimization process, we use the terminology defined in Appendix D and define a sensor-plane based, least squares error function in a form that is similar to Eq. (3.30). It is noted that the form given in Eqs. (3.34) and (3.35) is for a single camera view. As outlined in the remainder of the chapter, summation over all M views is performed using derivatives in Eqs. (3.36)–(3.39) to complete the Hessian matrix and perform the camera calibration process.

$$
E = \sum_{n=1}^{N} \mathbf{F}_n \cdot \mathbf{F}_n = \sum_{n=1}^{N} ((x_{sn} - X_{sn}(\boldsymbol{\beta}))^2 + (y_{sn} - Y_{sn}(\boldsymbol{\beta}))^2) \tag{3.33}
$$

$$
\begin{aligned}
\mathbf{F}_n &= (F_{n1}, F_{n2}) = ((x_{sn} - X_s(\beta_i, \mathbf{p_1})), (y_{sn} - Y_s(\beta_i, \mathbf{p_1})))\\
\boldsymbol{\beta} &= \text{vector of model parameters} = (\mathbf{Int}, \mathbf{Ext}, \mathbf{p_1})\\
&= (c_x, c_y, f_x, f_y, f_s, \kappa_1, \kappa_2, \kappa_3;\ n_x, n_y, n_z, t_x, t_y, t_z;\ X_{w1}, Y_{w1}, Z_{w1})
\end{aligned}
$$

Using Eqs. (3.32) and (3.33) with the developments in Appendix D, the gradient and Hessian for the vector metric function, $\boldsymbol{F}$, can be obtained. These expressions are used to define increment equations for updating all parameters. The form is as follows:

$$
\begin{aligned}
&\nabla_{\beta}\mathbf{E} = 2\mathbf{F^T} \cdot \frac{\partial \mathbf{F}}{\partial \boldsymbol{\beta}} = -2\mathbf{F^T} \cdot \frac{\partial \mathbf{X}}{\partial \boldsymbol{\beta}} = -2(x_{sn} - X_{sn}(\boldsymbol{\beta}))\frac{\partial X_{sn}}{\partial \boldsymbol{\beta}} - 2(y_{sn} - Y_{sn}(\boldsymbol{\beta}))\frac{\partial Y_{sn}}{\partial \boldsymbol{\beta}}\\
&(\boldsymbol{\nabla}_{\boldsymbol{\beta}}\boldsymbol{E})_j = -2(x_{sn} - X_{sn}(\boldsymbol{\beta}))\frac{\partial X_{sn}}{\partial \beta_j} - 2(y_{sn} - Y_{sn}(\boldsymbol{\beta}))\frac{\partial Y_{sn}}{\partial \beta_j}\\
&\boldsymbol{\nabla\nabla E} \cong 2\boldsymbol{\nabla F^T} \cdot \boldsymbol{\nabla F} = -2\frac{\partial \mathbf{X_n}}{\partial \boldsymbol{\beta}}\frac{\partial \mathbf{X_n}}{\partial \boldsymbol{\beta}} - 2\frac{\partial \mathbf{Y_n}}{\partial \boldsymbol{\beta}}\frac{\partial \mathbf{Y_n}}{\partial \boldsymbol{\beta}}
\end{aligned}
$$

After the kth iteration, the improved estimate can be written

$$
\boldsymbol{\beta}_{k+1} = \boldsymbol{\beta}_k + \boldsymbol{\Delta\beta} = \boldsymbol{\beta}_k - \left([\boldsymbol{\nabla F}(\boldsymbol{\beta}_k)^T \bullet \boldsymbol{\nabla F}(\boldsymbol{\beta}_k)]^{-1}\right) \bullet \left[\boldsymbol{\nabla F}(\boldsymbol{\beta}_k)^T \bullet \boldsymbol{F}(\boldsymbol{\beta}_k)\right] \tag{3.34}
$$

In index notation, the equations have the following form:

$$
\begin{aligned}
&\boldsymbol{\nabla F}(\boldsymbol{\beta})_{ij} = \frac{\partial(\mathbf{F})_i}{\partial(\boldsymbol{\beta})_j}(\boldsymbol{\beta}) = \frac{\partial F_i}{\partial \beta_j}(\boldsymbol{\beta}) = \frac{\partial[(\mathbf{x})_i - (\boldsymbol{X})_i]}{\partial \beta_j}(\boldsymbol{\beta}) = -\frac{\partial X_i}{\partial \beta_j}(\boldsymbol{\beta})\\
&\left[\boldsymbol{\nabla F}(\boldsymbol{\beta})^T \cdot \mathbf{F}\right]_i = -\sum_{j=1}^{2} \frac{\partial X_j}{\partial \beta_i}(\boldsymbol{\beta})(x_j - X_j(\boldsymbol{\beta}))\\
&\left[\boldsymbol{\nabla F}(\boldsymbol{\beta})^T \cdot \boldsymbol{\nabla F}(\boldsymbol{\beta})\right]_{ij} = \sum_{k=1}^{2} \frac{\partial X_k}{\partial \beta_i}(\boldsymbol{\beta})\frac{\partial X_k}{\partial \beta_j}(\boldsymbol{\beta})
\end{aligned}
\tag{3.35}
$$

As shown in Eq. (3.2.3), gradients in the 2D vector function **F** are required. Gradients for the key vector function, **X** (corresponds to the 2D distorted position in the sensor plane for our studies) are as follows:

$$
\begin{aligned}
&\frac{\partial X_1}{\partial c_x} = 1 + \frac{\partial D_x}{\partial c_x} && \frac{\partial X_2}{\partial c_x} = \frac{\partial D_y}{\partial c_x} \\
&\frac{\partial X_1}{\partial c_y} = \frac{\partial D_x}{\partial c_y} && \frac{\partial X_2}{\partial c_y} = 1 + \frac{\partial D_y}{\partial c_y} \\
&\frac{\partial X_1}{\partial f_x} = \frac{R_{11}X_W + R_{12}Y_W + R_{13}Z_W + t_x}{R_{31}X_W + R_{32}Y_W + R_{33}Z_W + t_z} + \frac{\partial D_x}{\partial f_x} && \frac{\partial X_2}{\partial f_x} = \frac{\partial D_y}{\partial f_x} \\
&\frac{\partial X_1}{\partial f_y} = \frac{\partial D_x}{\partial f_y} && \frac{\partial X_2}{\partial f_y} = \frac{R_{21}X_W + R_{22}Y_W + R_{23}Z_W + t_y}{R_{31}X_W + R_{32}Y_W + R_{33}Z_W + t_z} \\
& && \qquad + \frac{\partial D_y}{\partial f_y} \\
&\frac{\partial X_1}{\partial f_s} = \frac{R_{21}X_W + R_{22}Y_W + R_{23}Z_W + t_y}{R_{31}X_W + R_{32}Y_W + R_{33}Z_W + t_z} + \frac{\partial D_x}{\partial f_s} && \frac{\partial X_2}{\partial f_s} = \frac{\partial D_y}{\partial f_s} \\
&\frac{\partial X_1}{\partial \kappa_1} = \frac{\partial D_x}{\partial \kappa_1} && \frac{\partial X_2}{\partial \kappa_1} = \frac{\partial D_y}{\partial \kappa_1} \\
&\frac{\partial X_1}{\partial \kappa_2} = \frac{\partial D_x}{\partial \kappa_2} && \frac{\partial X_2}{\partial \kappa_2} = \frac{\partial D_y}{\partial \kappa_2} \\
&\frac{\partial X_1}{\partial \kappa_3} = \frac{\partial D_x}{\partial \kappa_3} && \frac{\partial X_2}{\partial \kappa_3} = \frac{\partial D_y}{\partial \kappa_3}
\end{aligned}
\tag{3.36}
$$

Derivatives of the model function with respect to the extrinsic parameters are also required with respect to (a) three components of translation vector and (b) three components of rotation tensor.

$$
\begin{aligned}
\frac{\partial X_1}{\partial t_x} &= \frac{f_x}{R_{31}X_W + R_{32}Y_W + R_{33}Z_W + t_z} + \frac{\partial D_x}{\partial t_x} \\
\frac{\partial X_2}{\partial t_x} &= \frac{\partial D_y}{\partial t_x} \\
\frac{\partial X_1}{\partial t_y} &= \frac{\partial D_x}{\partial t_y} + \frac{f_s}{R_{31}X_W + R_{32}Y_W + R_{33}Z_W + t_z} \\
\frac{\partial X_2}{\partial t_y} &= \frac{f_y}{R_{31}X_W + R_{32}Y_W + R_{33}Z_W + t_z} + \frac{\partial D_y}{\partial t_y} \\
\frac{\partial X_1}{\partial t_z} &= -f_x \frac{R_{11}X_W + R_{12}Y_W + R_{13}Z_W + t_x}{(R_{31}X_W + R_{32}Y_W + R_{33}Z_W + t_z)^2} - f_s \frac{R_{21}X_W + R_{22}Y_W + R_{23}Z_W + t_y}{(R_{31}X_W + R_{32}Y_W + R_{33}Z_W + t_z)^2} \\
&\quad + \frac{\partial D_x}{\partial t_z} \\
\frac{\partial X_2}{\partial t_z} &= -f_y \frac{R_{21}X_W + R_{22}Y_W + R_{23}Z_W + t_y}{(R_{31}X_W + R_{32}Y_W + R_{33}Z_W + t_z)^2} + \frac{\partial D_y}{\partial t_z}
\end{aligned}
\tag{3.37}
$$

Using the minimal parameter definition given in Appendix G for the rotation tensor, where $\mathbf{n} = (n_1, n_2, n_3) = \Theta(\bar{n}_x, \bar{n}_y, \bar{n}_z)$, the derivatives of each component in the rotation tensor are with respect to n_i and can be written in the following form:

$$\frac{\partial \mathbf{R}}{\partial \mathbf{n}} = \frac{\partial \mathbf{R}}{\partial \tilde{\mathbf{n}}}\frac{\partial \tilde{\mathbf{n}}}{\partial \mathbf{n}} + \frac{\partial \mathbf{R}}{\partial \Theta}\frac{\partial \Theta}{\partial \mathbf{n}} = \sum_{k=1}^{3} \frac{\partial R_{ij}}{\partial \tilde{n}_k}\frac{\partial \tilde{n}_k}{\partial n_L} + \frac{\partial R_{ij}}{\partial \Theta}\frac{\partial \Theta}{\partial n_L}$$

$$\mathbf{n} = (n_1, n_2, n_3) = (\Theta \tilde{n}_x, \Theta \tilde{n}_y, \Theta \tilde{n}_z)$$

$$\Theta = \left(n_1^2 + n_2^2 + n_3^2\right)^{\frac{1}{2}}$$

$$\frac{\partial R_{ij}}{\partial \tilde{n}_x} = (1 - \cos\Theta)\begin{pmatrix} 2\tilde{n}_x & 0 & 0 \\ \tilde{n}_y & \tilde{n}_x & -\frac{\sin\Theta}{1-\cos\Theta} \\ \tilde{n}_z & \frac{\sin\Theta}{1-\cos\Theta} & \tilde{n}_x \end{pmatrix} \qquad \frac{\partial \Theta}{\partial n} = \left\{\tilde{n}_x, \tilde{n}_y, \tilde{n}_z\right\}$$

$$\frac{\partial R_{ij}}{\partial \tilde{n}_y} = (1 - \cos\Theta)\begin{pmatrix} \tilde{n}_y & \tilde{n}_x & \frac{\sin\Theta}{1-\cos\Theta} \\ 0 & 2\tilde{n}_y & 0 \\ \frac{\sin\Theta}{1-\cos\Theta} & \tilde{n}_z & \tilde{n}_y \end{pmatrix}$$

$$\frac{\partial R_{ij}}{\partial \tilde{n}_z} = (1 - \cos\Theta)\begin{pmatrix} \tilde{n}_y & \tilde{n}_x & \frac{\sin\Theta}{1-\cos\Theta} \\ 0 & 2\tilde{n}_y & 0 \\ \frac{\sin\Theta}{1-\cos\Theta} & \tilde{n}_z & \tilde{n}_y \end{pmatrix}$$

$$\frac{\partial \tilde{\mathbf{n}}}{\partial \mathbf{n}} = \frac{\partial \tilde{n}_i}{\partial n_j} = \begin{pmatrix} (1-\tilde{n}_x^2) & -\tilde{n}_x\tilde{n}_y & -\tilde{n}_x\tilde{n}_z \\ & (1-\tilde{n}_y^2) & -\tilde{n}_y\tilde{n}_z \\ \text{SYMMETRIC} & & (1-\tilde{n}_z^2) \end{pmatrix}$$

$$\frac{\partial R_{ij}}{\partial \Theta} = \begin{pmatrix} \sin\Theta(\tilde{n}_x^2 - 1) & \tilde{n}_x\tilde{n}_y\sin\Theta - \tilde{n}_z\cos\Theta & \tilde{n}_x\tilde{n}_z\sin\Theta + \tilde{n}_y\cos\Theta \\ \tilde{n}_x\tilde{n}_y\sin\Theta + \tilde{n}_z\cos\Theta & \sin\Theta(\tilde{n}_y^2 - 1) & \tilde{n}_y\tilde{n}_z\sin\Theta - \tilde{n}_x\cos\Theta \\ \tilde{n}_x\tilde{n}_z\sin\Theta - \tilde{n}_y\cos\Theta & \tilde{n}_y\tilde{n}_z\sin\Theta + \tilde{n}_x\cos\Theta & \sin\Theta(\tilde{n}_z^2 - 1) \end{pmatrix} \tag{3.38}$$

To account for small variations in the three-dimensional positions of points on the target plane, derivatives are required with respect to (X_W, Y_W, Z_W) for each object point. Using the pinhole model position in Eq. (3.9), the derivatives are given in Eq. (3.39).

$$
\begin{aligned}
\frac{\partial X_1}{\partial X_W} &= f_x \left[\frac{R_{11}}{R_{31}X_W + R_{32}Y_W + R_{33}Z_W + t_z} \right] \\
&\quad - f_x R_{31} \left[\frac{R_{11}X_W + R_{12}Y_W + R_{13}Z_W + t_x}{(R_{31}X_W + R_{32}Y_W + R_{33}Z_W + t_z)^2} \right] \\
&\quad - f_s \left[\frac{R_{21}}{R_{31}X_W + R_{32}Y_W + R_{33}Z_W + t_z} \right] \\
&\quad - f_s R_{31} \left[\frac{R_{21}X_W + R_{22}Y_W + R_{23}Z_W + t_y}{(R_{31}X_W + R_{32}Y_W + R_{33}Z_W + t_z)^2} \right] + \frac{\partial D_x}{\partial X_W} \\
\frac{\partial X_2}{\partial X_W} &= f_y \left[\frac{R_{21}}{R_{31}X_W + R_{32}Y_W + R_{33}Z_W + t_z} \right] \\
&\quad - f_y R_{31} \left[\frac{R_{21}X_W + R_{22}Y_W + R_{23}Z_W + t_y}{(R_{31}X_W + R_{32}Y_W + R_{33}Z_W + t_z)^2} \right] + \frac{\partial D_y}{\partial X_W} \\
\frac{\partial X_1}{\partial Y_W} &= f_x \left[\frac{R_{12}}{R_{31}X_W + R_{32}Y_W + R_{33}Z_W + t_z} \right] \\
&\quad - f_x R_{32} \left[\frac{R_{11}X_W + R_{12}Y_W + R_{13}Z_W + t_x}{(R_{31}X_W + R_{32}Y_W + R_{33}Z_W + t_z)^2} \right] \\
&\quad + f_s \left[\frac{R_{22}}{R_{31}X_W + R_{32}Y_W + R_{33}Z_W + t_z} \right] \\
&\quad - f_s R_{32} \left[\frac{R_{21}X_W + R_{22}Y_W + R_{23}Z_W + t_y}{(R_{31}X_W + R_{32}Y_W + R_{33}Z_W + t_z)^2} \right] + \frac{\partial D_x}{\partial Y_W} \qquad (3.39) \\
\frac{\partial X_2}{\partial Y_W} &= f_y \left[\frac{R_{22}}{R_{31}X_W + R_{32}Y_W + R_{33}Z_W + t_z} \right] \\
&\quad - f_y R_{32} \left[\frac{R_{21}X_W + R_{22}Y_W + R_{23}Z_W + t_y}{(R_{31}X_W + R_{32}Y_W + R_{33}Z_W + t_z)^2} \right] + \frac{\partial D_y}{\partial Y_W} \\
\frac{\partial X_1}{\partial Z_W} &= f_x \left[\frac{R_{13}}{R_{31}X_W + R_{32}Y_W + R_{33}Z_W + t_z} \right] \\
&\quad - f_x R_{33} \left[\frac{R_{11}X_W + R_{12}Y_W + R_{13}Z_W + t_x}{(R_{31}X_W + R_{32}Y_W + R_{33}Z_W + t_z)^2} \right]
\end{aligned}
$$

$$+ f_s \left[\frac{R_{23}}{R_{31}X_W + R_{32}Y_W + R_{33}Z_W + t_z} \right]$$

$$- f_s R_{33} \left[\frac{R_{21}X_W + R_{22}Y_W + R_{23}Z_W + t_y}{(R_{31}X_W + R_{32}Y_W + R_{33}Z_W + t_z)^2} \right] + \frac{\partial D_x}{\partial Z_W}$$

$$\frac{\partial X_2}{\partial Z_W} = f_y \left[\frac{R_{23}}{R_{31}X_W + R_{32}Y_W + R_{33}Z_W + t_z} \right]$$

$$- f_y R_{33} \left[\frac{R_{21}X_W + R_{22}Y_W + R_{23}Z_W + t_y}{(R_{31}X_W + R_{32}Y_W + R_{33}Z_W + t_z)^2} \right] + \frac{\partial D_y}{\partial Z_W}$$

The derivatives given in Eqs. (3.36–3.39) are used to define the terms in the Jacobian matrix, $[\mathbf{J}]$, shown in Fig. 3.14. The approximate Hessian matrix is defined to be $[\mathbf{J}]^{\mathbf{T}}[\mathbf{J}]$ in Eq. (3.2.3). The structure of the resulting Hessian, $[\mathbf{H}]$, is shown in Fig. 3.15. The block structure of $[\mathbf{H}]$ is shown more clearly in Fig. 3.16.

3.2.4 Solution Details

The normal equations for updating the solution set can be written in the form given in Eq. (3.40), where x_i^j is defined as the ith ($i = 1,2$) component of the measured sensor position for the jth calibration grid point ($j = 1$ to N).

$$[\mathbf{H}]\{\xi\} = \{\mathbf{b}\} \quad ; \quad \{\mathbf{b}\} = [\mathbf{J}]^{\mathbf{T}}\{\mathbf{F}\}$$

$$\{\mathbf{F}\} = \left\{ \begin{array}{l} \left. \begin{array}{l} \left\{ \begin{array}{l} x_1^1 - X_1^1 \\ x_2^1 - X_2^1 \end{array} \right\} \\ \left\{ \begin{array}{l} x_1^2 - X_1^2 \\ x_2^2 - X_2^2 \end{array} \right\} \\ \vdots \\ \left\{ \begin{array}{l} x_1^N - X_1^N \\ x_2^N - X_2^N \end{array} \right\} \end{array} \right\} \text{first view} \\ \vdots \\ \vdots \\ \left. \begin{array}{l} \left\{ \begin{array}{l} x_1^1 - X_1^1 \\ x_2^1 - X_2^1 \end{array} \right\} \\ \left\{ \begin{array}{l} x_1^2 - X_1^2 \\ x_2^2 - X_2^2 \end{array} \right\} \\ \vdots \\ \left\{ \begin{array}{l} x_1^N - X_1^N \\ x_2^N - X_2^N \end{array} \right\} \end{array} \right\} M^{\text{th}} \text{ view} \end{array} \right\}_{2NM \times 1}$$

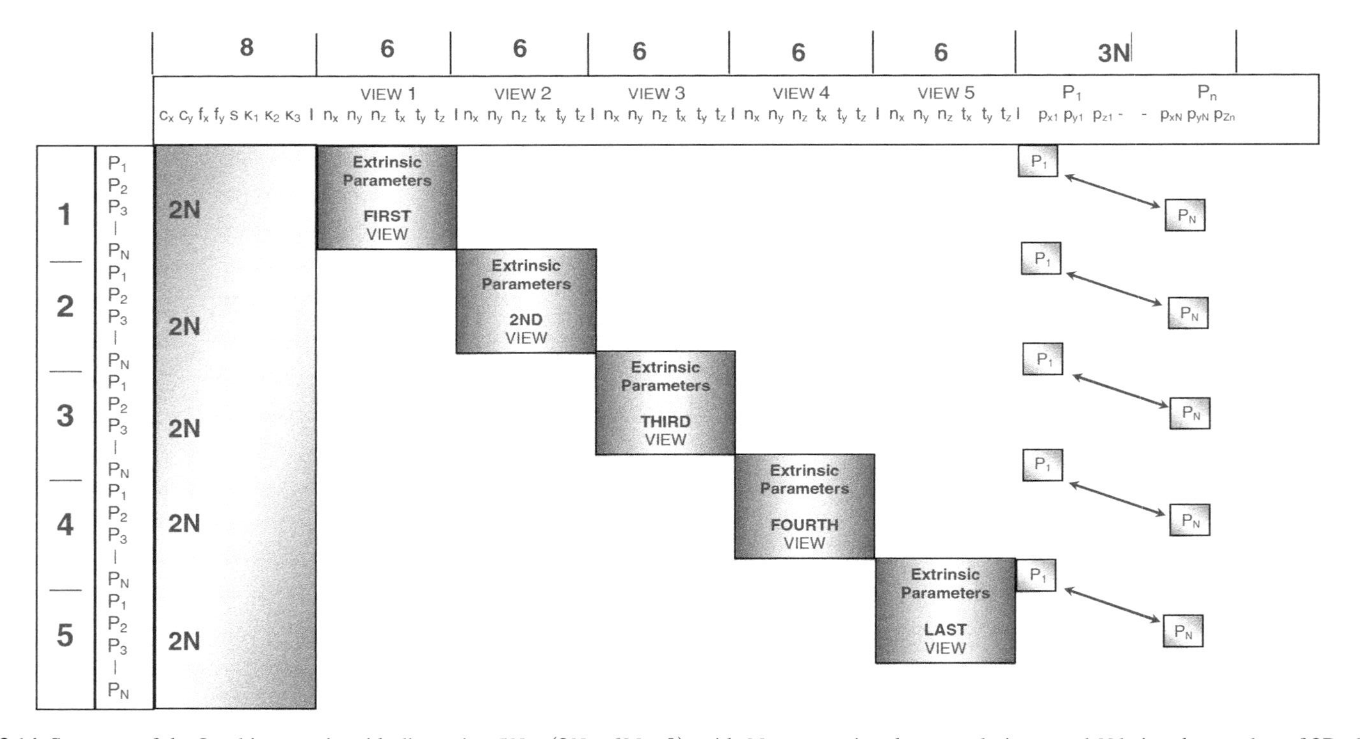

Fig. 3.14 Structure of the Jacobian matrix with dimension $5N \times (3N + 6M + 8)$, with M representing the several views, and N being the number of 3D object points. Here, $M = 5$. The target views are needed to generate equations and estimate 8 intrinsic parameters, 30 extrinsic parameters and the locations of N 3D positions

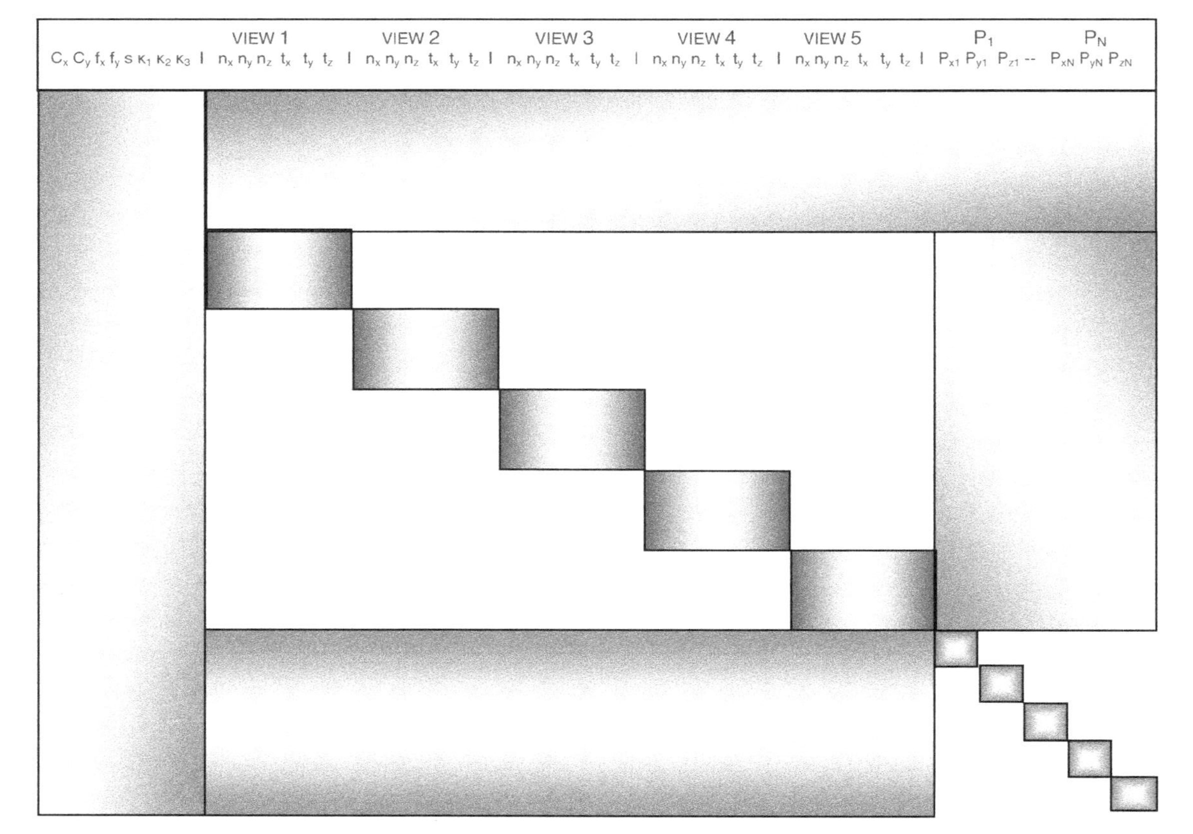

Fig. 3.15 Structure of the symmetric approximate Hessian matrix $[\mathbf{J^T}][\mathbf{J}]$ with dimension $3N + 6M + 8$. In this example, $M = 5$, with Hessian dimension of 38

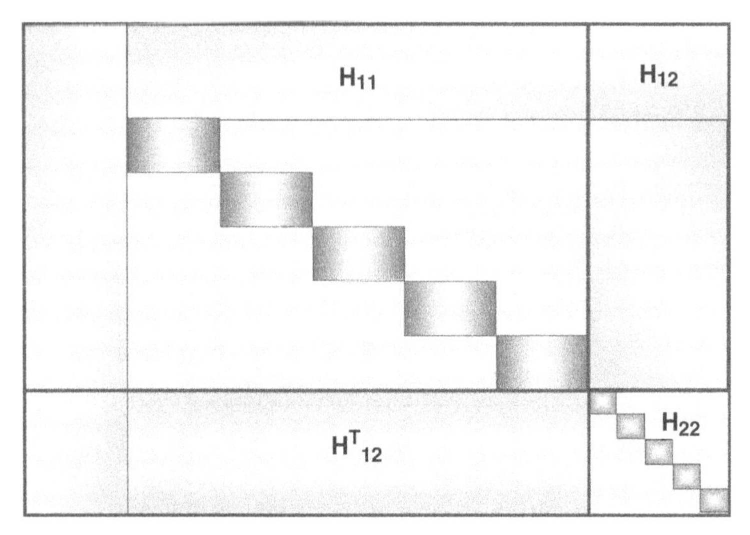

Fig. 3.16 Block submatrices within approximate Hessian matrix. $[\mathbf{H_{11}}]$ is a $(6M+8)$ square matrix. $[\mathbf{H_{22}}]$ is a $3N$ square matrix. $[\mathbf{H_{12}}]$ is a $6M+8 \times 3N$ rectangular matrix

$$[\mathbf{J}]_{ij} = \boldsymbol{\nabla} \boldsymbol{F}(\boldsymbol{\beta})_{ij} = \left[\frac{\partial X_i}{\partial \beta_j}(\boldsymbol{\beta})\right]_{(3N+6M+8)\times(2NM)}$$

$$\{\xi\} = \left\{ \begin{array}{l} \left.\begin{array}{l} c_x \\ c_y \\ f_x \\ f_y \\ f_s \\ \kappa_1 \\ \kappa_2 \\ \kappa_3 \end{array}\right\} \text{intrinsic} \\ \left.\begin{array}{l} t_x \\ t_y \\ t_z \\ n_x \\ n_y \\ n_z \end{array}\right\} \text{extrinsic, first view} \\ \vdots \\ \left.\begin{array}{l} t_x \\ t_y \\ t_z \\ n_x \\ n_y \\ n_z \end{array}\right\} \text{extrinsic, } M^{\text{th}} \text{ view} \\ \left.\begin{array}{l} X_W \\ Y_W \\ Z_W \end{array}\right\} \text{first point} \\ \vdots \\ \left.\begin{array}{l} X_W \\ Y_W \\ Z_W \end{array}\right\} N \text{ point} \end{array} \right\}_{(3N+6M+8)\times 1} \tag{3.40}$$

In the normal equations, **F** is the difference between the measured and predicted sensor positions for all views and is typically designated the residual or residue vector. A direct approach for solving the normal equations requires inversion of $[\mathbf{H}]$ each time a new update of the solution vector is required. For the cases shown in Table 3.1, the matrices are less than 350×350 and can be inverted relatively quickly. However, suppose a denser target is used with 2,500 target points (50×50 grid) and 50 views. With six extrinsic parameters per view and eight intrinsic parameters, the matrix is 7808. In such a case, the entire inversion and iteration process can be extremely time-consuming taking several hours to complete using a direct solution method. A more efficient scheme that reduces the solution time to $O(10^2)$ s can be developed using the block structure and the extremely sparse nature of the largest block in $[\mathbf{H}]$ that is primarily related to the re-estimation of the target points.

Consider the structure of the Hessian matrix shown in Figs. 3.14–3.16. We can write the following:

$$\begin{bmatrix} [\mathbf{H}_{11}]_{(6M+8)\times(6M+8)} & [\mathbf{H}_{12}]_{(6M+8)\times(3N)} \\ [\mathbf{H}_{12}]^{T}_{(3N)\times(6M+8)} & [\mathbf{H}_{22}]_{(3N)\times(3N)} \end{bmatrix}$$

$$\{\xi\} = \left\{ \begin{array}{l} \left. \begin{array}{c} c_x \\ c_y \\ \vdots \\ t_x \\ t_y \\ t_z \\ n_x \\ n_y \\ n_z \end{array} \right|_{(6M+8)\times 1} \quad \xi_1 \text{ Intrinsic and extrinsic} \\ \left. \begin{array}{c} X_{W_1} \\ Y_{W_1} \\ Z_{W_1} \\ \vdots \\ X_{W_N} \\ Y_{W_N} \\ Z_{W_N} \end{array} \right|_{3N\times 1} \quad \xi_2 \text{ 3D positions re-estimated} \end{array} \right\} \tag{3.41}$$

In this case, $[\mathbf{H}_{11}]$ and $[\mathbf{H}_{22}]$ are square matrices, with $[\mathbf{H}_{11}]$ being partially banded (section involving the extrinsic factors for all views) and $[\mathbf{H}_{22}]$ a sparse, banded matrix. The matrix $[\mathbf{H}_{22}]$ can be readily and efficiently inverted since it is a banded matrix with bandwidth of 3. Given this situation, pre-multiplication of Eq. (3.41) by a specific block matrix will reduce the solution process to a relatively straightforward procedure.

$$\begin{bmatrix} [\mathbf{I}] & -[\mathbf{H}_{12}][\mathbf{H}_{22}]^{-1} \\ [\mathbf{0}] & [\mathbf{I}] \end{bmatrix} \begin{bmatrix} [\mathbf{H}_{11}] & [\mathbf{H}_{12}] \\ [\mathbf{H}_{12}]^T & [\mathbf{H}_{22}] \end{bmatrix} \begin{Bmatrix} \xi_1 \\ \xi_2 \end{Bmatrix} = \begin{bmatrix} [\mathbf{I}] & -[\mathbf{H}_{12}][\mathbf{H}_{22}]^{-1} \\ [\mathbf{0}] & [\mathbf{I}] \end{bmatrix} \begin{Bmatrix} \mathbf{b}_1 \\ \mathbf{b}_2 \end{Bmatrix}$$

$$\Rightarrow \begin{bmatrix} [\mathbf{H}_{11}] - [\mathbf{H}_{12}][\mathbf{H}_{22}]^{-1}[\mathbf{H}_{12}]^T & [\mathbf{0}] \\ [\mathbf{H}_{12}]^T & [\mathbf{H}_{22}] \end{bmatrix} \begin{Bmatrix} \xi_1 \\ \xi_2 \end{Bmatrix} = \begin{Bmatrix} \{\mathbf{b}_1\} - [\mathbf{H}_{12}][\mathbf{H}_{22}]^{-1}\{\mathbf{b}_2\} \\ b_2 \end{Bmatrix}$$

$$[\overline{\mathbf{H}}] = [\mathbf{H}_{11}] - [\mathbf{H}_{12}][\mathbf{H}_{22}]^{-1}[\mathbf{H}_{12}]^T$$

$$\Rightarrow [\overline{\mathbf{H}}]\{\xi_1\} = \left\{\{\mathbf{b}_1\} - \left\{[\mathbf{H}_{12}][\mathbf{H}_{22}]^{-1}\{\mathbf{b}_2\}\right\}\right\}$$

$$\Rightarrow \{\xi_1\} = [\overline{\mathbf{H}}]^{-1}\left\{\{\mathbf{b}_1\} - \left\{[\mathbf{H}_{12}][\mathbf{H}_{22}]^{-1}\{\mathbf{b}_2\}\right\}\right\}$$

and

$$[\mathbf{H}_{12}]^T\{\xi_1\} + [\mathbf{H}_{22}]\{\xi_2\} = \{\mathbf{b}_2\}$$

$$\Rightarrow \{\xi_2\} = [\mathbf{H}_{22}]^{-1}\left\{\{\mathbf{b}_2\} - [\mathbf{H}_{12}]^T\{\xi_1\}\right\} \tag{3.42}$$

The matrix $[\overline{\mathbf{H}}]$ can be shown to be a square, symmetric matrix of size $6M+8$. Typically, this matrix is far smaller than $[\mathbf{H}]$ and can be readily inverted using algorithms designed for inversion of symmetric, positive definite matrices (e.g. Cholesky decomposition).

The values $\{\xi_1\}$ and $\{\xi_2\}$ computed in Eq. (3.42) are the first set of parameter increments. The values can be used to perform the first update and define the first set of revised parameter values. The updating process for the bundle adjustment problem defined in Eqs. (3.30–3.42) is repeated until the parameter values converge and further updating is not required.

3.2.5 Discussion

The camera calibration procedure described in the previous sections can be modified in a variety of ways to accomodate changes in the camera model, distortion correction procedures or measurement objective function.

For example, Fig. 3.17 shows both a typical grid target and a deformed image of the grid target.

If a priori distortion removal is performed as described in Section 3.1.2.3 to correct all measurement positions in the images, then the corrected positions would be as shown in Fig. 3.18.

In this case, the grid spacing in the corrected image is a close approximation to the actual spacing, with only a shift in the image to be determined. Thus, when using images corrected for distortions, then the distortion parameters, e.g. $(\kappa_1, \kappa_2, \kappa_3)$ in

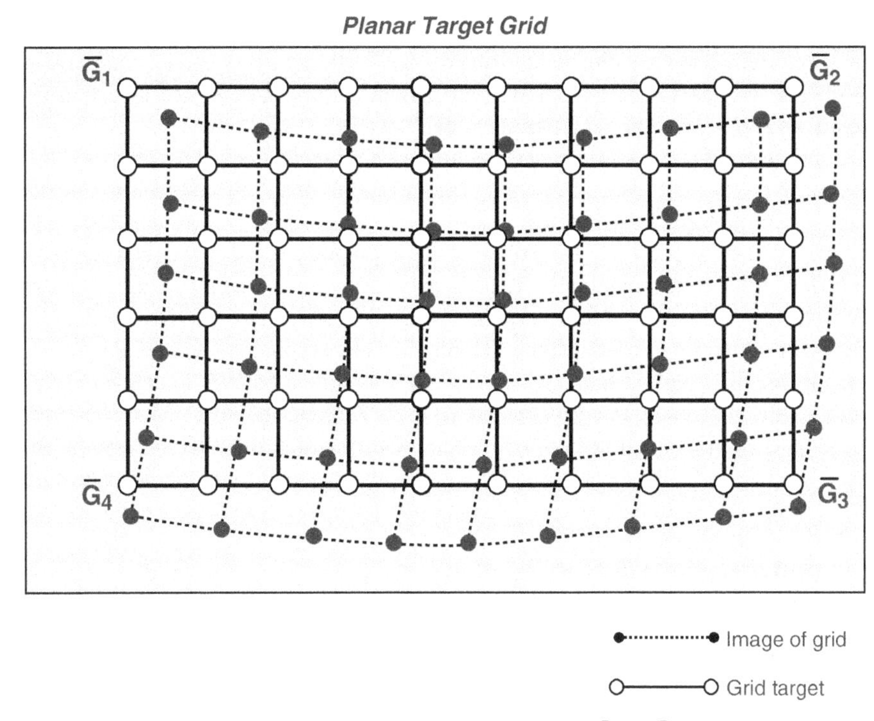

Fig. 3.17 Undistorted and distorted images of target grid. Points $\bar{G}_1 \rightarrow \bar{G}_4$ define the corner grid points in undistorted target image

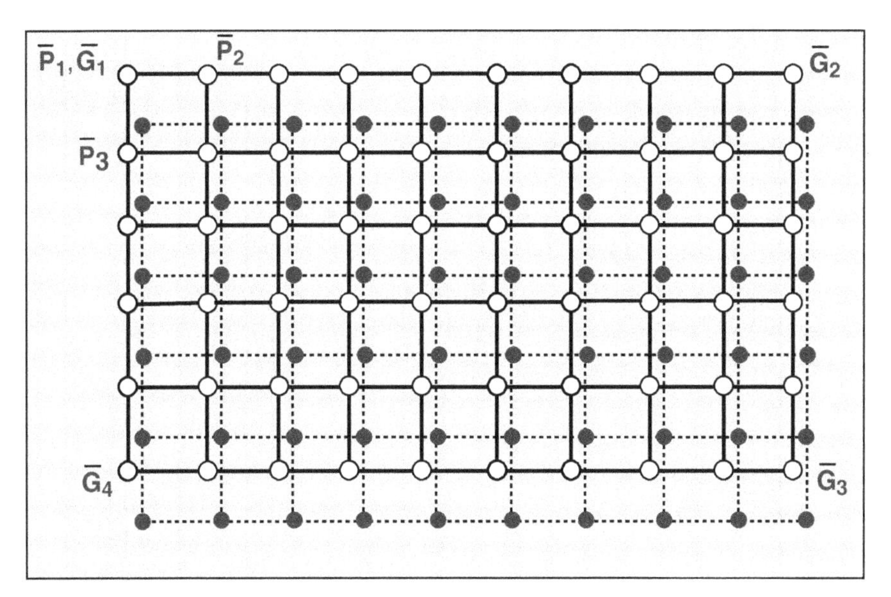

Fig. 3.18 Undistorted and distortion-corrected images of target grid. Points $\bar{G}_1$, $\bar{G}_2$ and $\bar{G}_3$ represent at least three non-collinear points required to constrain solution

Eq. (3.31), are not used in the projection model. In this case, only five intrinsic camera parameters are estimated duiring the calibration process and all comparisons in the least squares object function are between distortion-corrected measurement positions and projection model predictions.

Furthermore, if the sensor plane has orthogonal row and column directions, then the skew factor is removed from the camera model, resulting in one less intrinsic camera parameter to be estimated.

As noted in Section 3.2.2.1, previous work has shown that seven constraints in one view of the calibration target are necessary to ensure uniqueness in the solutions. For typical applications where a nominally planar grid target is used for calibration, the constraints imposed are (a) the 3D position of two points, $(X_W, Y_W, Z_W)|_{P_1}$ and $(X_W, Y_W, Z_W)|_{P2}$, and (b) the Z-position of a non-collinear point P_3 to define the rigid body rotation about the line formed by P_1 and P_2. Representative locations for points P_1, P_2 and P_3 are shown in Fig. 3.18. Since it is common practice to consider that P_1, P_2 and P_3 lie in the same plane, it is generally assumed that the planar grid lies in the plane $Z = 0$. In this case, the Z-location of all three points is set equal to zero.

To impose these constraints on the parameter solutions, let us assume that $P_1 = (\mathbf{0}, \mathbf{0}, \mathbf{0})$, $P_2 = (\mathbf{\Delta x}, \mathbf{0}, \mathbf{0})$ and $P_3 = (x_3, y_3, \mathbf{0})$, where the bold terms denote known values. Since the seven known values should not be re-estimated during the calibration process, the matrix formulation $[\mathbf{H}]\{\xi\} = \{\mathbf{b}\}$ would be modified to have a form given as follows:

$$[\mathbf{H}]\{\xi\} = \{\mathbf{b}\}$$

$$H_{22}\left\{ \underbrace{\begin{bmatrix} \cdot & & & & & & & & & & & \\ & \cdot & & & & & & & & & & \\ & & \ddots & & & & & & & & & \\ & & & 1 & 0 & 0 & & & & & & \\ & & & 0 & 1 & 0 & & & & & & \\ & & & 0 & 0 & 1 & & & & & & \\ & & & & & & 1 & 0 & 0 & & & \\ & & & & & & 0 & 1 & 0 & & & \\ & & & & & & 0 & 0 & 1 & & & \\ & & & & & & & & & \cdot & \cdot & \cdot \\ & & & & & & & & & \cdot & \cdot & \cdot \\ & & & & & & & & & \cdot & \cdot & 1 \\ & & & & & & & & & & & \ddots \end{bmatrix}}_{H_{22}} \right. \begin{Bmatrix} \xi_1 \\ \xi_2 \\ \vdots \\ P_{X1} \\ P_{Y1} \\ P_{Z1} \\ P_{X2} \\ P_{Y2} \\ P_{Z2} \\ P_{X3} \\ P_{Y3} \\ P_{Z3} \\ \vdots \end{Bmatrix} = \begin{Bmatrix} b_1 \\ b_2 \\ \vdots \\ 0 \\ 0 \\ 0 \\ \Delta x \\ 0 \\ 0 \\ x_3 \\ y_3 \\ 0 \\ \vdots \end{Bmatrix} \tag{3.43}$$

Though the procedure described above does not reduce the size of the matrix to be inverted, a reduction in the size of the matrix by seven is not significant relative to a typical size $O(5 \times 10^2)$. The procedure does eliminate the need to include additional unknowns into the optimization process to ensure that the constraints are enforced (e.g., Lagrange multipliers).

The single camera calibration procedure forms the basis for calibrating a multi-camera stereo vision system. For example, a stereo-vision system can be considered to be a series of independent cameras that view the same target grid during the calibration process. When the cameras are viewed as independent vision devices, then each camera is calibrated separately using the procedures defined in this section. In this case, the extrinsic parameters of each camera are estimated relative to a common coordinate system, typically defined by the rows and columns that compose the target grid. Details for stereo system calibration are discussed in Section 4.2.3.

Chapter 4
Two-Dimensional and Three-Dimensional Computer Vision

Two-dimensional (2D) and three-dimensional (3D) computer vision employ the pinhole camera model, distortion models and general optimization procedures described in Chapter 3. For two-dimensional computer vision, it is assumed that the motions of a planar object occur within the object plane. In most 2D cases, the object plane is nominally parallel to the camera sensor plane. In 3D computer vision,[1] the only restrictions placed on the motion of a curvilinear object are (a) the object remains in focus during the motion and (b) points of interest on the object are imaged onto two or more camera sensor planes. In the following sections, details regarding models and calibration procedures for 2D and 3D computer vision are provided.

4.1 Two-Dimensional Computer Vision

Figure 4.1 illustrates the projection of points on a planar object to positions in a parallel sensor plane. For this case, all points on the object are located at the same distance, Z, from the sensor plane. Thus, each object vector O_oP is in a scaled version of each image plane vector, $O_I p$. In this case, the projection of object points to image points is reduced to an affine projection.

4.1.1 Model

To simplify Eq. (3.9) for this application, the following assumptions are used. First, the world coordinate system is assumed to be aligned with a planar object coordinate system that has its origin at the intersection of the optical axis with the object. Hence, $[\mathbf{R}] = [\mathbf{I}]$, the identity matrix.

[1] Extensive additional background in stereo-vision is available in books such as Faugeras [69, 70] or Hartley [105].

M.A. Sutton et al., *Image Correlation for Shape, Motion and Deformation Measurements: Basic Concepts, Theory and Applications,* DOI: 10.1007/978-0-387-78747-3_4,

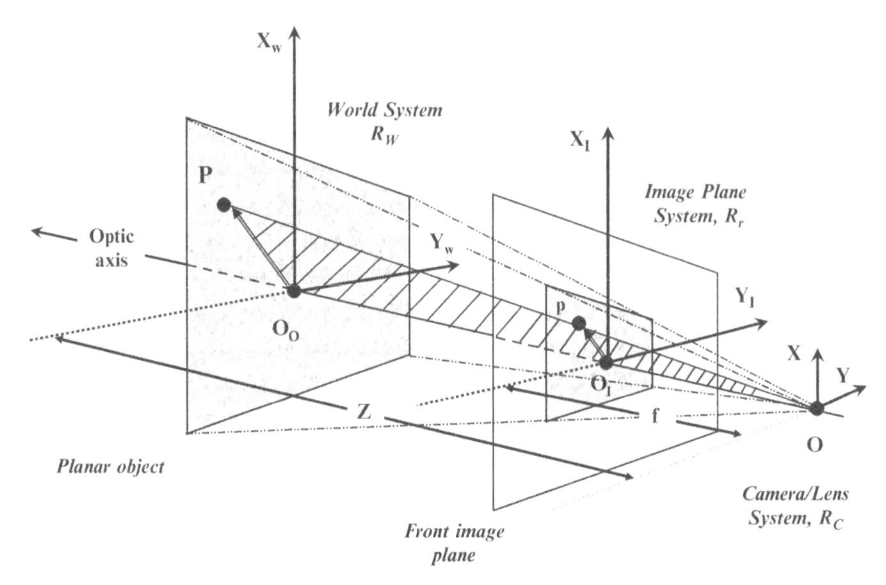

Fig. 4.1 2D imaging of planar object using front image plane model. Object point P and corresponding image point p are related by an affine transformation. Position vectors for both P and p lie in a plane containing the optical axis. World coordinates of P are $(X_w, Y_W, 0)$ in the $X_w - Y_W - Z_W$-system. Sensor position of the corresponding image point p is (x_s, y_s) in the sensor system (not shown)

Second, the sensor plane is assumed to be $\theta = \pi/2$ with $\theta = 0$. Combining these, we can write the relationship between planar object coordinates and sensor positions in the following form:

$$\begin{Bmatrix} x_s \\ y_s \end{Bmatrix} = \begin{Bmatrix} \left(\frac{fS_x}{Z}\right) X_W \\ \left(\frac{fS_y}{Z}\right) Y_W \end{Bmatrix} + \begin{Bmatrix} c_x \\ c_y \end{Bmatrix} = \begin{Bmatrix} s_x X_W \\ s_y Y_W \end{Bmatrix} + \begin{Bmatrix} c_x \\ c_y \end{Bmatrix} \tag{4.1}$$

Inspection of Figs. 3.4 and 4.1 shows that f/Z defines a transverse image magnification factor, M_T, where $f > Z$ corresponds to images that are physically larger than the object and $f < Z$ corresponds to image de-magnification. It is noted that the variable t_z defined in Eq. (3.9) satisfies the equation $Z = t_Z$ for the 2D construction shown in Fig. 2.1. Thus, for the same scale factor, higher magnification corresponds to fewer imaging sensors per unit length on the object. Defining (s_x, s_y) as the number of sensors per unit object length on the image plane, Eq. (4.1) indicates that there are four intrinsic parameters to define the position of an image point for 2D computer vision; s_x, s_y, c_x and c_y.

If it is assumed that Eq. (3.31) is appropriate for defining image distortions, then there are three radial distortion parameters, $(\kappa_1, \kappa_2, \kappa_3)$, to be determined. Hence, seven intrinsic camera parameters are to be determined during the calibration process for 2D imaging.

4.1.2 Bundle Adjustment for Plane-to-Plane Imaging

Since in-plane motions are assumed to occur in 2D computer vision, calibration is performed with planar motions. Thus, each view corresponds to a specific set of two-dimensional motions that are a combination of 2D translations $(t_x, t_y)_i$ and/or in-plane rotations, θ_i, where $i = 1, \ldots, M$. In this case, each calibration view results in three additional parameters to be estimated as part of the calibration process.

To obtain optimal estimates for the intrinsic and extrinsic camera parameters, we again minimize the form shown in Eq. (3.33) using a Levenberg-Marquardt algorithm such as described in Appendix D. The structure of the Jacobian matrix for 2D imaging is shown in Fig. 4.2.

For 2D imaging, **Int** = $(s_x, s_y, c_x, c_y, \kappa_1, \kappa_2, \kappa_3)$. Assuming that there are N target grid points to be re-estimated during the optimization process and M views with three rigid body motions per view, for this application there are $2MN$ equations relating predicted and measured 2D sensor positions to optimally (a) estimate all intrinsic and extrinsic camera parameters and (b) re-estimate all 2D object target points.

- Seven intrinsic parameters
- $3M$ extrinsic parameters positioning the camera relative to the grid
- $2N$ coordinates of the target grid points

For plane-to-plane transformations, it is common to specify the location of two points on the target (e.g., the 2D coordinates of two points). If this is done, there are $2MN$ equations to solve for $[2N+3M+7]-4$ unknown parameters. Table 4.1 shows results for three bundle adjustment examples that mimic those shown in Table 3.1 for imaging 3D points.

4.1.3 Solution Details

The general procedures outlined in Section 3.2 are applicable for updating and optimizing the various calibration parameters.

- Equations (3.29), (3.30) and (3.33) are used to define an error function and estimate the various parameters.
- Equation (3.31) may be used to define radial distortion error.
- Figure 3.15 and Eq. (3.2.3) with associated partial derivatives are used to define the Jacobian matrix with the structure shown in Fig. 4.2.
- The general form of the normal equations for updating the solution set is given in Eqs. (3.40) and (3.41). The size of the various block structures in the Hessian matrices is shown in Eq. (4.2).
- The procedure described in Sections 3.2.3 and 3.2.4 and shown in Eq. (3.43) is used to incorporate constraints on the solution set.
- General procedure defined in Eq. (3.2.3) and Appendix D is used to update the solution parameter set and obtain optimal parameters.

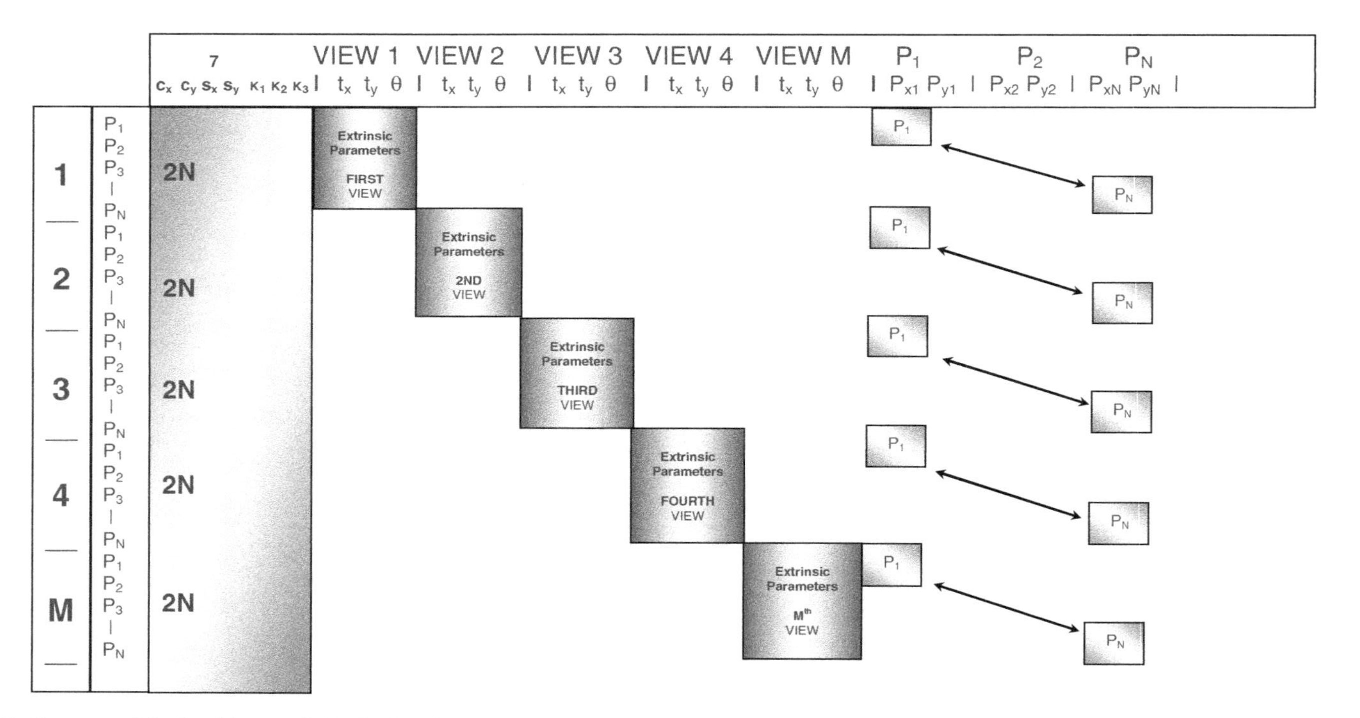

Fig. 4.2 Structure of the Jacobian matrix for 2D imaging model. There are *M* views and a total of *N* points in the target to generate equations and estimate seven intrinsic parameters, 3*N* extrinsic parameters and the locations of *N* 2D points on calibration target

Table 4.1 Bundle adjustment examples with unknowns, equations and constraints in 2D camera calibration

Parameters and equations	Example 1	Example 2	Example 3
Number of 2D points (N)	$8 \times 8 = 64$	$10 \times 10 = 100$	$6 \times 6 = 36$
Number of views (M)	10	7	5
Intrinsic parameters (7)	7	7	7
Extrinsic parameters ($3M$)	$10 \times 3 = 30$	$7 \times 3 = 21$	$5 \times 3 = 15$
2D components re-estimated ($2N$)	$64 \times 2 = 128$	$100 \times 2 = 200$	$36 \times 2 = 72$
Number of constraints	4	4	4
Total number of unknowns $((2N+3M+7)-4)$	161	224	101
Equations ($2NM$)	$2 \times 64 \times 10 = 1{,}280$	$2 \times 100 \times 7 = 1{,}400$	$2 \times 36 \times 5 = 360$

4.1.4 Discussion

The full camera calibration procedure outlined above can be used to correct for modest inaccuracies in a camera lens or target grid. However, if high quality lenses are used in a scientific grade digital image acquisition system, then it is likely that the effects of image distortion are small and the calibration process can be greatly simplified.

Furthermore, if qualitative estimates for s_x and s_y are sufficient for a specific application (e.g., measurement of other camera parameters is not required), then a target containing a measurement standard can be attached to the surface of a planar object and imaged to estimate the scale factors. To simplify the process of converting target data into scale factors, the edges of the measurement standard are aligned with the row and column directions in the sensor plane and several images of the grid standard are recorded and measurements averaged to optimally estimate the pixel positions of several of the marks having a known spacing along both the horizontal and row directions. Using these values, s_x and s_y have been determined with accuracy better than 0.2% in several applications.

$$\begin{bmatrix} [\mathbf{H_{11}}]_{(3M+7)\times(3M+7)} & [\mathbf{H_{12}}]_{(3M+7)\times(2N)} \\ [\mathbf{H_{12}}]^T_{(2N)\times(3M+7)} & [\mathbf{H_{22}}]_{(2N)\times(2N)} \end{bmatrix}$$

$$\{\xi\} = \left\{ \begin{array}{l} \left. \begin{array}{c} c_x \\ c_y \\ \vdots \\ t_x \\ t_y \\ \theta \end{array} \right|_{(3M+7)\times 1} \quad \xi_1 \text{ Intrinsic and extrinsic} \\ \left. \begin{array}{c} X_{w1} \\ Y_{w1} \\ \vdots \\ X_{wN} \\ Y_{wN} \end{array} \right|_{2N\times 1} \quad \xi_2 \text{ 2D positions re-estimated} \end{array} \right\} \tag{4.2}$$

4.2 Three-Dimensional Computer Vision

As discussed in Chapter 3, a pinhole camera performs a perspective projection. As shown algebraically in Eq. (3.9), such a projection transforms a 3D object point into a 2D image-point, thereby removing the third dimension in an irreversible manner. As shown in the top image of Fig. 4.3, the two 3D points Q and R are imaged onto the same image-point p as they are lying on the same projective ray (C, p), demonstrating that there exists an infinity of 3D points that correspond to the image point p. As shown in the bottom image in Fig. 4.3, it is possible to recover the three-dimensional position of the true object points by using two cameras to record simultaneous image points of the same object. If two image points are (p, q'), then the unique 3D point is Q. If the corresponding image points are (p, r'), then the unique 3D point is R.

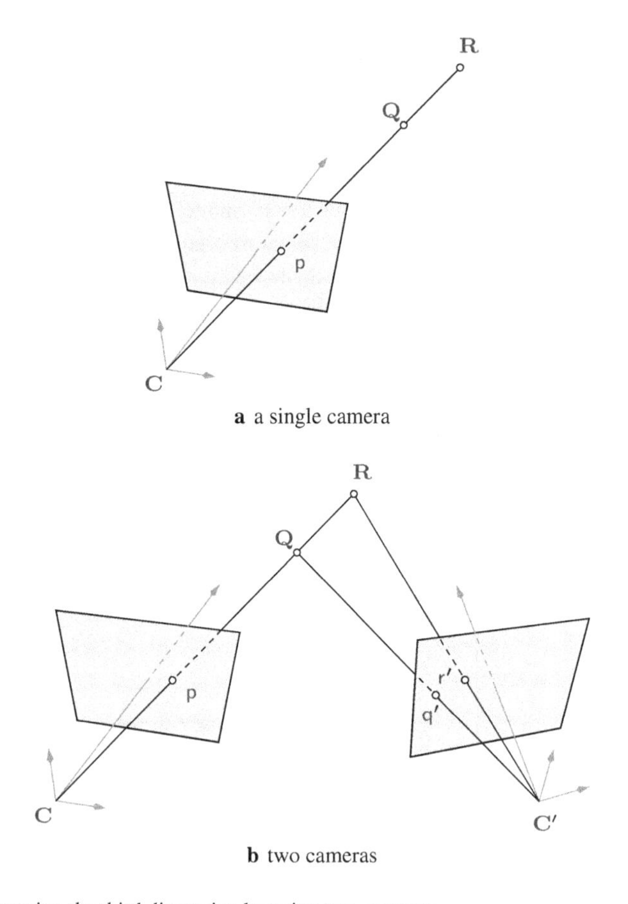

Fig. 4.3 Recovering the third dimension by using two cameras

4.2.1 Geometry of a Stereovision System

Consistent with the single-camera geometry shown in Fig. 3.1 and the associated terminology in Eqs. (3.1)–(3.9), Fig. 4.4 presents a schematic of the geometry for a typical two-camera stereovision system.

The extrinsic transformations can be written as follows, where both the rigid body rotation tensor and the translation vector are combined into a 3×4 matrix.

$$
\begin{aligned}
[\mathbf{T}]_{\mathbf{R_W}-\mathbf{R_C}} &= [\mathbf{T}] \doteq \begin{bmatrix} R & t \\ 0 & 1 \end{bmatrix} \\
[\mathbf{T}]_{\mathbf{R_C}-\mathbf{R_{C'}}} &= [\mathbf{T_S}] \doteq \begin{bmatrix} R & t \\ 0 & 1 \end{bmatrix}_{\mathbf{S}} \\
[\mathbf{T}]_{\mathbf{R_W}-\mathbf{R_{C'}}} &= [\mathbf{T'}] \doteq \begin{bmatrix} R & t \\ 0 & 1 \end{bmatrix}_{\mathbf{C'}}
\end{aligned}
\tag{4.3}
$$

Here, $[\mathbf{T}]$ transforms vectors in the world system, R_W, into the pinhole system, R_C, located in the left camera. Similarly, $[\mathbf{T}]'$ transforms vectors in the world system, R_W, into the pinhole system, $R_{C'}$, located in the right camera. Finally, $[\mathbf{T_S}]$ transforms vectors in R_C into vectors in $R_{C'}$. The transformations are related as follows:

$$
[\mathbf{T_S}]\,[\mathbf{T}] = [\mathbf{T}]' \tag{4.4}
$$

Following the notation in Chapter 3, consider a general three-dimensional point $\mathbf{M}$. Using the corresponding homogeneous forms for the 3D position, $\tilde{\mathbf{M}}$, the scaled sensor location in the left camera, $\tilde{\mathbf{m}}$ and the corresponding scaled sensor location in right camera, $\tilde{\mathbf{m}}'$, Eq. (3.7) gives the following:

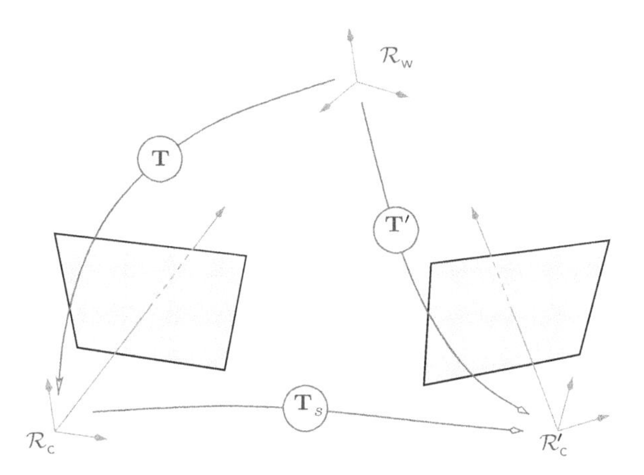

Fig. 4.4 The reference frames associated with the 3D scene and to each camera are linked by three rigid body transformations

$$\begin{aligned}\tilde{\mathbf{m}} &= [\mathbf{K}] \bullet [\mathbf{T}] \bullet \{\tilde{\mathbf{M}}\} \\ \tilde{\mathbf{m}}' &= [\mathbf{K}'] \bullet [\mathbf{T}'] \bullet \{\tilde{\mathbf{M}}\}\end{aligned} \tag{4.5}$$

Applying Eqs. (4.4) and (4.5), we have the following relationship between the two image points:

$$\begin{aligned}\tilde{\mathbf{m}}' &= [\mathbf{K}'] \bullet [\mathbf{T}'] \bullet \{\tilde{\mathbf{M}}\} \\ &= [\mathbf{K}'] \bullet [\mathbf{T_S}] \bullet [\mathbf{T}] \bullet \{\tilde{\mathbf{M}}\}\end{aligned} \tag{4.6}$$

As shown in Eq. (4.6), the sensor positions for a common 3D point are directly related through the transformation matrices.

4.2.2 Epipolar Constraint

Given an object point **M** and the homogeneous form for the 2D projections $\tilde{\mathbf{m}}$ and $\tilde{\mathbf{m}}'$ on both image planes in a stereo-vision system, it has been shown [69, 70, 105, 178] that any stereo-imaging system consisting of cameras that can be modeled by Eqs. (3.1)–(3.9) has a geometric constraint between $\tilde{\mathbf{m}}$ and $\tilde{\mathbf{m}}'$. Known as the **epipolar constraint**, the relationship can be written in the form:

$$\{\tilde{\mathbf{m}}'\}^{\mathbf{T}} \bullet [\mathbf{F}]_{3\times 3} \bullet \{\tilde{\mathbf{m}}\} = \mathbf{0} \tag{4.7}$$

The square fundamental matrix, $[\mathbf{F}]$, has size 3×3 and rank 2. Conceptually, Eq. (4.7) associates (a) to $\tilde{\mathbf{m}}$ a line in the right image called its associated epipolar line and (b) to $\tilde{\mathbf{m}}'$ an epipolar line in the left image [20]. Specifically, if $\{\tilde{\mathbf{m}}\}$ is the homogeneous representation for an image point in the left image, then $[\mathbf{F}] \cdot \{\tilde{\mathbf{m}}\}$ is its associated epipolar line in the right image. As shown in Eq. (4.7), the homogeneous representation for the corresponding image-point $\{\tilde{\mathbf{m}}'\}$ associated with $\{\tilde{\mathbf{m}}\}$

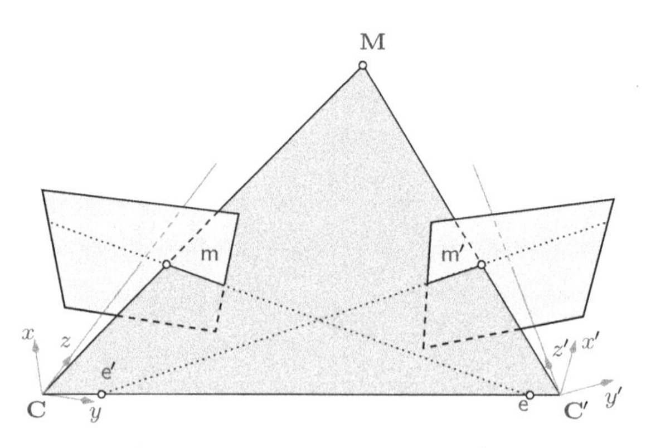

Fig. 4.5 Epipolar geometry: the camera optical centers **C** and **C**$'$, the 3D point **M** and its images m and m$'$ lie in a common plane. This plane intersects each image plane in an epipolar line

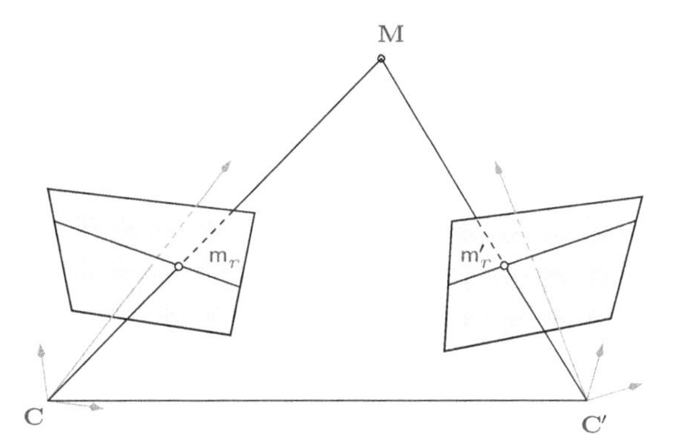

Fig. 4.6 The epipolar constraint reduces the matching problem to a single scan along the epipolar line

necessarily lies along its epipolar line. Figure 4.5 presents a schematic of the stereo correspondences and epipolar lines. Since the calibration process provides sufficient information to define the fundamental matrix, Eq. (4.7) can be combined with the known position of a point in the reference image (e.g., $\tilde{\mathbf{m}}$) to define the corresponding line in the second camera's image plane along which the matching point, $\tilde{\mathbf{m}}'$, is located. Therefore, as shown in Fig. 4.6, the search process to locate the matching point, $\tilde{\mathbf{m}}'$, is confined to a limited region along the epipolar line defined by Eq. (4.7) in the image plane of the second camera.

4.2.3 Calibration in Stereovision

There are two approaches commonly used for calibration of a stereovision system. The first approach is to independently calibrate each camera in a stereovision arrangement using motions of the same target pattern. As shown in Fig. 4.7, stereo views of a common grid pattern may be used to define a common world coordinate system for all cameras, thereby providing the basis for relating image locations in all cameras to a common 3D position.

The second calibration approach is to consider both cameras as a single "measurement system". Oftentimes known as a stereo-rig or stereo-system, the optimization procedure for this system is constructed to determine the relative orientation and position of an arbitrarily selected camera with respect to the remaining camera.

4.2.3.1 Stereovision-Independent Camera Calibration

The calibration procedures described in Chapter 3 are used to calibrate each camera in the stereo system. To perform calibration, simultaneous images are acquired of the same target grid in *M* different views by both cameras.

To relate the image positions in each separate camera for every 3D point, the calibration process is performed in a manner that ensures a common world coordinate

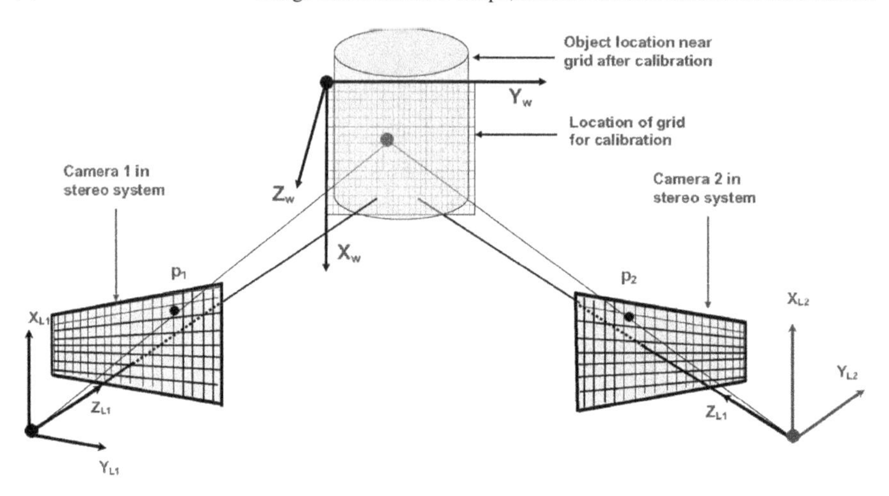

Fig. 4.7 Schematic of two camera stereo-vision system during imaging of a target grid. World system is generally aligned with the first grid location. Origin is at an arbitrary grid intersection point

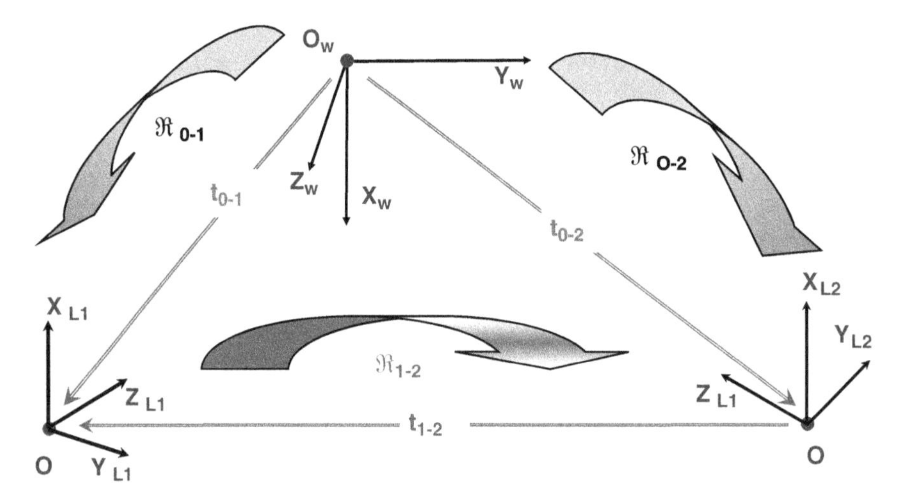

Fig. 4.8 Schematic describing transformations between cameras in a stereo-vision system. World coordinate system is assumed to be the same for both cameras

system is used to define the extrinsic parameters. One way to ensure that the same world coordinate system is specified for all cameras is to use (a) a specific target grid point as the origin of the world coordinate system, $\mathbf{O_W}$, (b) a specific line of points in the target grid as the X_W axis and (c) a perpendicular line of points in the target grid as the Y_W axis. This situation is shown schematically in Fig. 4.7.

As shown in Fig. 4.8, $\mathbf{R_{0-1}}$ and $\mathbf{R_{0-2}}$ rotate the world system to align with the left camera (camera 1) and the right camera (camera 2), respectively, and the vectors $\mathbf{t_{0-1}}$ and $\mathbf{t_{0-2}}$ translate the world system to the origin (pinhole) of the camera 1 and camera 2 systems, respectively. Then the form of the transformation can be written as follows:

$$\begin{Bmatrix} X_{L1} \\ Y_{L1} \\ Z_{L1} \end{Bmatrix} = [\mathbf{R_{0-1}}] \begin{Bmatrix} X_W \\ Y_W \\ Z_W \end{Bmatrix} + \begin{Bmatrix} t_{X_{L1}} \\ t_{Y_{L1}} \\ t_{Z_{L1}} \end{Bmatrix}$$

$$\begin{Bmatrix} X_{L2} \\ Y_{L2} \\ Z_{L2} \end{Bmatrix} = [\mathbf{R_{0-2}}] \begin{Bmatrix} X_W \\ Y_W \\ Z_W \end{Bmatrix} + \begin{Bmatrix} t_{X_{L2}} \\ t_{Y_{L2}} \\ t_{Z_{L2}} \end{Bmatrix} \tag{4.8}$$

$$\begin{Bmatrix} X_W \\ Y_W \\ Z_W \end{Bmatrix} = \text{world coordinates for common 3D point, } P$$

Independent calibration of each camera is performed using procedures outlined in Chapter 3 to determine (a) the extrinsic parameters shown in Eq. (4.8) and (b) the intrinsic parameters (e.g., distortion parameters, center location, f_x and f_y) for both cameras. To ensure a common global coordinate system, image positions in both cameras for the same common point are used.

4.2.3.2 Stereovision-Combined System Calibration

The procedure described in Chapter 3 is modified slightly to calibrate a stereovision system. To construct a "system calibration" procedure, we begin with the following assumptions.

- One of the cameras is selected arbitrarily to be the "master" camera (e.g., camera 1).
- The orientation and position of the world coordinate system is determined relative to the master camera system.
- The orientation and position of all other cameras in the stereovision system are defined relative to the master camera system. The relationship between the individual cameras (e.g., orientation and pinhole position) is assumed to be invariant during the calibration and measurement processes. Thus, the set of rotations and translations required to relate the master camera system to each additional system is considered to be part of the set of **intrinsic parameters**.
- Calibration motions of the object are determined relative to the master camera system, with the relative position of each camera used to estimate sensor positions in other cameras.

Conceptually, the relationships required for combined system calibration are also shown in Fig. 4.8. If the master camera is assumed to be camera 1, only the orientation ($\mathbf{R_{0-1}}$) and the position vector ($\mathbf{t_{0-1}}$) are considered to be **extrinsic parameters**. Thus, the intrinsic parameters include (a) the six independent terms in $\mathbf{R}_{1-2}$ and vector $\mathbf{t}_{1-2}$, (b) the internal parameters for camera 1 and (c) the internal parameters for camera 2.

In equation form, the relationship between the several coordinate systems can be written as follows:

$$\begin{Bmatrix} X_{L1} \\ Y_{L1} \\ Z_{L1} \end{Bmatrix} = [\mathbf{R}_{0-1}] \begin{Bmatrix} X_W \\ Y_W \\ Z_W \end{Bmatrix} + \begin{Bmatrix} t_{X_{0-1}} \\ t_{Y_{0-1}} \\ t_{Z_{0-1}} \end{Bmatrix}$$

$$\begin{Bmatrix} X_{L2} \\ Y_{L2} \\ Z_{L2} \end{Bmatrix} = [\mathbf{R}_{1-2}] \begin{Bmatrix} X_W \\ Y_W \\ Z_W \end{Bmatrix} + \begin{Bmatrix} t_{X_{1-2}} \\ t_{Y_{1-2}} \\ t_{Z_{1-2}} \end{Bmatrix} \tag{4.9}$$

$$\begin{Bmatrix} X_W \\ Y_W \\ Z_W \end{Bmatrix} = \text{world coordinates for common 3D point, } P$$

$$\begin{Bmatrix} t_{X_{1-2}} \\ t_{Y_{1-2}} \\ t_{Z_{1-2}} \end{Bmatrix} = \text{position of } O_2 \text{ relative to } O_1 \text{ as defined in camera 2 system}$$

In this formulation, all 3D positions of an object are relative to the master camera view. Thus, the orientation and origin for each stereo view are defined relative to the master camera system.

4.2.3.3 Bundle Adjustment for Modified Calibration Procedure

Consider a common object point, **P**. Let the image of point **P** be (x_{s1}, y_{s1}) in camera 1 and (x_{s2}, y_{s2}) in camera 2. The least squares metric relating measured image positions and model predictions can be written as follows:

$$\begin{aligned} E &= \sum_{n=1}^{N} (\mathbf{f_{n1}} \cdot \mathbf{f_{n1}} + \mathbf{f_{n2}} \cdot \mathbf{f_{n2}}) \\ &= \sum_{n=1}^{N} \left((x_{n1} - X_{n1}(\boldsymbol{\beta}))^2 + (y_{n1} - Y_{n1}(\boldsymbol{\beta}))^2 \right) \\ &\quad + \sum_{n=1}^{N} \left((x_{n2} - X_{n2}(\boldsymbol{\beta}))^2 + (y_{n2} - Y_{n2}(\boldsymbol{\beta}))^2 \right) \end{aligned}$$

$N =$ number of points on target grid being imaged

$\mathbf{f_{n1}} = \mathbf{x_{n1}} - \mathbf{X_{n1}}(\beta) =$ 2D vector defined in sensor plane of camera 1

$$\mathbf{f_{n2}} = \mathbf{x_{n2}} - \mathbf{X_{n2}}(\boldsymbol{\beta}) = \text{2D vector defined in sensor plane of camera 2}$$

$$\begin{aligned}\boldsymbol{\beta} &= \text{vector of model parameters}\\ &= \{\{\mathbf{Int}\},\{\mathbf{Ext}\},\{\mathbf{P}\}\}\end{aligned}$$

$$\begin{aligned}\mathbf{Int} = \{ &\{c_x, c_y, f_x, f_y, f_s, \kappa_1, \kappa_2, \kappa_3\}_1\\ &\{c_x, c_y, f_x, f_y, f_s, \kappa_1, \kappa_2, \kappa_3\}_2\\ &\{n_x, n_y, n_z, t_x, t_y, t_z\}_{1-2}\}\end{aligned}$$

$$\mathbf{Ext} = \{\{n_x, n_y, n_z, t_x, t_y, t_z\}_{\text{view }1}, \cdots, \{n_x, n_y, n_z, t_x, t_y, t_z\}_{\text{view }M}\}$$

$$\mathbf{P} = \{\{P_x, P_y, P_z\}_1, \{P_x, P_y, P_z\}_2, \cdots, \{P_x, P_y, P_z\}_N\} \tag{4.10}$$

As shown in Eq. (4.10), the orientation and position of camera 2 relative to camera 1 and the internal parameters for cameras 1 and 2 are considered to be intrinsic parameters.

Equation (4.2.3.3) shows the relationship between the three-dimensional position in the two camera systems. The orientation and position of the grid relative to camera 1 for all M views are considered to be extrinsic parameters. The 3D positions of all N points in the calibration target as seen in each view are considered to be parameters for re-estimation.

To obtain optimal estimates for the intrinsic and extrinsic camera parameters, we minimize the form shown in Eq. (4.10) using a Levenberg–Marquardt algorithm and a modified form of the procedures described in Sections 3.2.3 and 3.2.4.

$$\begin{Bmatrix} X \\ Y \\ Z \end{Bmatrix}_{\text{Cam1}} = [\mathbf{R_{0-1}}] \begin{Bmatrix} X_W \\ Y_W \\ Z_W \end{Bmatrix} + \begin{Bmatrix} t_X \\ t_Y \\ t_Z \end{Bmatrix}_{0-1}$$

$$\Rightarrow \begin{Bmatrix} X \\ Y \\ Z \\ 1 \end{Bmatrix}_{\text{Cam1}} = \begin{bmatrix} \{[\mathbf{R_{0-1}}]\}_{3\times 3} & \{\{\mathbf{t_{0-1}}\}\}_{3\times 1} \\ \{\{\mathbf{0}\}^{\mathbf{T}}\}_{1\times 3} & \{1\}_{1\times 1} \end{bmatrix} \begin{Bmatrix} X_W \\ Y_W \\ Z_W \\ 1 \end{Bmatrix}$$

$$\begin{Bmatrix} X \\ Y \\ Z \end{Bmatrix}_{\text{Cam2}} = [\mathbf{R_{1-2}}] \begin{Bmatrix} X_W \\ Y_W \\ Z_W \end{Bmatrix} + \begin{Bmatrix} t_X \\ t_Y \\ t_Z \end{Bmatrix}_{1-2}$$

$$\Rightarrow \begin{Bmatrix} X \\ Y \\ Z \\ 1 \end{Bmatrix}_{\text{Cam2}} = \begin{bmatrix} \{[\mathbf{R_{1-2}}]\}_{3\times 3} & \{\{\mathbf{t_{1-2}}\}\}_{3\times 1} \\ \{\{\mathbf{0}\}^{\mathbf{T}}\}_{1\times 3} & \{1\}_{1\times 1} \end{bmatrix} \begin{Bmatrix} X \\ Y \\ Z \\ 1 \end{Bmatrix}_{\text{Cam1}}$$

$$
\begin{aligned}
&= \begin{bmatrix} \{[\mathbf{R}_{1-2}]\}_{3\times3} & \{\{\mathbf{t}_{1-2}\}\}_{3\times1} \\ \{\{\mathbf{0}\}^{\mathbf{T}}\}_{1\times3} & \{1\}_{1\times1} \end{bmatrix} \begin{bmatrix} \{[\mathbf{R}_{0-1}]\}_{3\times3} & \{\{\mathbf{t}_{0-1}\}\}_{3\times1} \\ \{\{\mathbf{0}\}^{\mathbf{T}}\}_{1\times3} & \{1\}_{1\times1} \end{bmatrix} \begin{Bmatrix} X_W \\ Y_W \\ Z_W \\ 1 \end{Bmatrix} \\
&= \begin{bmatrix} \{[\mathbf{R}_{1-2}][\mathbf{R}_{0-1}]\}_{3\times3} & \{[\mathbf{R}_{1-2}]\{\mathbf{t}_{1-2}\}+\{\mathbf{t}_{0-1}\}\}_{3\times1} \\ \{\{\mathbf{0}\}^{\mathbf{T}}\}_{1\times3} & \{1\}_{1\times1} \end{bmatrix}_{4\times4} \begin{Bmatrix} X_W \\ Y_W \\ Z_W \\ 1 \end{Bmatrix} \\
&= \begin{bmatrix} \{[\Re]\}_{3\times3} & \{\{\Im\}\}_{3\times1} \\ \{\{\mathbf{0}\}^{\mathbf{T}}\}_{1\times3} & \{1\}_{1\times1} \end{bmatrix}_{4\times4} \begin{Bmatrix} X_W \\ Y_W \\ Z_W \\ 1 \end{Bmatrix}
\end{aligned}
\tag{4.11}
$$

$$
\alpha_2 \begin{Bmatrix} x_s \\ y_s \\ 1 \end{Bmatrix}_{\text{Cam2}} = \begin{bmatrix} \{[\mathbf{A}_2][\Re]\}_{3\times3} & \{[\mathbf{A}_2][\Im]\}_{3\times1} \end{bmatrix} \begin{Bmatrix} X_W \\ Y_W \\ Z_W \\ 1 \end{Bmatrix} = \left[\{\overline{\mathbf{P}}\}_{3\times4}\right] \begin{Bmatrix} X_W \\ Y_W \\ Z_W \\ 1 \end{Bmatrix}
$$

$$
[\mathbf{A}_2] = \begin{bmatrix} f_{x_2} & f_{s_2} & c_{x_2} \\ 0 & f_{y_2} & c_{y_2} \\ 0 & 0 & 1 \end{bmatrix}
$$

where:

$$
\begin{aligned}
f_{x_2} &= (fS_x)_{\text{Cam2}} \\
f_{y_2} &= (fS_y/\sin\theta)_{\text{Cam2}} \\
f_{s_2} &= (-fS_x\cot\theta)_{\text{Cam2}} \\
c_{x_2} &= (-S_x(\hat{c}_x - \hat{c}_y\cot\theta))_{\text{Cam2}} \\
c_{y_2} &= (-S_y\hat{c}_y/\sin\theta)_{\text{Cam2}}
\end{aligned}
$$

Table 4.2 shows results for three examples that mimic those shown in Tables 3.1 and 4.1 for imaging points with single cameras. In these examples, **Int** has 20 parameters to be determined. Assuming that there are N target grid points to be re-estimated during the optimization process and M views with three rigid body motions per view, for this application there are $4MN$ equations relating predicted and measured 2D sensor positions in **both** cameras for optimal estimation of all intrinsic and extrinsic camera parameters, as well as re-estimation of all 3D object target points. Specifying two 3D points and one Z-direction as constraints, the total number of unknowns to be determined is $3N+6M+13$.

Table 4.2 Bundle adjustment examples with unknowns, equations and constraints for combined stereovision system

Parameters and equations	Example 1	Example 2	Example 3
Number of 3D points (N)	$8 \times 8 = 64$	$10 \times 10 = 100$	$6 \times 6 = 36$
Number of views (M)	10	7	5
Intrinsic parameters (20)	20	20	20
Extrinsic parameters ($6M$)	$10 \times 6 = 60$	$7 \times 6 = 42$	$5 \times 6 = 30$
3D components re-estimated ($3N$)	$64 \times 3 = 192$	$100 \times 3 = 300$	$36 \times 3 = 108$
Number of constraints	7	7	7
Total number of unknowns $((3N+6M+20)-7)$	265	355	151
Equations ($4NM$)	$4 \times 64 \times 10 = 2{,}560$	$4 \times 100 \times 7 = 2{,}800$	$4 \times 36 \times 5 = 720$

4.2.4 Discussion

Independent camera calibration is the direct application of the principles in Chapter 3 to two cameras that are oriented to view a common region. Since the cameras are to be used to convert each measured 2D sensor position into a common 3D position, the relative position of all cameras must be unaltered during calibration. After calibration has been completed, the relative position must either remain unaltered or be known using additional information.

Since the initial position of the world system is used to define R_{0-1} and the vector t_{0-1}, **any change in position of all cameras in either an independently calibrated or combined calibration stereovision system that does not affect their relative orientation and position** will add only a rigid body motion to the 3D measurements. Thus, calibration can be performed either (a) at a different location than the actual experiment and moved to the experiment location or (b) using a calibration object that is offset from the true object position, without affecting the measured shape, strains or distortions of the object.

Comparison of Tables 3.1 and 4.2 shows the following for a two-camera stereovision system. Combined stereovision system calibration requires that a few more parameters be determined than needed for calibration of one camera. Specifically, a stereo-rig calibration process solves $4MN$ equations to estimate $3N + 6M + 13$ unknowns, whereas a single camera calibration solves $2MN$ to estimate $3N+6M+1$ parameters. The corresponding reduced computation time is not a major advantage for combined stereo-vision system calibration since current computational capabilities are sufficient to complete calibration of either system in a few seconds.

To populate the Jacobian matrix, it is necessary to obtain the partial derivatives of the sensor position model functions. For the combined stereovision system model, there are four separate model functions relating sensor positions in each camera to the various camera parameters. Since positions in the master camera are only a function of extrinsic parameters, intrinsic parameters in camera 1, and target positions, the functions defining distorted sensor positions in the master camera, (x_{s_1}, y_{s_1}) have derivatives that are the same as given previously, e.g., Eq. (3.39).

For the second camera, the partial derivatives $\partial x_{s_2}/\partial \beta_k$ and $\partial y_{s_2}/\partial \beta_k$ are required, where the β_i coefficients are given in Eq. (4.12).

$$
\begin{aligned}
&\{\beta_1, \beta_2, \cdots, \beta_{22}\} \\
&\quad = \left\{ \{c_x, c_y, f_x, f_y, f_s, \kappa_1, \kappa_2, \kappa_3\}_1 \{c_x, c_y, f_x, f_y, f_s, \kappa_1, \kappa_2, \kappa_3\}_2 \{n_x, n_y, n_z, t_x, t_y, t_z\}_{1-2} \right\} \\
&\{\beta_{23}, \beta_{24}, \cdots, \beta_{28}; \cdots; \beta_{6M+17}, \beta_{6M+18}, \cdots, \beta_{6M+22}\} \\
&\quad = \{\{n_x, n_y, n_z, t_x, t_y, t_z\}_{\text{view } 1}, \cdots, \{n_x, n_y, n_z, t_x, t_y, t_z\}_{\text{view } M}\} \\
&\{\beta_{6M+23}, \beta_{6M+24}, \beta_{6M+25}; \beta_{6M+26}, \beta_{6M+27}, \beta_{6M+28}; \cdots; \\
&\quad \beta_{3N+6M+20}, \beta_{3N+6M+21}, \beta_{3N+6M+22}\} = \{\{P_x, P_y, P_z\}_1, \{P_x, P_y, P_z\}_2, \cdots, \{P_x, P_y, P_z\}_N\}
\end{aligned}
\tag{4.12}
$$

The functions defining sensor positions in camera 2, (x_{s_2}, y_{s_2}), are given in Eq. (4.2.3.3), with the definition of the components of $[\mathbf{R}]$ provided in Appendix G. Derivatives of the functions in Eq. (4.2.3.3) follow procedures similar to those given in Chapter 3.

Chapter 5
Digital Image Correlation (DIC)

5.1 Introduction to Image Matching

Image matching is a discipline of computer vision that is of central importance to a large number of practical applications. To name just a few, image matching is used to solve problems in industrial process control, automatic license plate recognition in parking garages, biological growth phenomena, geological mapping, stereo vision, video compression and autonomous robots for space exploration. Since the applications are so varied, there are a wide variety of approaches and algorithms in use today, many specialized to a given task. For instance, highly specialized algorithms exist to determine motion vectors of small tracer particles used in the study of fluid flows. Digital image correlation is no exception, and algorithms are employed that take the physics of the underlying deformation processes into account. In one regard, however, digital image correlation is somewhat unique. Due to the miniscule motions that are often of interest in engineering applications, the resolution requirements are much higher than for most other applications. To accurately measure the stress-strain curve for many engineering materials, length changes on the order of 10^{-5} m/m have to be resolved. These requirements have led to the development of many algorithms targeted towards providing high resolution with minimal systematic errors.

This chapter discusses the fundamental problems in image matching with a focus on resolving the motions on the surface of deforming structures. Various fundamental concepts for digital image correlation are presented, and the most commonly used approaches are explored in detail.

5.1.1 The Aperture Problem

It is generally not possible to find the correspondence of a single pixel in one image in a second image. Typically, the gray value of a single pixel can be found at thousands of other pixels in the second image, and there is no unique correspondence.

M.A. Sutton et al., *Image Correlation for Shape, Motion and Deformation Measurements: Basic Concepts, Theory and Applications*, DOI: 10.1007/978-0-387-78747-3_5,

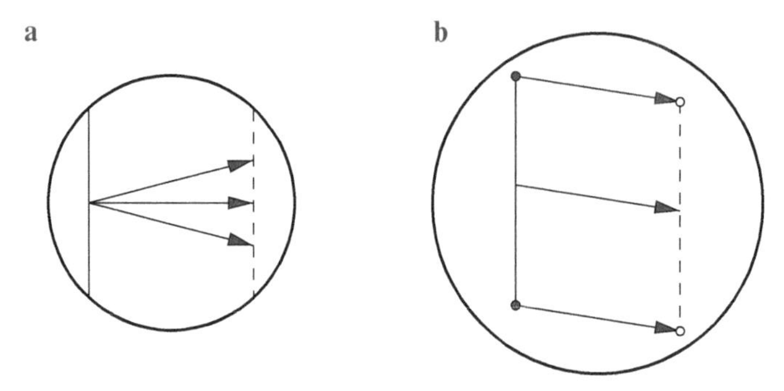

Fig. 5.1 The aperture problem in image matching. **a** A point on the line can match arbitrary points on the displaced line. **b** The aperture has been enlarged to include the end points of the line. The motion vector is now unique

We therefore consider finding the correspondence of a small neighborhood around the pixel of interest. While the neighborhood provides additional information, the matching problem still may not be unique. This can be illustrated by placing an aperture over the image.

Figure 5.1 illustrates this problem for a line in an image. Even though the component of the motion vector that is perpendicular to the line can be resolved, the motion along the line cannot. A point on the line in the first image can be matched to an arbitrary point on the line in the second image. This ambiguity is commonly referred to as the aperture problem. In the right part of the figure, the aperture has been increased revealing the end points of the line. In this case, the motion vector can be uniquely determined.

5.1.2 The Correspondence Problem

The aperture problem is a special case of the more general correspondence problem. There are many situations in which a unique correspondence between features in two images cannot be established. Two cases are illustrated in Fig. 5.2. For a repeating structure such as a grid of small dots, motion can only be resolved up to an unknown multiple of the grid constant, as illustrated in Fig. 5.2a. When the entire grid is considered by increasing the aperture size, the correspondence problem again becomes unique and the motion vectors can be resolved.

The correspondence problem becomes even more difficult to resolve if we do not restrict ourselves to rigid motion, but consider cases where the object undergoes deformation. This is illustrated in Fig. 5.2b. For a textureless structure undergoing deformation, we cannot obtain any motion information inside its boundaries, since no features are present. But due to the deformation, it is even impossible to determine motion vectors on the boundary of the structure.

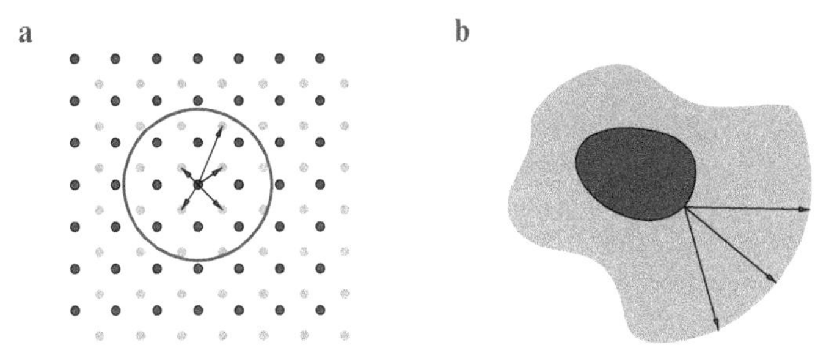

Fig. 5.2 Correspondence problem for **a** a repeating structure, where a unique correspondence can only be found if the edge of the grid is included in the aperture and **b** a textureless deforming structure, where no correspondence can be established without further assumptions

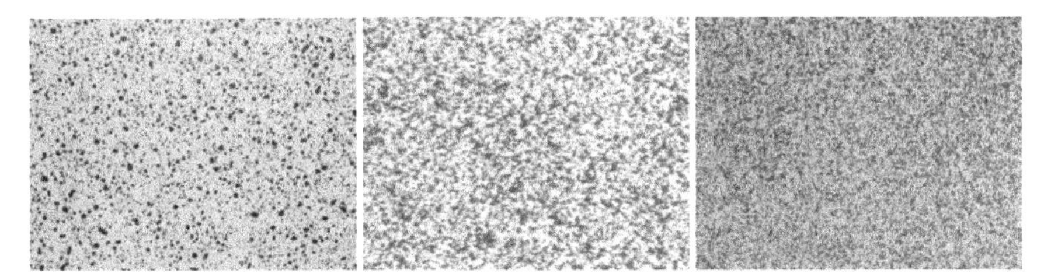

Fig. 5.3 Examples of typical speckle patterns

5.1.3 Speckle Pattern

To solve the correspondence problem uniquely, the object surface has to exhibit certain properties. The discussion of the aperture problem above has shown that oriented structures such as lines limit the determination of motion vectors to the component orthogonal to the structure. The ideal surface texture should therefore be isotropic, i.e., it should not have a preferred orientation. Furthermore, we have shown that repeating textures can lead to misregistration problems. The preferred surface texture should therefore be non-periodic. These requirements naturally lead to the use of random textures, such as the speckle pattern formed when a coherently illuminated surface is viewed through an aperture. The patterns commonly applied typically resemble laser speckle patterns to some degree. However, the patterns used in digital image correlation adhere to the surface and deform with the surface, and therefore no loss of correlation occurs even under large translations and deformations. Some examples of speckle patterns are shown in Fig. 5.3. One of the key features of good speckle patterns is their high information content. Since the entire surface is textured, information for pattern matching is available everywhere on the surface, and not only on a comparatively sparse grid. This permits the use of a relatively small aperture for pattern matching, commonly referred to as a subset or window.

5.2 Image Matching Methods

While it is relatively simple for a human observer to identify motion in successive images, it is not straightforward to formulate the problem in mathematical terms, and indeed, many different approaches exist. In this section, we derive a simple method based on optical flow and show how it relates to the classic Lucas–Kanade tracker. We then extend the simple template matching approach to permit matching in varying lighting conditions and take object deformations into account.

5.2.1 Differential Methods

To illustrate the problem of motion estimation, we consider a one-dimensional problem, as illustrated in Fig. 5.4. Let $G(x,t)$ be the intensity distribution on the object as a function of time. If the motion is sufficiently small, we can approximate the gray values around a point of interest by a first order Taylor expansion

$$G(x+\Delta x,t) = G(x,t) + \frac{\partial G}{\partial x}\Delta x \tag{5.1}$$

If the object moves with a constant velocity $\dot{u}$, a gray value will be displaced by an amount $\Delta x = \dot{u}\Delta t$ in a time interval Δt, and the gray value distribution after a time step Δt is merely a shifted copy of the original gray value distribution

$$\begin{aligned} \Delta G &= G(x,t+\Delta t) - G(x,t) \\ &= G(x-\dot{u}\Delta t,t) - G(x,t) \\ &= G(x-\Delta x,t) - G(x,t) \end{aligned} \tag{5.2}$$

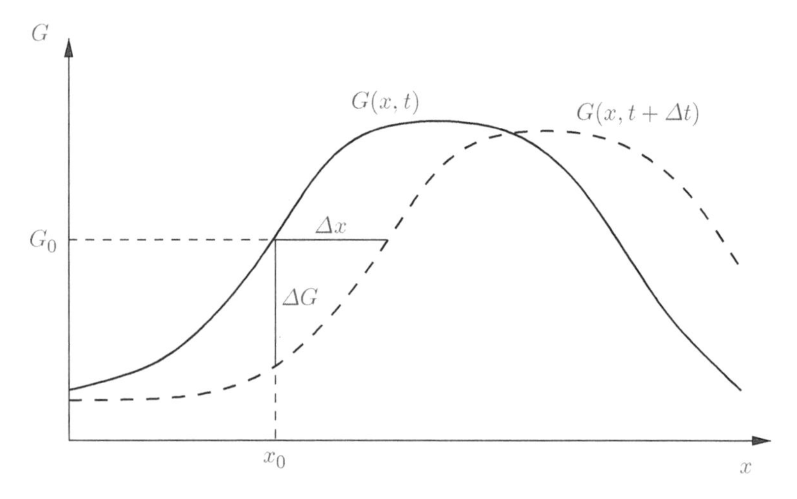

Fig. 5.4 Illustration of differential motion estimation for a 1-d problem

By substituting Eq. (5.1) into Eq. (5.2), we can express the change in gray value ΔG as a function of the slope of the intensity as follows:

$$\begin{aligned} G(x,t+\Delta t) - G(x,t) &= -\frac{\partial G}{\partial x}\Delta x \\ \Delta G &= -\frac{\partial G}{\partial x}\Delta x \end{aligned} \tag{5.3}$$

and thus obtain an estimate for the motion

$$\Delta x = -\frac{\Delta G}{\frac{\partial G}{\partial x}} \tag{5.4}$$

For small motions, the displacement is thus given by the change in intensity divided by the slope of the intensity pattern. By substituting $\Delta x = \dot{u}\Delta t$ into Eq. (5.4), we can obtain an estimate for the velocity

$$\dot{u}\Delta t = -\frac{\Delta G}{\frac{\partial G}{\partial x}} \tag{5.5}$$

and after taking the limit $\Delta t \rightarrow 0$, we arrive at

$$\frac{\partial G}{\partial t} + \dot{u}\frac{\partial G}{\partial x} = 0 \tag{5.6}$$

For the two-dimensional velocity $\mathbf{v}$, the same derivation can be followed using the Taylor expansion $G(\mathbf{x}+\boldsymbol{\Delta}\mathbf{x}) = G(\mathbf{x}) + \boldsymbol{\Delta}\mathbf{x}\cdot\nabla G$, and one arrives at the following equation

$$\frac{\partial G}{\partial t} + \mathbf{v}\cdot\nabla G = 0 \tag{5.7}$$

This equation is commonly known as the *brightness change constraint equation* for the *optical flow*, and has been at the center of a large body of research over the past decades. While a thorough discussion of optical flow methods is beyond our scope, we will discuss a simple solution method for determining displacements based on Eq. (5.7). We rewrite Eq. (5.7) for the discrete case and multiply with the time step between images to obtain

$$\boldsymbol{\Delta}\mathbf{x}\cdot\nabla G = -\Delta G \tag{5.8}$$

Inspection of Eq. (5.8) reveals that it is generally not possible to determine two-dimensional motion without further information. First, if the intensity gradients are zero, the motion cannot be determined at all. But even for non-zero gradients, Eq. (5.8) provides only one equation for two unknowns. Since the dot product $\boldsymbol{\Delta}\mathbf{x}\cdot\nabla G$ can be written as the magnitude of the gradient multiplied by the component of $\boldsymbol{\Delta}\mathbf{x}$ oriented in the direction of the local gray value gradient, i.e., the component $\Delta_{\perp}$ perpendicular to the local edge, we can resolve the motion perpendicular to the local gray value edge

$$\Delta_{\perp} = -\Delta G/|\nabla G| \tag{5.9}$$

This limitation is a mathematical expression of the aperture problem discussed in Section 5.1.1, and can be resolved by using a small neighborhood, e.g., a square subset of the image, instead of a single point. Assuming that the motion is approximately constant over a small neighborhood, we can write Eq. (5.8) for a collection of N points within the vicinity of the point of interest to give the following

$$\begin{bmatrix} \frac{\partial G^1}{\partial x} & \frac{\partial G^1}{\partial y} \\ \frac{\partial G^2}{\partial x} & \frac{\partial G^2}{\partial y} \\ \vdots & \vdots \\ \frac{\partial G^N}{\partial x} & \frac{\partial G^N}{\partial y} \end{bmatrix} \begin{bmatrix} \Delta \bar{x} \\ \Delta \bar{y} \end{bmatrix} = - \begin{bmatrix} \Delta G^1 \\ \Delta G^2 \\ \vdots \\ \Delta G^N \end{bmatrix} , \tag{5.10}$$

For more than two points, this is an over-determined equation system of the form

$$\mathbf{G} \boldsymbol{\Delta} \bar{\mathbf{x}} = -\mathbf{g} , \tag{5.11}$$

which can be solved for the average motion $\boldsymbol{\Delta}\bar{\mathbf{x}}$ in a least squares sense by

$$\boldsymbol{\Delta}\bar{\mathbf{x}} = -(\mathbf{G}^T\mathbf{G})^{-1}\mathbf{G}^T\mathbf{g} \tag{5.12}$$

Equation (5.12) can be written in the following summation form

$$\begin{bmatrix} \Delta \bar{x} \\ \Delta \bar{y} \end{bmatrix} = - \begin{bmatrix} \sum \left(\frac{\partial G}{\partial x}\right)^2 & \sum \left(\frac{\partial G}{\partial x}\frac{\partial G}{\partial y}\right) \\ \sum \left(\frac{\partial G}{\partial x}\frac{\partial G}{\partial y}\right) & \sum \left(\frac{\partial G}{\partial y}\right)^2 \end{bmatrix}^{-1} \begin{bmatrix} \sum \frac{\partial G}{\partial x}\Delta g \\ \sum \frac{\partial G}{\partial y}\Delta g \end{bmatrix} \tag{5.13}$$

Equations (5.12) and (5.13) show that the motion can be estimated as long as the matrix $\mathbf{G}^T\mathbf{G}$ is non-singular, i.e.,

$$\det(\mathbf{G}^T\mathbf{G}) = \sum \left(\frac{\partial G}{\partial x}\right)^2 \sum \left(\frac{\partial G}{\partial y}\right)^2 - \left(\sum \frac{\partial G}{\partial x}\frac{\partial G}{\partial y}\right)^2 \neq 0 \tag{5.14}$$

Trivially, this means that not all gray value derivatives must be zero. In other words, motion estimation is not possible in regions of constant gray values. Furthermore, the determinant will also vanish if all gray value gradients are aligned in the same direction. In this case, the partial derivatives in the two coordinate directions are related by a constant factor, and it can readily be shown that this implies that the determinant in Eq. (5.13) will vanish. Again, this result mathematically describes the correspondence problem discussed in Section 5.1.2.

5.2.1.1 Concluding Remarks

The relatively simple method outlined above for determining local motions that occur between two images was derived from simple geometric considerations that naturally result in the established concepts of *optical flow*. In its derivation, two underlying assumptions have been made. First, the motion has to be sufficiently small for the first-order Taylor's series expansion to be valid. Second, the motion has to be approximately constant throughout the neighborhood used for motion estimation. Aside from the limitation to small motions, a less obvious disadvantage of the method is embedded in Eqs. (5.12) and (5.13). These equations show that the metric used for defining an optimal solution is the algebraic distance $||\mathbf{g}+\mathbf{G}\mathbf{\Delta}\bar{\mathbf{x}}||^2$, a quantity that has no intuitive meaning. In the following sections, an approach is described that employs an intuitive metric and overcomes the limitation to small motions.

5.2.2 Template Matching

In this section, we derive a motion estimation method based on minimizing the gray value difference between a small subset from one image (template) and a displaced copy in another image. We assume that between the two images no lighting changes occur, i.e., the template and its displaced copy only differ by Gaussian random noise. We denote the reference image from which the template is taken by F, and the image after displacement by G. We are now trying to minimize the squared difference in gray values, known as a sum of squares deviation (SSD), over a neighborhood

$$\bar{\mathbf{d}}_{\text{opt}} = \text{argmin} \sum \left| G(\mathbf{x}+\bar{\mathbf{d}}) - F(\mathbf{x}) \right|^2 \tag{5.15}$$

To solve for the optimal displacement vector $\bar{\mathbf{d}}_{\text{opt}}$, we can use a simple iterative algorithm by expanding our cost function into a first-order Taylor series

$$\chi^2(\bar{d}_x+\Delta_x, \bar{d}_y+\Delta_y) = \sum \left| G(\mathbf{x}+\bar{\mathbf{d}}) - \frac{\partial G}{\partial x}\Delta_x - \frac{\partial G}{\partial y}\Delta_y - F(\mathbf{x}) \right|^2 \tag{5.16}$$

Here, $\bar{d}_x$ and $\bar{d}_y$ are the current estimates for the average motion of the subset, and Δ_x and Δ_y are the incremental motion updates sought in the current iteration. Taking partials of Eq. (5.16) with respect to Δ_x and Δ_y and setting them to zero results in the following linear equation system for the incremental updates in each iteration

$$\begin{bmatrix} \Delta_x \\ \Delta_y \end{bmatrix} = \begin{bmatrix} \sum \left(\frac{\partial G}{\partial x}\right)^2 & \sum \frac{\partial G}{\partial x}\frac{\partial G}{\partial y} \\ \sum \frac{\partial G}{\partial x}\frac{\partial G}{\partial y} & \sum \left(\frac{\partial G}{\partial y}\right)^2 \end{bmatrix}^{-1} \begin{bmatrix} \sum \frac{\partial G}{\partial x}(F-G) \\ \sum \frac{\partial G}{\partial y}(F-G) \end{bmatrix} \tag{5.17}$$

Equation (5.17) can be used to iteratively improve the estimate for the average motion in the pth iteration using $\bar{\mathbf{d}}^{p+1} = \bar{\mathbf{d}}^p + \mathbf{\Delta}$ until convergence to the optimal motion vector $\bar{\mathbf{d}}_{\text{opt}}$ is reached. This iterative image registration technique is the well-known Lucas–Kanade tracker algorithm [170]. The Lucas–Kanade tracker is an extension of the differential motion estimator derived in the previous section. As can be seen by comparing Eqs. (5.13) and (5.17), the differential approach is equivalent to a single iteration of the Lucas–Kanade method carried out for an initial motion estimate of zero, i.e., at the integer location. However, the Lucas–Kanade algorithm is not restricted to small motion vectors, and indeed, arbitrarily large motions can be found as long as the initial estimate is within the convergence radius of the method.

5.3 Subset Shape Functions

The image matching algorithms discussed so far are limited to the determination of the average in-plane displacement of a typically square subset between two images. In many engineering applications, however, the measurement of complex displacement fields is of interest, and the specimen might experience elongation, compression, shear or rotation. In other words, an initially square reference subset might assume a considerably distorted shape in a later image after deformation. Consider an image that is slowly rotated around its center. As the rotation angle increases, the similarity between the original subset and the rotated subset decreases. This phenomenon is often-times referred to as decorrelation, and is illustrated in Fig. 5.5. The graph illustrates the sum of square residuals χ^2 between the original and rotated subsets as a function of rotation angle. As can be seen in the graph, decorrelation occurs rapidly even for small rotation angles, and at 10°, the rotated pattern is almost entirely uncorrelated, i.e., the value of the χ^2 function at the correct displacement is not significantly lower than that for an arbitrary location. One of the key advantages of the iterative matching algorithm is that it is not limited to determining pure translations, and it can easily be extended to account for deformations. This is accomplished by introducing a subset shape function $\xi(\mathbf{x}, \mathbf{p})$ that transforms pixel coordinates in the reference subset into coordinates in the image after deformation. The SSD cost function is then written as

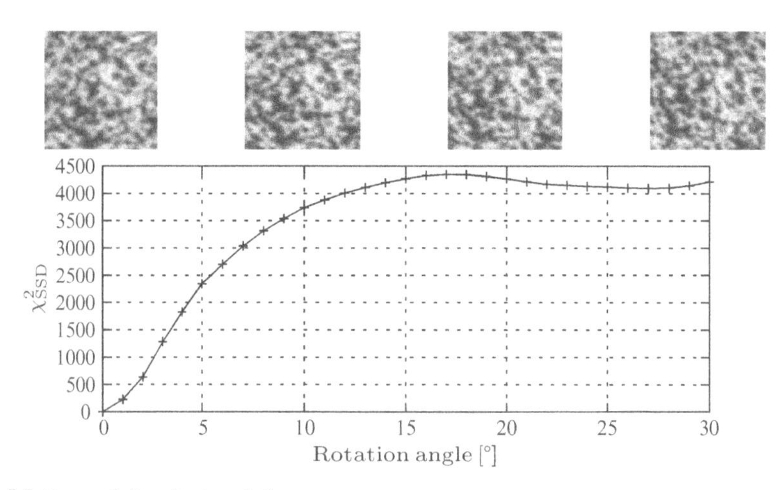

Fig. 5.5 Decorrelation due to rotation

$$\chi^2(\mathbf{p}) = \sum (G(\xi(\mathbf{x},\mathbf{p})) - F(\mathbf{x}))^2 , \tag{5.18}$$

and we optimize for the parameter vector $\mathbf{p}$ of the shape function. For the simple case of pure translation, the shape function can be written as

$$\xi(\mathbf{x},\mathbf{p}) = \mathbf{x} + \begin{bmatrix} p_0 \\ p_1 \end{bmatrix} , \tag{5.19}$$

where p_0 and p_1 represent the mean subset displacements in the x and y-direction, respectively. Similarly, one can write a shape function that permits the subset to undergo an affine transformation in the following manner

$$\xi(\mathbf{x},\mathbf{p}) = \begin{bmatrix} p_0 \\ p_1 \end{bmatrix} + \begin{bmatrix} 1+p_2 & p_3 \\ p_4 & 1+p_5 \end{bmatrix} \mathbf{x} \tag{5.20}$$

The minimization of Eq. (5.18) requires computation of derivatives of the cost function with respect to all parameters $\mathbf{p}$. In the case of affine shape functions, the derivatives can readily be computed as products of the gray value derivatives with the (subset) coordinates, and the parameter update can be computed from

$$\Delta\mathbf{p} = \mathbf{H}^{-1}\mathbf{q} , \tag{5.21}$$

with the symmetric Hessian matrix

$$\mathbf{H} = \begin{bmatrix} \sum G_x^2 & \sum G_x G_y & \sum G_x^2 x & \sum G_x^2 y & \sum G_x G_y x & \sum G_x G_y y \\ & \sum G_y^2 & \sum G_x G_y x & \sum G_x G_y y & \sum G_y^2 x & \sum G_y^2 y \\ & & \sum G_x^2 x^2 & \sum G_x^2 xy & \sum G_x G_y x^2 & \sum G_x G_y xy \\ & & & \sum G_x^2 y^2 & \sum G_x G_y xy & \sum G_x G_y y^2 \\ & & & & \sum G_y^2 x^2 & \sum G_y^2 xy \\ & & & & & \sum G_y^2 y^2 \end{bmatrix} \tag{5.22}$$

and

$$\mathbf{q} = \begin{bmatrix} \sum G_x(F-G) \\ \sum G_y(F-G) \\ \sum G_x x(F-G) \\ \sum G_x y(F-G) \\ \sum G_y x(F-G) \\ \sum G_y y(F-G) \end{bmatrix} , \tag{5.23}$$

where partial differentiation is denoted by subscripts, i.e., $G_x = \partial G/\partial x$. It is noted that the introduction of higher-order shape functions further complicates the correspondence problem. For instance, the in-plane rotation component cannot be determined for a pattern of concentric circles, and the use of higher-order shape functions is not possible in this case. Whether a particular gray value pattern is suitable for a particular shape function can mathematically be determined by whether the Hessian matrix can be inverted or not.

5.3.1 Polynomial Shape Functions

A simple approach to account for increasingly complex subset deformations is to use shape functions that are polynomials in the subset coordinates. This family of shape functions includes pure displacements (zero-order polynomial) and affine transformations (first-order polynomial), but can be extended to allow for quadratic and higher-order polynomial functions. The computation of derivatives of the cost function for polynomial shape functions is accomplished using the chain rule, and derivatives with respect to the shape function parameters are easily found as products of gray value derivatives and powers of subset coordinates.

Polynomial shape functions have another advantage. It is particularly simple to analyze what answer a digital image correlation algorithm will yield for a given displacement field. The image correlation algorithm does not directly fit a displacement field to measured displacement data, but accomplishes this goal through minimizing an error function defined on a secondary measure, namely the intensity distribution. It is, however, the central assumption that the true displacement field will be best approximated by the local shape function when the error function defined in terms of the gray values attains its minimum. In this case, the subset used for image correlation defines a polynomial low-pass filter applied to the displacement field encoded in the images. For zero-order polynomial shape functions this is intuitive. Rather than measuring the displacement value at the center of the subset, the digital image correlation method computes the average displacement of the subset. Thus, the subset shape function acts as a box filter for the underlying displacement field. Mathematically, the shape function parameters obtained by image correlation can directly be predicted by minimizing the difference between the shape function $\mathbf{\Phi}(\mathbf{p})$ and the underlying displacement field $\mathbf{u}$ by solving

$$\mathbf{p}_{\text{opt}} = \operatorname{argmin} \sum |\mathbf{\Phi}(\mathbf{x},\mathbf{p}) - \mathbf{u}(\mathbf{x})|^2 \tag{5.24}$$

For simplicity, we limit the following discussion to the one-dimensional case and focus on the x-component of the displacement vector. The shape function is given by a polynomial in the x-coordinate of the subset

$$\phi(x) = p_0 + p_1 x + \cdots + p_n x^N \tag{5.25}$$

and minimization of Eq. (5.24) gives rise to a linear equation system

$$\begin{bmatrix} \sum x^0 & \sum x & \sum x^2 & \ldots & \sum x^N \\ \ldots & \sum x^2 & \sum x^3 & \ldots & \sum x^{N+1} \\ \ldots & \ldots & \sum x^4 & \ldots & \sum x^{N+2} \\ \vdots & \vdots & \vdots & \vdots & \vdots \\ \ldots & \ldots & \ldots & \ldots & \sum x^{2N} \end{bmatrix} \begin{Bmatrix} p_0 \\ p_1 \\ p_2 \\ \vdots \\ p_N \end{Bmatrix} = \begin{Bmatrix} \sum u(x) \\ \sum u(x)x \\ \sum u(x)x^2 \\ \vdots \\ \sum u(x)x^N \end{Bmatrix} \tag{5.26}$$

with a symmetric matrix. If the subset has $2M+1$ points, i.e., the summations over x-coordinates are carried out from $x=-M$ to $x=M$, all odd powers in the subset coordinates vanish due to symmetry, and one obtains the simplified equation system

$$\begin{bmatrix} \sum x^0 & 0 & \sum x^2 & \dots & \sum x^N \\ \dots & \sum x^2 & 0 & \dots & \sum x^{N+1} \\ \dots & \dots & \sum x^4 & \dots & \sum x^{N+2} \\ \vdots & \vdots & \vdots & \vdots & \vdots \\ \dots & \dots & \dots & \dots & \sum x^{2N} \end{bmatrix} \begin{Bmatrix} p_0 \\ p_1 \\ p_2 \\ \vdots \\ p_N \end{Bmatrix} = \begin{Bmatrix} \sum u(x) \\ \sum u(x)x \\ \sum u(x)x^2 \\ \vdots \\ \sum u(x)x^N \end{Bmatrix} \tag{5.27}$$

For a known displacement field, Eq. (5.27) permits the direct computation of the subset center displacements $u_0 = p_0$ for polynomial shape functions. Furthermore, inspection of Eq. (5.27) reveals that the parameters p_0 and p_1, and p_1 and p_2 are uncorrelated due to the symmetric placement of data points on either side of the subset center. In other words, zero and first-order shape functions yield identical subset center displacements $u_0 = p_0$, while first and second-order shape functions yield identical derivatives $du/dx = p_1$, irrespective of the order of the actual displacement $u(x)$. Equation (5.27) can also be interpreted as a Savitzky–Golay [249] low-pass filter that takes input values $u(x)$ and yields a smoothed value $u_0 = p_0$ for the displacement at the subset center. For a given order, Eq. (5.27) can be solved for p_0 in terms of the unknown $u(x)$. This yields an equation of the form

$$\begin{aligned} p_0 = {} & h(M)u(-M) + h(M-1)u(-M+1) + \cdots \\ & + h(0)u(0) + \cdots + h(-M+1)u(M-1) + h(-M)u(M) \end{aligned} \tag{5.28}$$

which represents a convolution of $u(x)$ with a filter kernel $h(x)$. The coefficients in $h(x)$ depend on the order of the shape function. For zero and first-order shape functions, this filter is a simple box filter with

$$h_{0/1}(x) = \frac{1}{2M+1}\,, \tag{5.29}$$

and for a second-order filter, the coefficients are found as

$$h_2(x) = A - Bx^2\,, \tag{5.30}$$

where

$$\begin{aligned} A &= \frac{1}{2M+1-\dfrac{\left(\sum_{x=-M}^{M} x^2\right)^2}{\sum_{x=-M}^{M} x^4}} \\ B &= A\frac{\sum_{x=-M}^{M} x^2}{\sum_{x=-M}^{M} x^4} \end{aligned} \tag{5.31}$$

The transfer functions for zero and second-order shape functions and a subset size of 25 ($M = 12$) are shown in Fig. 5.6. It can be seen that the second-order filter significantly extends the cut-off toward higher wave numbers. In practice, this means that second-order shape functions will introduce less systematic error in measuring

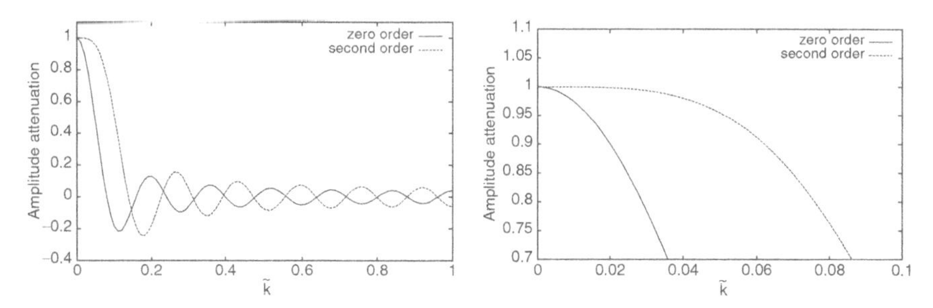

Fig. 5.6 Transfer function of the low-pass filter corresponding to a subset size of 25 pixels for zero and second order shape functions. The right graph shows a detail for small wave numbers

displacement fields that vary quickly over short distances, i.e., displacement fields containing higher spatial frequencies. The higher cut-off can also be interpreted as an increase in spatial resolution, since the spatial resolution of the measured displacement is given by the highest spatial wave number. While we refer to [253] for a more detailed discussion of polynomial shape functions and the systematic errors associated with them, several points are worth mentioning. First and foremost, shape functions always have to be chosen such that they are capable of accurately representing the underlying displacement field over the size of the subset. Otherwise, decorrelation will occur and Eq. (5.27) is no longer valid. Even though zero and first order polynomial shape functions theoretically yield the same center point displacement, zero order shape functions suffer from decorrelation in many practical applications, e.g., when rotations or appreciable strains are present. While lower order shape functions execute faster and have better noise properties (see [253]), they cannot be used in many applications due to decorrelation. Second, the low-pass filtering effect of the subset shape functions can be limited by minimizing the size of the subset. This underscores the need for sufficiently dense speckle patterns that have a large information content even in a small subset. Third, the filter properties of the low-pass filters that result from polynomial shape functions are not very desirable, as evidenced in the oscillations shown in Fig. 5.6. Furthermore, the two-dimensional box filters also show significant anisotropy. This problem can be remedied by the introduction of a weighting function $w(\mathbf{x})$ in the cost function in Eq. (5.18):

$$\chi^2(\mathbf{p}) = \sum w(\mathbf{x})\left(\mathbf{G}(\xi(\mathbf{x},\mathbf{p})) - \mathbf{F}(\mathbf{x})\right)^2 \tag{5.32}$$

For the weighting function, the Gaussian distribution is a natural choice, as it provides the best compromise between spatial and displacement resolution.

5.3.2 Shape Functions for Stereo Matching

The displacements that arise between two views of the same object from different viewing angles are generally not described by affine transformations, even if the object is a plane. This is due to the non-linear nature of the perspective projection described in Chapter 2. Using Eq. (3.7), we can write the relationship between sensor coordinates and world coordinates as

$$\alpha \begin{Bmatrix} x_s \\ y_s \\ 1 \end{Bmatrix} = \begin{bmatrix} \Lambda \end{bmatrix}_{3\times 4} \begin{Bmatrix} X_W \\ Y_W \\ Z_W \\ 1 \end{Bmatrix} \tag{5.33}$$

For a planar object, we can always align the global coordinate system such that the object is in the $X_w Y_w$-plane, i.e., the Z-component of the object coordinates vanishes ($Z_w = 0$). In this case, we can omit the Z_w-coordinate from the equation and obtain

$$\alpha \begin{Bmatrix} x_s \\ y_s \\ 1 \end{Bmatrix} = \begin{bmatrix} \mathbf{H} \end{bmatrix}_{3\times 3} \begin{Bmatrix} X_W \\ Y_W \\ 1 \end{Bmatrix}, \tag{5.34}$$

where

$$\mathbf{H} = \begin{bmatrix} \lambda_{11} & \lambda_{12} & \lambda_{14} \\ \lambda_{21} & \lambda_{22} & \lambda_{24} \\ \lambda_{31} & \lambda_{32} & \lambda_{34} \end{bmatrix} \tag{5.35}$$

The 3×3-matrix $\mathbf{H}$ is a so-called plane-to-plane homography that relates image coordinates to coordinates on a plane in space. Since Eq. (5.34) is defined up to a scale, the homographic mapping only contains eight independent parameters, and homographies are typically normalized such that $H_{33} = 1$. If two cameras are viewing the same planar scene, there will be a homography for each camera that relates its image coordinates to coordinates in the plane. Since homographies can be inverted, Eq. (5.34) can be used to relate the image coordinates between two cameras viewing a planar object

$$\begin{aligned} \alpha \mathbf{x}_{s1} &= \mathbf{H}_1 \mathbf{H}_2^{-1} \mathbf{x}_{s2} \\ &= \mathbf{H}_{12} \mathbf{x}_{s2} \end{aligned} \tag{5.36}$$

Thus, the image coordinates between the two cameras are themselves related by a homography. For a digital image correlation algorithm to accurately reconstruct planar objects, or objects that are locally well approximated by a plane, the shape functions used should therefore be capable of representing homographies. This can be accomplished in several ways. A straightforward approach is to use a homography as the shape function

$$\mathbf{\Phi}(\mathbf{x}, \mathbf{p}) = \begin{Bmatrix} \dfrac{p_{11}x + p_{12}y + p_{13}}{p_{31}x + p_{32}y + 1} \\ \dfrac{p_{21}x + p_{22}y + p_{23}}{p_{31}x + p_{32}y + 1} \end{Bmatrix} \tag{5.37}$$

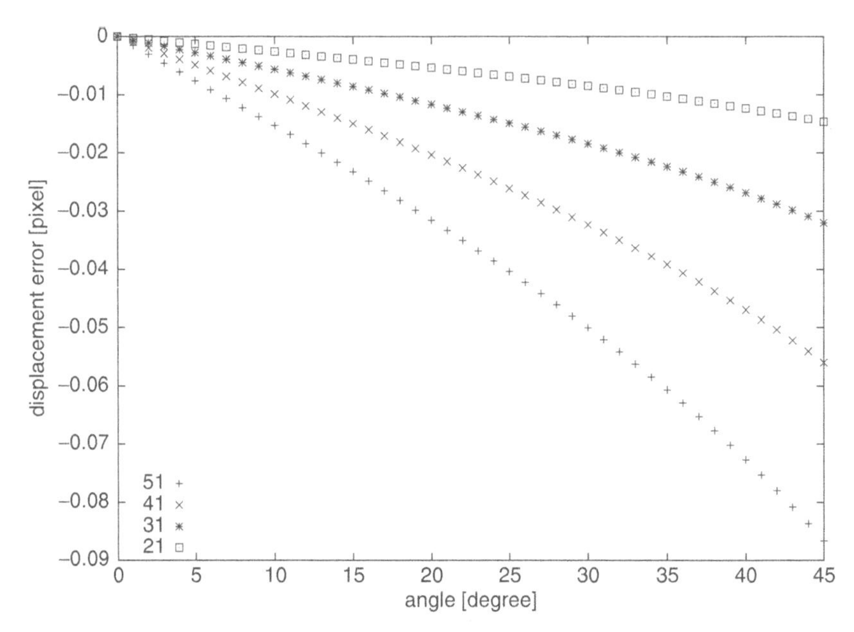

Fig. 5.7 Error in center point coordinate computed by digital image correlation when affine shape functions are used for stereo matching

Due to the non-linear nature of this shape function, the computation of the Hessian matrix becomes somewhat more complicated, and it has been our experience that homographic shape functions can suffer from convergence problems in some implementations. Another approach is to use shape functions that can approximate homographies sufficiently well. Inspection of Eq. (5.37) reveals that if p_{31} and p_{32} are small, the homographic shape function approaches an affine transformation. This poses the question whether affine shape functions can approximate homographies well enough to be useful for stereo matching in practical applications. In many situations, this is actually the case. Figure 5.7 shows the center-point error if affine shape functions are used to approximate the homographic mapping as a function of the stereo angle and different subset sizes. The focal lengths of the two cameras were chosen to approximately correspond to a 25 mm focal length on a typical CCD camera. As can be seen, the bias can be kept fairly small for small subset sizes and stereo angles below 25°. However, substantial error arises for large subset sizes and large stereo angles. In such cases, it is best to avoid affine shape functions for stereo matching.

For a calibrated stereo system, a different approach can be used. As shown in Chapter 4, point correspondences are constrained to the epipolar line in stereo imaging. If the calibration of the two cameras and their relation in space is known, it is possible to find homographic transformations that can be applied to the images such that all epipolar lines become horizontal and are aligned on the same scanline in each of the two stereo images. This process is normally referred to as *rectification*, and is a widely used method in computer vision. In *rectified* images, it is possible to use much simpler shape functions, for example

$$\Phi(\mathbf{x},\mathbf{p}) = \begin{Bmatrix} p_0 + p_1 x + p_2 y \\ y \end{Bmatrix} \tag{5.38}$$

While image rectification has the advantage that fairly simple shape functions can be used with good success, there are some drawbacks to the method. First, any error in the calibration of the stereo system will lead to some error in the rectified images, and the digital image correlation algorithm is constrained to find an erroneous match instead of the best match. This issue can be addressed by not constraining the subset to the epipolar line via the introduction of an additional shape function parameter

$$\Phi(\mathbf{x},\mathbf{p}) = \begin{Bmatrix} p_0 + p_1 x + p_2 y \\ y + p_3 \end{Bmatrix} \tag{5.39}$$

Second, image rectification requires interpolation, which can introduce positional error into the rectified images. Furthermore, the resampling process can introduce aliasing if the rectifying homographies compress all or part of an image. This situation should be avoided during the computation of the rectifying homographies.

5.3.3 Summary

One of the primary advantages of the iterative digital image correlation algorithm discussed in this chapter is that it can easily cope with complex deformation fields by incorporating subset shape functions into the matching algorithm. The algorithm is therefore the preferred choice for deformation measurements and stereo matching. The analysis of polynomial shape functions has shown that subset shape functions are equivalent to low-pass filters applied to the displacement fields encoded in the images. This limits the spatial resolution of the method, and the experimentalist should take care to apply appropriate speckle patterns such that small subsets can be used for matching, particularly when affine shape functions are used to approximate non-linear displacement fields, e.g., the displacement fields between stereo views.

5.4 Optimization Criteria for Pattern Matching

Even under near ideal experimental conditions, there will be differences between the intensity of images recorded at different times. These changes may be due to a variety of reasons, e.g., changes in lighting, changes in specimen reflectivity due to strain or changes in the orientation of the specimen. Such changes may be localized and not affect the entire image equally. This is of particular concern when trying to match subsets from stereo images, as the difference in camera angle to the object results in localized brightness differences in most cases. It is therefore critical to develop matching algorithms that can accurately measure the correct correspondence between subsets even if the intensity values undergo significant changes. The template matcher discussed so far was derived by minimizing the squared gray value differences between the reference subset and the subset after motion. This simple optimization criterion cannot be expected to yield accurate results if, for example, one image is twice as bright as the other. The squared sum of differences (SSD), however, is just one of many optimization criteria that can be used for template matching, and indeed, the digital image correlation method owes its name to the use of the normalized cross-correlation criterion

$$\chi^2_{NCC} = \frac{\sum FG}{\sqrt{\sum F^2 \sum G^2}} \tag{5.40}$$

The normalized cross-correlation criterion is bounded in the interval $[0,1]$, and attains its maximum for perfectly matching patterns. An attractive feature of the normalized cross-correlation criterion is that it is independent of scale. For practical applications, this means that matching two patterns that differ in a multiplicative change in intensity due to varying lighting conditions becomes possible. The disadvantage of the cross-correlation criterion, however, is that its derivatives are more complex to implement and more time consuming to evaluate. We therefore seek to derive efficient optimization criteria that can be used if the pattern undergoes changes in lighting. The lighting variations are accounted for by substituting the gray values G by a photometric transformation $\Phi(G)$ in Eq. (5.15)

$$\chi^2 = \sum |\Phi(G) - F|^2 \tag{5.41}$$

Different criteria can now be derived for different functional forms of Φ.

5.4.1 Offset in Lighting

If the lighting during image acquisition changes only by an offset, the photometric transformation becomes

$$\Phi(G) = G + b \tag{5.42}$$

and the cost function can be written as

$$\chi^2 = \sum (G_i + b - F_i)^2 \tag{5.43}$$

The offset b can be solved for explicitly as an additional parameter in the optimization problem. Alternatively, a criterion can be developed that does not explicitly depend on b, but implicitly minimizes Eq. (5.43). At a given iteration, the optimal value for b can be found by minimizing the cost function with respect to b, i.e.,

$$b_{\text{opt}} = \text{argmin}_b \sum (G_i + b - F_i)^2 \tag{5.44}$$

The optimal estimate for b can be found by

$$\frac{\partial \chi^2}{\partial b} = 2\sum(G_i + b - F_i)$$
$$\frac{\partial \chi^2}{\partial b} = 0 \quad \Longrightarrow \quad b_{\text{opt}} = \bar{F} - \bar{G} \tag{5.45}$$
$$\bar{F} = \frac{\Sigma F}{n}; \; \bar{G} = \frac{\Sigma G}{n}$$

By substituting the best estimator b_{opt} in Eq. (5.43), we obtain the zero-mean sum of square difference criterion (ZSSD)

$$\chi^2_{\text{ZSSD}} = \sum\left((G_i - \bar{G}) - (F_i - \bar{F})\right)^2 \tag{5.46}$$

5.4.2 *Scale in Lighting*

In a similar manner, one can obtain a scale-invariant criterion by defining the photometric transformation

$$\Phi(G) = aG \tag{5.47}$$

and the cost function

$$\chi^2 = \sum(aG_i - F_i)^2 \tag{5.48}$$

The optimal estimate for a is found by

$$\frac{\partial \chi^2}{\partial a} = 2\sum(aG_i - F_i)G_i$$
$$\frac{\partial \chi^2}{\partial a} = 0 \quad \Longrightarrow \quad a_{\text{opt}} = \frac{\sum F_i G_i}{\sum G_i^2} \tag{5.49}$$

This leads to the so-called normalized sum of squared difference criterion (NSSD)

$$\chi^2_{\text{NSSD}} = \sum\left(\frac{\sum F_i G_i}{\sum G_i^2} G_i - F_i\right)^2 \tag{5.50}$$

5.4.3 *Offset and Scale in Lighting*

We can combine the two photometric transformations above and derive an optimization criterion that allows for an offset and a scale in lighting by defining our photometric transformation as

$$\Phi(G) = aG + b \tag{5.51}$$

and the cost function

$$\chi^2 = \sum (aG_i + b - F_i)^2 \tag{5.52}$$

We can derive two equations to solve for the optimal estimators for a and b by writing

$$\begin{aligned}
\frac{\partial \chi^2}{\partial a} &= 2\sum (aG_i + b - F_i)G_i \\
\frac{\partial \chi^2}{\partial a} &= 0 \quad \Longrightarrow \quad a_{\text{opt}} = \frac{\sum (F_i - b)G_i}{\sum G_i^2} \\
\frac{\partial \chi^2}{\partial b} &= 2\sum (aG_i + b - F_i) \\
\frac{\partial \chi^2}{\partial b} &= 0 \quad \Longrightarrow \quad b_{\text{opt}} = \frac{\sum F_i - aG_i}{\sum 1} = \frac{\sum F_i - aG_i}{n}
\end{aligned} \tag{5.53}$$

Introducing $\bar{F}_i = F_i - \bar{F}$ and $\bar{G}_i = G_i - \bar{G}$ and after some algebraic manipulations, one arrives at

$$\begin{aligned}
a_{\text{opt}} &= \frac{\sum \bar{F}_i \bar{G}_i}{\sum \bar{G}_i^2} \\
b_{\text{opt}} &= \bar{F} - \bar{G}\frac{\sum \bar{F}_i \bar{G}_i}{\sum \bar{G}_i^2}
\end{aligned} \tag{5.54}$$

and finally the so-called zero-mean normalised sum of squared difference (ZNSSD) criterion

$$\chi^2_{\text{ZNSSD}} = \sum \left(\left(\frac{\sum \bar{F}_i \bar{G}_i}{\sum \bar{G}_i^2} G_i - \bar{G}\frac{\sum \bar{F}_i \bar{G}_i}{\sum \bar{G}_i^2} \right) - F_i + \bar{F} \right)^2 \tag{5.55}$$

5.4.4 Concluding Remarks

In the previous sections, we have derived optimization criteria that permit accurate template matching even if the images differ by localized changes in brightness. The underlying assumption behind the approach is that all lighting changes are approximately constant over the size scale of the subset. The use of these criteria is critical for many practical applications, as changes in lighting are difficult to control. This is a particular problem for stereo matching, as the difference in viewing angles between the two stereo cameras oftentimes results in substantial differences in brightness of corresponding image regions. The equations for the different criteria are summarized in Table 5.1. The table also contains equations for the less used normalized cross-correlation (NCC) as well as the sum of absolute differences (SAD) criterion. While the criteria that can accommodate changes in lighting, particularly the ZNSSD criterion, appear rather complex and difficult to implement, it is noted that they can be evaluated in a single pass just like the SSD criterion and introduce virtually no computational overhead compared to the SSD criterion.

Table 5.1 Summary of common optimization criteria

Name	Formula	Intensity changes	$\Phi(G)$
SSD	$\sum_i (G_i - F_i)^2$	none	$\Phi = G$
ZSSD	$\sum ((G_i - \bar{G}) - (F_i - \bar{F}))^2$	offset	$\Phi = G + b$
NSSD	$\sum \left(\frac{\sum F_i G_i}{\sum G_i^2} G_i - F_i \right)^2$	scale	$\Phi = aG$
ZNSSD	$\sum \left(\left(\frac{\sum \bar{F}_i \bar{G}_i}{\sum \bar{G}_i^2} G_i - \bar{G} \frac{\sum \bar{F}_i \bar{G}_i}{\sum \bar{G}_i^2} \right) - (F_i - \bar{F}) \right)^2$	scale + offset	$\Phi = aG + b$
NCC	$1 - \frac{\sum_i F_i G_i}{\sqrt{\sum_i F_i^2 \sum_i G_i^2}}$	scale	$\Phi = aG$
SAD	$\sum_i \lvert F_i - G_i \rvert$	none	$\Phi = G$

5.5 Efficient Solution Methods

As a full-field method, digital image correlation is typically used to compute a large number of displacement vectors throughout the image. For example, if one wishes to analyze displacement vectors on 5 pixel centers in a typical $1{,}024 \times 1{,}024$ pixel image, approximately 40,000 displacement vectors have to be computed. With the ongoing trend towards higher resolution cameras, the number of data points that need to be analyzed keeps increasing. Furthermore, the analysis is typically not limited to a single image, but rather applied to a sequence of images. It is not uncommon to capture several hundred images during dynamic deformation events, resulting in the need to repeat the basic image matching algorithm several million times. For this reason, the computational efficiency of digital image correlation algorithms is of central importance. In the following sections, we discuss how the Lucas–Kanade algorithm can be modified to improve its efficiency.

5.5.1 *Efficient Update Rules for Planar Motion*

As discussed above, the minimization of the cost function

$$\chi^2(\mathbf{p}) = \sum (G(\xi(\mathbf{x}, \mathbf{p})) - F(\mathbf{x}))^2 \tag{5.56}$$

is accomplished by iteratively computing parameter updates $\Delta \mathbf{p}$ until convergence is reached. In each step, the gray values and the gradients of the gray values in the deformed image G have to be evaluated at the current locations given by the shape function $\xi(\mathbf{p})$. The approximation of the Hessian matrix then needs to be computed and inverted, and multiplied by a vector assembled from the gray value differences multiplied by gradient terms. To minimize the computational cost of the algorithm, it would be advantageous to arrive at a formulation of the problem that does not require the recomputation of the Hessian matrix in every step of the iteration. To explore this possibility, we first examine the problem of pure translation, and rewrite the cost function in Eq. (5.56) as

$$\chi^2(\mathbf{p}) = \sum (G(\mathbf{x}+\mathbf{p}) - F(\mathbf{x}))^2 \tag{5.57}$$

The gray values $F(\mathbf{x})$ and $G(\mathbf{x}+\mathbf{p})$ each form an $N \times N$ matrix and can be thought of as images. In each iteration, the algorithm discussed so far computes an incremental motion vector $\boldsymbol{\Delta p}$ that moves image G to bring it in better alignment with F. By swapping the role of F and G, it should be possible to compute an incremental motion vector $\boldsymbol{\Delta u}$ that moves image F to bring it in closer alignment with G. Conceptually, these motion vectors should differ only in their sign, i.e.,

$$\boldsymbol{\Delta u} = -\boldsymbol{\Delta p} \, , \tag{5.58}$$

and we can compute the update to the parameter vector $\mathbf{p}$ as

$$\mathbf{p}^{n+1} = \mathbf{p}^n - \boldsymbol{\Delta u} \tag{5.59}$$

This new update rule will be referred to as the *inverse* update rule as opposed to the *forwards* update rule discussed in the previous sections. By swapping the role of F and G for the purpose of the parameter update computation, the derivatives that appear in the Hessian matrix are now derivatives of the original image F, i.e., they remain constant in each iteration:

$$\mathbf{H} = \begin{bmatrix} \sum \left(\frac{\partial F}{\partial x}\right)^2 & \sum \frac{\partial F}{\partial x}\frac{\partial F}{\partial y} \\ \sum \frac{\partial F}{\partial x}\frac{\partial F}{\partial y} & \sum \left(\frac{\partial F}{\partial y}\right)^2 \end{bmatrix} \tag{5.60}$$

Aside from the obvious advantage that the Hessian matrix only has to be computed and inverted once, this has other, more subtle benefits. In the *forwards* formulation, gray value derivatives had to be evaluated at subpixel locations, whereas the *inverse* update rule requires evaluation of derivatives at integer pixel locations only. This greatly reduces the computational cost of the algorithm since it eliminates the computation of interpolating derivative filter coefficients and the computation of the corresponding convolution. This operation is particularly expensive for higher-order shape functions, since the subpixel position is generally different for each pixel in the subset. Another benefit resulting from the evaluation of derivatives at integer locations in the inverse update scheme is that it is much easier to design good derivative filters. Whether the subpixel derivatives are computed using interpolation of pre-computed gradient images or using interpolating derivative filters, e.g., derivatives of B-splines, the resulting derivatives generally suffer from a positional bias, which is eliminated by the inverse scheme.

While the derivation of the inverse update rule is very intuitive and the benefits are obvious, it is unclear whether this algorithm indeed minimizes the cost function (5.57). Fortunately, the equivalence of the two algorithms has been mathematically proven in [25], and empirical evidence has shown the two algorithms to have the same convergence properties.

5.5.2 Extension to General Shape Functions

In the previous section, we have derived an inverse update algorithm for digital image correlation that is applicable to planar motion only. Since most practical applications require the use of at least affine shape functions, we now extend the concept to general shape functions. Again, we swap the role of the reference image and the deformed image for the purpose of computing parameter updates. We introduce a shape function $\xi(\mathbf{x},\mathbf{u})$ that is applied to the reference image F and compute an incremental parameter update $\Delta\mathbf{u}^n$ in the n-th iteration from

$$\Delta\mathbf{u}^n = \operatorname{argmin} \sum (G(\xi(\mathbf{x},\mathbf{p}^n)) - F(\xi(\mathbf{x},\mathbf{u})))^2 \tag{5.61}$$

It is important to note that the shape function used for the reference image $\xi(\mathbf{x},\mathbf{u})$ is always the identity transformation, i.e., the shape of the reference subset is never changed. The shape function is merely used to compute a parameter update $\Delta\mathbf{u}$ that *would* bring the reference image in better alignment with the deformed image, but the actual updates are applied to the shape function $\xi(\mathbf{x},\mathbf{p})$ for the deformed image. It is clear that the success of this strategy depends on the ability to compute an updated parameter vector $\mathbf{p}^{n+1}$ from $\mathbf{p}^n$ and $\Delta\mathbf{u}^n$. Since we now compute the update for the reference image and we are seeking the update to the shape function parameters in the deformed image, it can be assumed that the update to $\Delta\mathbf{p}$ must somehow be the opposite of the update $\Delta\mathbf{u}$. If $\Delta\mathbf{u}$ implies a motion to the right, $\Delta\mathbf{p}$ should result in a motion to the left. Similarly, a clock-wise rotation in $\Delta\mathbf{u}$ should result in a counter-clockwise rotation in $\Delta\mathbf{p}$, or a contraction should result in an expansion. In other words, we are looking for a parameter update $\Delta\mathbf{u}'$ that inverts the effect of $\xi(\mathbf{x},\Delta\mathbf{u})$, or

$$\xi(\mathbf{x},\Delta\mathbf{u}') = \xi^{-1}(\mathbf{x},\Delta\mathbf{u}) \tag{5.62}$$

This implies that our shape function must be invertible, which is the case for most shape functions of interest. For affine shape functions, for instance, a simple matrix inversion is required. Once the parameters $\Delta\mathbf{u}'$ of the inverse update have been found, the question of how to compute $\Delta\mathbf{p}$ still remains. At this point, we re-examine the update computation more closely to understand what the parameter updates $\Delta\mathbf{u}$ and $\Delta\mathbf{u}'$ describe. The subset in the reference image is typically an $N \times N$ sub-image I of gray values. The current shape function $\xi(\mathbf{x},\mathbf{p}^n)$ can be used to compute a corresponding gray value in the deformed image for each pixel in the reference sub-image, and the gray values from the deformed image can themselves be organized in an $N \times N$ image J. We now break down the step of computing parameter updates into two separate steps:

1. Compute an $N \times N$ sub-image J by evaluating $G(\xi, \mathbf{p}^n)$.
2. Compute a parameter update $\Delta \mathbf{u}^n$ that brings the square reference subset in better alignment with the square sub-image J.

Mathematically, the second step is accomplished by computing the update from

$$\Delta \mathbf{u}^n = \operatorname{argmin} \sum (J(\mathbf{x}) - I(\xi(\mathbf{x}, \mathbf{u})))^2 , \tag{5.63}$$

which no longer depends on the current parameter estimate $\mathbf{p}^n$. In other words, the parameter updates are always computed for square sub-images and only depend on the gray values of J, but not on the shape function parameters that were used to generate J. Since the update computation is independent of $\mathbf{p}$, it is clear that the update to the parameter vector $\Delta \mathbf{p}$ is not equal to the inverse update $\Delta \mathbf{u}'$, i.e., $\Delta \mathbf{p} \neq \Delta \mathbf{u}'$. To further illustrate this, and to determine the correct way to compute the parameter update $\Delta \mathbf{p}$, consider the example illustrated in Fig. 5.8. The deformed image corresponds to a 90° counter-clock wise rotation. The reference subset used for matching is indicated as a square box in the reference image, and the top-left corner is marked by a triangle. Suppose the current parameter vector correctly captures the 90° rotation, but is 1 pixel too low, as indicated by the box in the deformed image. The square sub-images I and J used for the update computation are also illustrated. Since the parameter update is computed for the reference sub-image I, the computed update will express a motion to the left of 1 pixel, which would make the reference subset look identical to the deformed subset J, as illustrated. The inverse update, $\Delta \mathbf{u}'$ corresponds to a motion of 1 pixel to the right, which in turn makes the deformed subset J look identical to the reference subset I, as illustrated. The correct motion vector, however, is a motion upwards, which can be found by applying the current shape function to the inverse update vector $\Delta \mathbf{u}'$. This means that the shape function parameters for the next iteration can no longer be found by adding a vector update $\Delta \mathbf{p}$ to the current parameter vector $\mathbf{p}$. Instead, they must be composed by applying the current shape function to the inverse of the shape function update computed in the iteration

$$\xi(\mathbf{x}, \mathbf{p}^{n+1}) \leftarrow \xi(\mathbf{x}, \mathbf{p}^n) \circ \xi^{-1}(\mathbf{x}, \Delta \mathbf{u}) \tag{5.64}$$

For this reason, the algorithm is commonly referred to as the *inverse compositional* algorithm and it was first introduced in [24]. The inverse compositional algorithm has been proven to be equivalent to the forwards additive algorithm to first order in [24], and a detailed analysis of the algorithms can be found in [25]. In our experience, the inverse compositional algorithm outperforms the forwards additive algorithm by a factor between two and five, depending on the interpolation method used.

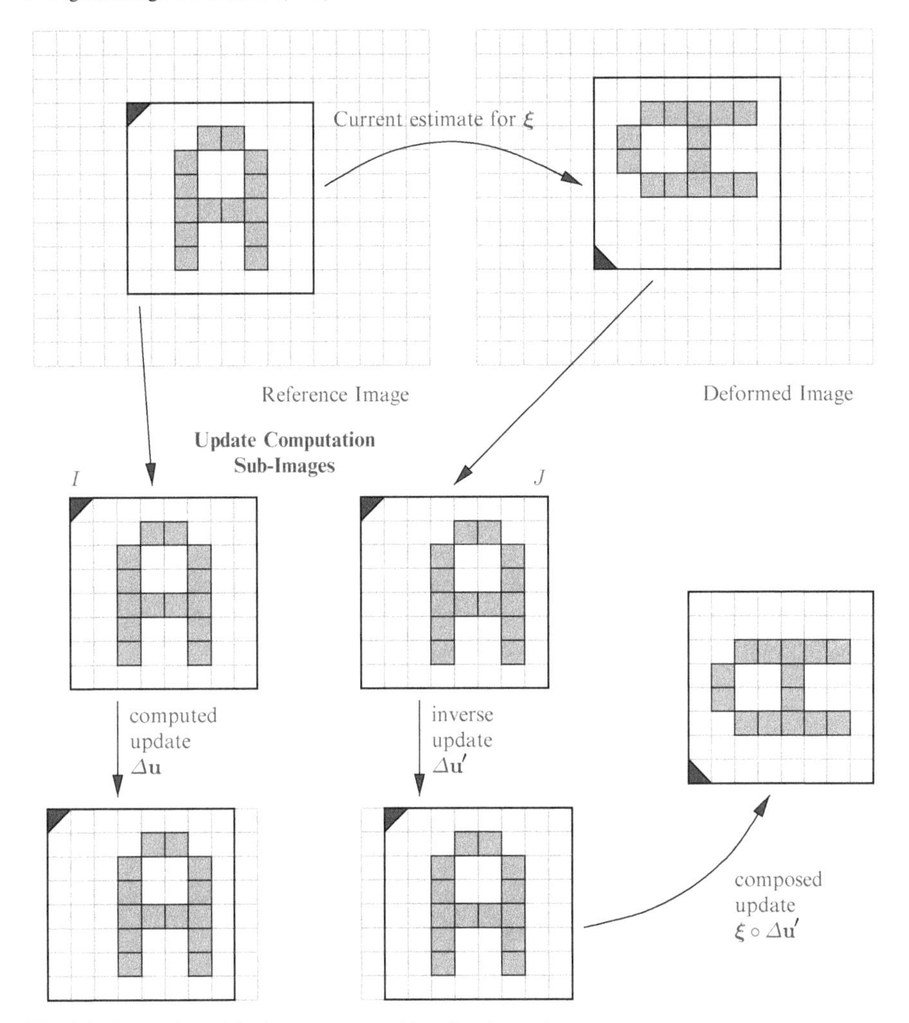

Fig. 5.8 Illustration of the inverse compositional update rule

5.6 Matching Bias

5.6.1 Interpolation Bias

In order to obtain sub-pixel accuracy, the cost function χ^2 has to be evaluated at non-integer locations. Therefore, gray values have to be interpolated between the sample points (pixels). As interpolation is a method of approximating a value between two samples, the use of gray value interpolation in the matching algorithm can be expected to introduce errors. However, it is not immediately obvious how gray value interpolation will affect the matching result. For simplicity, consider the case of matching two monochromatic waves

$$f(x) = \cos(kx)$$
$$g(x) = \cos(kx + \Delta\phi) \tag{5.65}$$

The best match between the two continuous waves is found for a displacement of $u_t = -\Delta\phi/k$. According to the sampling theorem, the two continuous waves can be represented by discrete samples if sampled more than twice per period. Consequently, the best match should still be found at $u_t = -\Delta\phi/k$, if the continuous waves are reconstructed from the samples. Herein, however, lies a practical problem. To reconstruct the original wave from the samples, a convolution of the samples with an infinite convolution kernel has to be performed. The convolution kernel for the ideal reconstruction is a sinc function that decreases only slowly with an envelope of $1/|x|$. Since this ideal reconstruction would involve a huge number of operations for each pixel within a subset, ideal reconstruction is impractical. Therefore, one has to resort to approximations of the ideal interpolation function. A classical approach to interpolation is the use of polynomials to represent the values in between samples. The simplest example is linear interpolation, which can be represented as a convolution operation with a convolution kernel

$$h(x) = [1 - x, x] \tag{5.66}$$

It is apparent that this operation significantly differs from the ideal interpolation represented by a sinc function. However, it is still not obvious how the matching result will be influenced by using this approximation. Since polynomial interpolation is a linear, shift-invariant operation, it will preserve the wave number of the original wave, but potentially alter both amplitude and phase. That is, the interpolated reconstruction of the original wave $g(x)$ will take on the form of

$$g_i(x) = a(x)\cos(kx + \Delta\phi + \phi_e(x)) \tag{5.67}$$

Here, $a(x)$ represents the amplitude attenuation and $\phi_e(x)$ the phase error due to interpolation. Both amplitude attenuation and phase shift are a function of the position. This is immediately apparent since at the sample points, the interpolation produces the original samples and thus does not introduce any error. The effect of interpolation on the matching result can now be predicted for the simple case of two monochromatic waves by examining at what measured displacement u_m the original wave $f(x)$ and the interpolated wave $g_i(x + u_m)$ minimize the cost function χ^2. For simplicity, let's consider a scale invariant cost functions such as the NSSD cost function. In this case, the amplitude attenuation will not affect the matching result, and the minimum of the cost function will be found at the displacement u_m that brings the original wave $f(x)$ and the interpolated wave $g_i(x + u_m)$ in phase. This is the case for

$$u_m = -\frac{\Delta\phi}{k} - \frac{\phi_e(x + u_m)}{k} \tag{5.68}$$

The matching error between the measured displacement and the true displacement of the two waves $\Delta u = u_m - u_t$ can therefore be expressed as

$$\Delta u = -\frac{\phi_e(x + u_m)}{k} \tag{5.69}$$

Equation (5.69) shows that the displacement between two sampled monochromatic waves measured by image correlation using gray value interpolation has a position-dependent, systematic bias of the sub-pixel information that depends on the phase error of the interpolator used. It is therefore important to understand the phase error of interpolation filters and choose filters that minimize the error.

5.6.1.1 Phase Error of Interpolation Filters

To understand the phase error of interpolation filters, it is convenient to analyze the transfer function of the filter in the Fourier domain. We can think of interpolation as applying a shift to the signal such that the sought subpixel value is moved to an integer location. To interpolate a subpixel value at a position ε, we seek to find the function $f'(x-\varepsilon)$ from our discrete samples and obtain the interpolated value as $f'(0)$. In the Fourier domain, we can accomplish this shift by applying the shift theorem and obtain

$$\hat{f}' = \exp(-i\varepsilon\pi\tilde{k})\hat{f} \tag{5.70}$$

Ideally, an interpolation filter should therefore have a transfer function

$$\hat{h}_{\text{ideal}}(\varepsilon) = \exp(-i\varepsilon\pi\tilde{k}) \tag{5.71}$$

This filter has a unit magnitude, i.e., it does not alter the amplitude, but introduces a linear phase shift $\varepsilon\pi\tilde{k}$ to bring the subpixel value to the integer position. The phase shift of an interpolation filter compared to the ideal transfer function can thus be found as

$$\Delta\phi = \arctan\left(\frac{\mathbf{Im}\,\hat{h}(\tilde{k},\varepsilon)}{\mathbf{Re}\,\hat{h}(\tilde{k},\varepsilon)}\right) - \varepsilon\pi\tilde{k}\,, \tag{5.72}$$

and the corresponding positional error as

$$\Delta = \Delta\phi\lambda/2\pi = \Delta\phi/(\pi\tilde{k}) \tag{5.73}$$

Here, $\tilde{k} = 2/\lambda$ is the normalized wave number, being equal to unity at the Nyquist frequency ($\lambda_N = 2$ pixels). As an example, we analyze the phase shift of cubic polynomial interpolation. Cubic polynomial interpolation can be expressed as a convolution operation with a convolution kernel

$$[h_0, h_1, h_2, h_3] \tag{5.74}$$

For convenience, we choose the subpixel location ε to have its origin at the midpoint between samples, and group the polynomial expressions for the filter coefficients into even and odd parts:

$$\begin{aligned}
h_0 &= +\frac{1}{4}\varepsilon^2 - \frac{1}{16} + \frac{1}{24}\varepsilon - \frac{1}{6}\varepsilon^3 \\
h_1 &= -\frac{1}{4}\varepsilon^2 + \frac{9}{16} - \frac{9}{8}\varepsilon + \frac{1}{2}\varepsilon^3 \\
h_2 &= -\frac{1}{4}\varepsilon^2 + \frac{9}{16} + \frac{9}{8}\varepsilon - \frac{1}{2}\varepsilon^3 \\
h_3 &= +\frac{1}{4}\varepsilon^2 - \frac{1}{16} - \frac{1}{24}\varepsilon + \frac{1}{6}\varepsilon^3
\end{aligned} \tag{5.75}$$

The corresponding transfer function is then found as

$$\begin{aligned}
\hat{h}_c(\tilde{k},\varepsilon) &= (-\frac{1}{2}\varepsilon^2 + \frac{9}{8})\cos(\frac{1}{2}\pi\tilde{k}) + (\frac{1}{2}\varepsilon^2 - \frac{1}{8})\cos(\frac{3}{2}\pi\tilde{k}) \\
&\quad + i\left((\frac{9}{4}\varepsilon - \varepsilon^3)\sin(\frac{1}{2}\pi\tilde{k}) - (\frac{1}{24}\varepsilon - \frac{1}{3}\varepsilon^3)\sin(\frac{3}{2}\pi\tilde{k})\right)
\end{aligned} \tag{5.76}$$

Using Eqs. (5.72) and (5.73), the positional error as a function of sub-pixel position and wave number is found as

$$\Delta(\tilde{k},\varepsilon) = \frac{1}{\pi\tilde{k}} \arctan \frac{((\frac{9}{4}\varepsilon - \varepsilon^3)\sin(\frac{1}{2}\pi\tilde{k}) - (\frac{1}{24}\varepsilon - \frac{1}{3}\varepsilon^3)\sin(\frac{3}{2}\pi\tilde{k}))}{(-\frac{1}{2}\varepsilon^2 + \frac{9}{8})\cos(\frac{1}{2}\pi\tilde{k}) + (\frac{1}{2}\varepsilon^2 - \frac{1}{8})\cos(\frac{3}{2}\pi\tilde{k})} - \varepsilon \tag{5.77}$$

The positional error Δ and the amplitude attenuation $|\hat{h}|$ are shown in Fig. 5.9 as a function of the fractional position ε and the wave number $\tilde{k}$. At the integer positions $\varepsilon = 1/2$ and $\varepsilon = -1/2$, both errors vanish. At the mid–point, $\varepsilon = 0$, the positional error is zero for symmetry reasons, but the amplitude error is highest. It is noted that the general shape of the error shown in Fig. 5.9 is the same for all polynomial and B-spline interpolators, only the magnitude of the errors changes.

The effect of the positional interpolation error on sub-pixel reconstruction is difficult to predict, as the error varies with frequency. For the case of reconstructing the displacement between two monochromatic waves, however, the effect can easily be imagined. For a scale-invariant cost function χ^2 the amplitude attenuation of

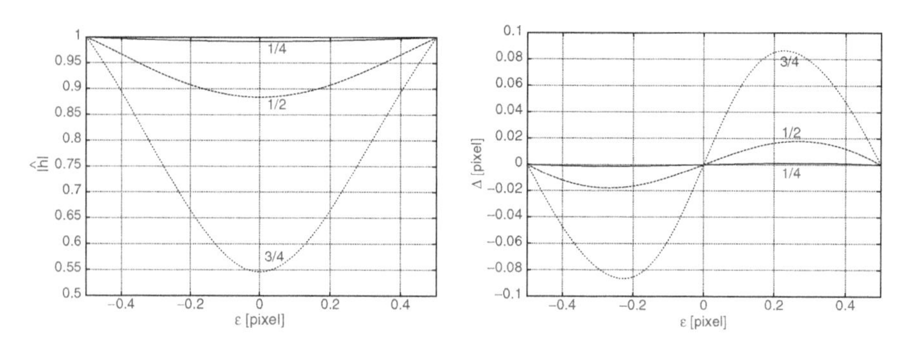

Fig. 5.9 Amplitude attenuation (*left*) and positional errors (*right*) of cubic polynomial interpolation as a function of the fractional position ε for wave numbers $\tilde{k} = 1/4, 1/2$ and $3/4$, as indicated

interpolation has no effect. The minimum of the cost function will occur when the original wave and the wave interpolated from the displaced copy are in phase. This will be the case if the measured displacement u^* plus the positional error $\Delta(u^*)$ is equal to the true displacement u_T. Therefore, the reconstruction error $\Delta u = u^* - u_T$ becomes

$$\Delta u = -\Delta(u^*) \tag{5.78}$$

For a first-order approximation of the error as a function of the true displacement u_T, one can assume that $\Delta(u^*) \approx \Delta(u_T)$, and the reconstruction error becomes the negative interpolation error [252].

5.6.1.2 Optimized Interpolation and Bias Comparison

In the previous section, we have shown that the accuracy of digital image correlation strongly depends on the phase accuracy of the interpolation filter used to reconstruct gray values at non-integer locations. As the left-hand graph in Fig. 5.9 indicates, cubic polynomial interpolation will introduce errors on the order of 1/50 pixel when matching structures with a period of 4 pixels, and errors of approximately 1/15 pixel can be expected for structures with a period of 3 pixels. Cubic polynomial interpolation is therefore clearly not well-suited for digital image correlation. The question arises as to what interpolation functions are suitable for image correlation, and how much error one can expect due to interpolation. The answer to that question lies in a compromise between bias and execution time. Generally, the more filter coefficients are used for interpolation, the better the result and the longer the execution time for the analysis. However, for a fixed number of filter coefficients, different interpolation schemes can provide vastly different results. For instance, cubic B-spline interpolation has much lower phase and amplitude errors than cubic polynomial interpolation with the same number of filter coefficients. The only additional computational expense for B-spline interpolation is a recursive pre-filter applied to the image [252,316,317]. If such variations between the performance of different interpolation methods exist, the question arises whether there is an optimal interpolation kernel for a given number of coefficients. Since the interpolation bias depends on the frequency content of the speckle pattern, it is not possible to construct one filter kernel that minimizes bias for all speckle patterns. However, it is possible to obtain optimized interpolation filters that outperform polynomial or B-spline interpolation and appreciably reduce bias. This is accomplished by a minimization process that minimizes the difference between the transfer function of the interpolation filter and the ideal transfer function from Eq. (5.71)

$$\int_{-0.5}^{0.5}\int_{0}^{1} w(\tilde{k})\left|\hat{h}(\tilde{k},\varepsilon) - \hat{h}_{\text{ideal}}(\tilde{k},\varepsilon)\right|^2 d\tilde{k}d\varepsilon \quad \longrightarrow \quad \min \tag{5.79}$$

Here, a weighting function $w(\tilde{k})$ is used that permits some flexibility in selecting the wave number range the filters are optimized for. To optimize a filter, one

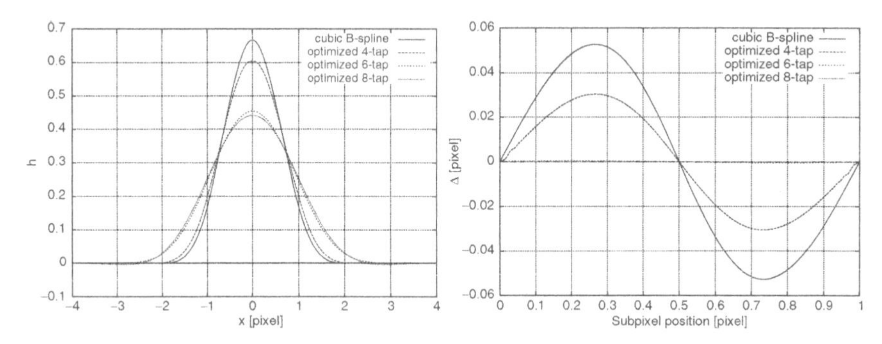

Fig. 5.10 Comparison of convolution kernels (*left*) for cubic B-spline and optimized four, six, and eight-tap interpolation filters. The graph on the right shows position errors Δ of the filters for a wave number of $\tilde{k} = 0.75$ (the errors for the optimized six-tap and eight-tap filters are very close to zero and practically coincide with the line $\Delta = 0$)

can parameterize its continuous convolution mask $h(x)$ in some form, e.g., using a spline, and find the spline coefficients that minimize (5.79). Even better results can be obtained by integrating recursive prefilters based on the B-spline transformation [316, 317]. This has been done for filters with a support of 4, 6 and 8 pixels, respectively, and the resulting convolution masks are shown in the left graph in Fig. 5.10. For reference, the convolution mask of the cubic B-spline interpolation filter is shown as well. The right graph in Fig. 5.10 shows the positional errors of the different interpolation methods for a normalized wave number $\tilde{k} = 0.75$. It can be seen that the optimized four-tap filter has roughly half the positional error of cubic B-spline interpolation at the same computational cost. The higher-order B-spline filters have been omitted, but a similar reduction in error is obtained using the filter optimization technique outlined above. The optimized six and eight-tap filters do not show any significant positional error at this wave number, while the four coefficient filters show errors of approximately 0.025 and 0.05 pixels, respectively.

To get a sense of how much error one can expect due to the phase errors in commonly used interpolation filters, we analyze the interpolation bias in a numerical study. To isolate the interpolation bias, it is important that the input data not contain any interpolation bias itself. The input images are therefore generated by applying a Fourier filter according to the shift theorem (5.70). A series of 20 images was generated for each pattern, corresponding to subpixel shift increments of 0.05 pixel between images. The shift of each image with respect to the original image was than determined using a digital image correlation algorithm that implements different interpolation filters. The results for two different images are shown in Fig. 5.11. The images were taken from specimens prepared for tension testing using black and white spray paint. The image on the left has a very fine-grained texture, while the speckle size in the right image is much larger. It can be seen that the interpolation bias is an order of magnitude higher for the image with the fine texture. For the fine texture, cubic B-spline interpolation produces interpolation bias of approximately 1/40 pixel, and the optimized four coefficient interpolation filter

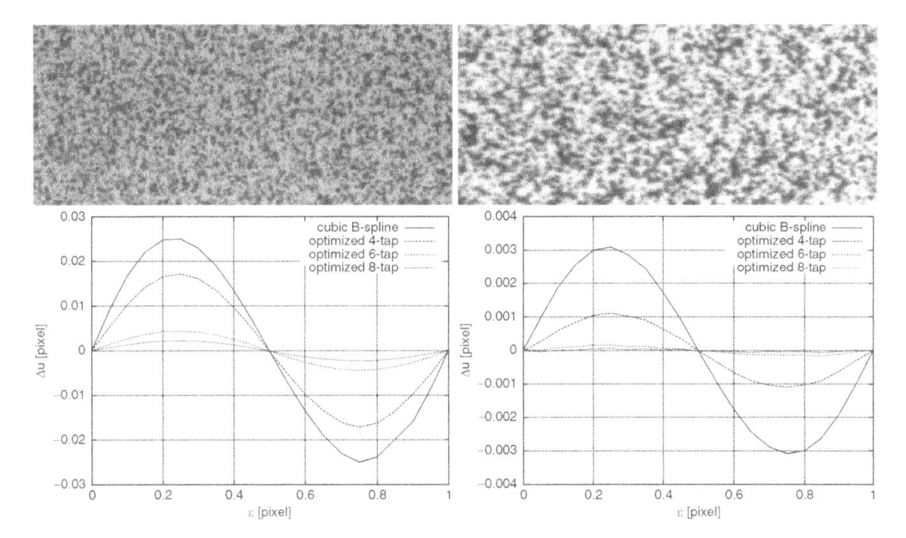

Fig. 5.11 Comparison of interpolation bias as a function of subpixel position for different interpolation filters. The pattern used is shown above the two plots

approximately 1/50 pixel. This amount of bias is clearly not acceptable, and for this reason, we strongly recommend against the use of four-tap filters despite their computational efficiency. The optimized six and eight coefficient filters exhibit bias well below 1/200 pixel even for the pattern with small speckles. The eight coefficient filter shows approximately half the error compared to the six coefficient filter, and the additional computational expense appears justified.

5.6.1.3 Strain Errors Due to Interpolation Errors

The interpolation bias discussed in the previous sections directly translates to a bias in strain measurements. If, instead of the true displacement $u_t(x)$, a biased displacement $u_b(x) = u_t(x) + \Delta u(x)$ is measured, the resulting strain becomes

$$\begin{aligned} \varepsilon_b &= \frac{\partial u_b(x)}{\partial x} = \varepsilon_t + \frac{\partial \Delta u}{\partial x} = \varepsilon_t + \frac{\partial \Delta u}{\partial u_t}\frac{\partial u_t}{\partial x} \\ &= \varepsilon_t\left(1 + \frac{\partial \Delta u}{\partial u_t}\right) \end{aligned} \tag{5.80}$$

The relative strain bias $\Delta\varepsilon_b = (\varepsilon_b - \varepsilon_t)/\varepsilon_t$ can now be found as

$$\Delta\varepsilon_b = \frac{\partial \Delta u}{\partial u}\,, \tag{5.81}$$

i.e., the relative bias in strain due to interpolation bias is proportional to the slope of the displacement bias shown in Fig. 5.11. Even though the absolute error of 1/40th of a pixel produced by cubic B-spline interpolation for the finer of the two speckle patterns might appear acceptable in some applications, the slope of the bias approaches 20% at the integer positions. This means that one must expect strain errors of 20% of the actual strain level for this particular speckle pattern when cubic B-spline interpolation is used. This again emphasizes the importance of proper interpolation filters in digital image correlation applications.

5.6.2 Bias Due to Noise

In practical applications, images are always contaminated by some noise. While imaging sensors can introduce various forms of noise into the signal, we limit ourselves to the study of Gaussian random noise and assume our reference and deformed images are composed of noise-free images with the addition of a noise term τ with a standard deviation σ

$$\begin{aligned} \bar{F} &= F + \tau_0 \\ \bar{G} &= G + \tau_1 \end{aligned} \tag{5.82}$$

We now formulate the sum of squared differences criterion for the noisy images, and limit our derivation to the one-dimensional case of determining a horizontal shift ξ

$$\chi^2(\xi) = \sum (G(\xi) - F)^2 + 2(G(\xi) - F)(\tau_1(\xi) - \tau_0) + (\tau_1(\xi) - \tau_0)^2 \tag{5.83}$$

We now study the case where $F = G$, i.e., there is no shift between the images, and attempt to determine the displacement ξ that minimizes Eq. (5.83). For the first term in Eq. (5.83), it is immediately apparent that the minimum is reached for a displacement $\xi = 0$, since the (uncontaminated) images are identical. We assume the second term can be neglected, and will come back to this assumption at a later point. The term of interest, $\sum(\tau_1(\xi) - \tau_0)^2$ can be further simplified by neglecting the term $\sum 2\tau_0\tau_1(\xi)$, since it is the product of two random variables with an expectation value of zero. These approximations simplify Eq. (5.83) to

$$\chi^2(\xi) = \sum (G(\xi) - F)^2 + \tau_1(\xi)^2 + \text{const} \tag{5.84}$$

The first term in Eq. (5.84) can be approximated in the vicinity of the minimum by a parabola. Whether and how much bias is introduced by the addition of noise will therefore be governed by the functional form of $\sum \tau_1(\xi)^2$. This term represents the variance of the noise multiplied by the number of data points. At a shift of $\xi = 0$, the noise is unaffected by the interpolation filters. However, since interpolation filters act as low-pass filters for non-integer locations, the variance of the filtered noise will necessarily be lower. This is explained by the fact that the low-pass filter introduces spatial correlation in the noise, i.e., the filtered noise values are no longer uncorrelated. It follows that the second term in Eq. (5.84) has its maximum at a shift $\xi = 0$ where the original noise signal is preserved, and its minimum at the mid-point

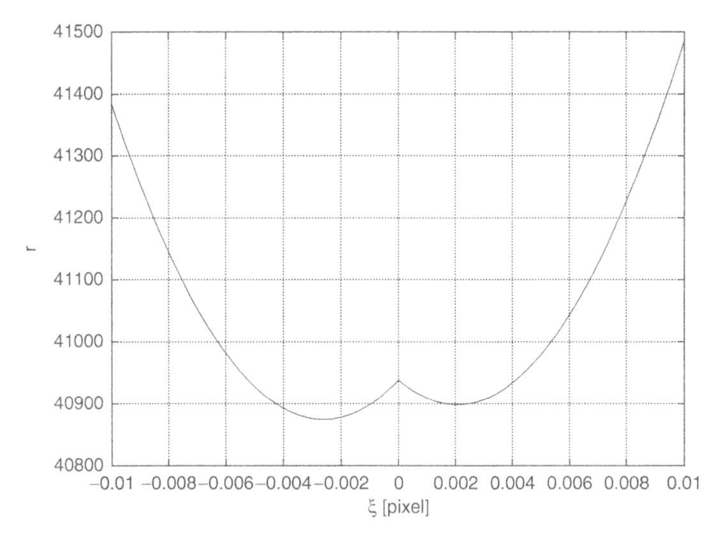

Fig. 5.12 Error function for two identical images corrupted by Gaussian noise with a standard deviation $\sigma = 2$ and cubic polynomial interpolation

$\xi = 0.5$, where the amplitude attenuation of the interpolation filter is highest. The superposition of this function with the parabolic first term in Eq. (5.84) will result in two minima, one on either side of the imposed shift of $\xi = 0$. This is illustrated in Fig. 5.12 for cubic polynomial interpolation and a noise level of two gray values. As can be seen in the figure, the graph is not symmetric with respect to $\xi = 0$, as the above discussion would indicate. This is due to the omission of the second term in Eq. (5.83), which was not omitted in the computation of the graph and causes a slight skew.

5.6.2.1 Mathematical Derivation

It is of interest to obtain a mathematical expression or bound on the bias that arises due to additive Gaussian noise. The bias will depend on the amount of noise, the signal content in the image, the sub-pixel location of the displaced subset, as well as the interpolation function used. The noise dependent term in Eq. (5.84) can be calculated if the transfer function of the interpolation filter $\hat{h}(\tilde{k}, \xi)$ is known as a function of the (normalized) wave number $\tilde{k}$ and the sub-pixel position ξ. Since Gaussian noise has a uniform power spectrum, the variance of the filtered signal can be computed from the integral of the filter's power spectrum along the wave number axis. Therefore, the noise dependent term in Eq. (5.84) can be written as

$$\sum_{1}^{N} \tau_1(\xi)^2 = N\sigma^2 \int_0^1 \hat{h}(\tilde{k}, \xi)^2 d\tilde{k} \tag{5.85}$$

The data dependent term in Eq. (5.84) can be approximated in the vicinity of the minimum by a quadratic function. By eliminating second-order terms, we can write

$$\sum (G(\xi) - F)^2 = \sum (\frac{dF}{d\xi})^2 \xi^2 \tag{5.86}$$

with the further assumption that the derivatives of the reference image are the same as those of the displaced image at the minimum. The constant term in Eq. (5.84) is the variance of the noise multiplied by the number of data points, so we can write

$$\chi^2(\xi) = \sum (\frac{dF}{d\xi})^2 \xi^2 + N\sigma^2 \left(1 + \int_0^1 \hat{h}(\tilde{k}, \xi)^2 d\tilde{k}\right) \tag{5.87}$$

We can further simplify this equation by assuming that the second term can be approximated by a linear function over the small range of the bias, and write

$$\chi^2(\xi) = \sum (\frac{dF}{d\xi})^2 \xi^2 + N\sigma^2 (1 + P + Q\xi) , \tag{5.88}$$

where P is the value of the noise dependent term and Q its derivative with respect to ξ. The biased position can be calculated by finding the minimum of the above parabola, which resolves to

$$\xi = -\frac{N\sigma^2 Q}{2\sum (\frac{dF}{d\xi})^2} \tag{5.89}$$

This equation incorporates the simple explanation for the cause of the bias mentioned earlier. The bias is proportional to the noise energy and inverse proportional to the energy in the derivative of the pattern, i.e., the strength of the edges required for matching. More importantly, the bias is also proportional to the slope of the squared amplitude attenuation of the interpolation filter. This has the immediate consequence that noise will not introduce any bias for a shift of 1/2 pixel, as interpolation filters are symmetric with respect to the mid-point position. Furthermore, Eq. (5.89) gives us a strategy to minimize bias due to noise. In Fig. 5.13, the integral of the squared amplitude attenuation and its derivatives are shown for various interpolation filters. The derivative corresponds to the factor Q in Eq. (5.89), and provides an assessment of how much bias is introduced by one interpolation filter compared to another. As can be seen, cubic polynomial interpolation introduces the highest amount of bias, and has the unfortunate property of a discontinuity at the integer location. This discontinuity is particularly detrimental in the context of strain measurements, as large errors in strain can be expected wherever the displacement crosses integer pixel boundaries. B-spline filters do not suffer from this discontinuity. The family of optimized interpolation filters shows the least amount of bias, but the four-tap filter shows small oscillations near the integer position. Overall, the optimized filters have significantly less bias than B-spline filters with the same support. This, and the substantially improved phase accuracy of the optimized filters, makes them the preferred choice for digital image correlation.

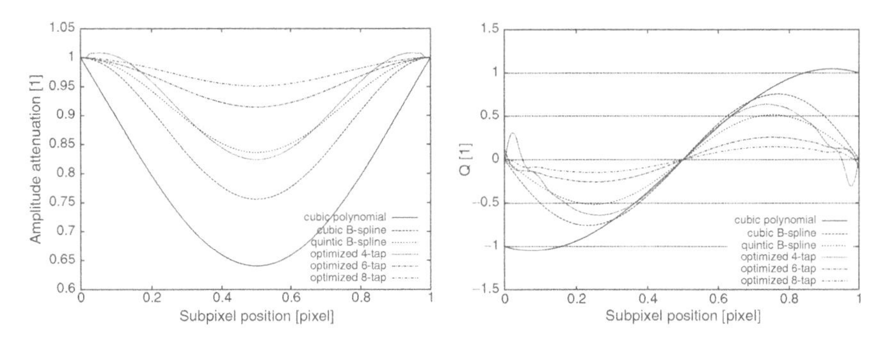

Fig. 5.13 Integral of the squared amplitude attenuation of various interpolation filters as a function of subpixel position (*left*) and the corresponding derivatives with respect to the subpixel position (*right*)

5.7 Statistical Error Analysis

5.7.1 Derivation for the One-Dimensional Case

The bias due to phase errors in interpolation filters as well as the noise induced bias can also be derived using statistical error analysis. The derivation also provides confidence margins for the displacements obtained with digital image correlation. For simplicity, we limit the derivation to the one-dimensional case of determining the uniform translation ξ in the horizontal direction. As previously, we use Eq. (5.82) and assume that the two images F and G are contaminated with Gaussian random noise τ_0 and τ_1, respectively, with the same distribution $N(0,\sigma)$. We write the SSD criterion as

$$\chi(t)^2 = \sum \left|\bar{G}_i(\xi) - \bar{F}_i\right|^2 , \tag{5.90}$$

and limit ourselves to sub-pixel motions ξ in the range $[0,1]$. To interpolate quantities $a_i(\xi)$ and their derivatives $\nabla a_i(\xi)$ (e.g., gray values or noise), we use cubic polynomial interpolation and define

$$\begin{aligned} a_i(\xi) &= \mathbf{c}(\xi)\cdot\mathbf{C}\cdot\hat{\mathbf{a}}_i \\ \nabla a_i(\xi) &= \mathbf{c}'(\xi)\cdot\mathbf{C}\cdot\hat{\mathbf{a}}_i , \end{aligned} \tag{5.91}$$

where

$$\begin{aligned} \mathbf{c}(\xi) &= \left[1\ \xi\ \xi^2\ \xi^3\right] \\ \mathbf{c}'(\xi) &= \left[0\ 1\ 2\xi\ 3\xi^2\right] \\ \hat{\mathbf{a}}_i &= \left[a_{i-1}\ a_i\ a_{i+1}\ a_{i+2}\right]^T \\ \mathbf{C} &= \begin{bmatrix} 0 & 1 & 0 & 0 \\ -1/3 & -1/2 & 0 & -1/6 \\ 1/2 & -1 & 0.5 & 0 \\ -1/6 & 1/2 & -1/2 & -1/6 \end{bmatrix} \end{aligned} \tag{5.92}$$

The exact translation encoded in the images is denoted by t_e, and the difference between the exact translation and the measured translation is denoted by t. The deviation of $E(t)$ represents the residual vector of errors in the estimated translation, with expectation $E(t)$ and variance $Var(t)$. The deviation of $E(t)$ from zero is a metric for bias in the image matching process. Similarly, $Var(t)$ is a measure of variability in the matching process [336].

We now develop the cost function into a Taylor series at the true displacement t_e

$$\begin{aligned}\chi(t)^2 &= \sum \left|\bar{G}_i(t_e) - \bar{F}_i\right|^2 \\ &= \sum \left|G_i(t_e) + \nabla G_i(t_e)t + \tau_{1i}(t_e) + \nabla\tau_{1i}(t_e)t - (F_i + \tau_{0i})\right|^2\end{aligned} \tag{5.93}$$

Letting $d\chi^2/dt = 0$, and neglecting all first-order products in $\tau_{0/1}$, we obtain the expectation value $E(t)$

$$\begin{aligned}E(t) &= -\frac{\sum(G_i(t_e) - F_i)\nabla G(t_e)}{\sum \nabla G_i(t_e)^2} - \frac{\sum \tau_{1i}(t_e)\nabla\tau_{1i}(t_e)}{\sum \nabla G_i(t_e)^2} \\ &= -\frac{\sum(G_i(t_e) - F_i)\nabla G(t_e)}{\sum \nabla G_i(t_e)^2} - \frac{\sum(\mathbf{c}(t_e)\cdot\mathbf{C}\cdot\hat{\tau}_{1i})(\mathbf{c}'(t_e)\cdot\mathbf{C}\cdot\hat{\tau}_{1i})}{\sum \nabla G_i(t_e)^2}\end{aligned} \tag{5.94}$$

After some algebraic manipulation, and with the further assumption that $\nabla G_i(t_e) \approxeq \nabla F_i$, we obtain

$$E(t) = -\frac{\sum\left(\mathbf{c}(t_e)\cdot\mathbf{C}\cdot\hat{\mathbf{G}}_i - F_i\right)\nabla F_i}{\sum(\nabla F_i)^2} - \frac{N\sigma^2 Q(t_e)}{2\sum(\nabla F_i)^2}, \tag{5.95}$$

where

$$Q(t_e) \approxeq 3.82106\cdot(t_e - 0.5) - 8.11528\cdot(t_e - 0.5)^3 \tag{5.96}$$

It is noted that the factor $Q(t_e)$ in this formulation is the same as in Eq. (5.89), and its functional form can be seen in the right-hand graph in Fig. 5.13. The variance, $Var(t)$ can be obtained using the following first order partial derivatives and Eq. (5.95).

$$\begin{aligned}Var(t) &= \sigma^2\cdot\left(\sum_i\left(\frac{\partial t}{\partial\tau_{1i}}\right)^2 + \sum_i\left(\frac{\partial t}{\partial\tau_{0i}}\right)^2\right) \\ &\cong \frac{2\sigma^2}{\sum_i\left[\nabla F(x_i)\right]^2}\end{aligned} \tag{5.97}$$

As shown in Eq. (5.95), the expectation value $E(t)$ of the bias t has two terms. The first term only depends on the difference between the interpolated gray values and the true gray values, and indicates that bias will occur due to interpolation errors. The second term is the already familiar expression from Eq. (5.89) that describes the bias due to noise. Equation (5.97) shows that the variance in the measurement of

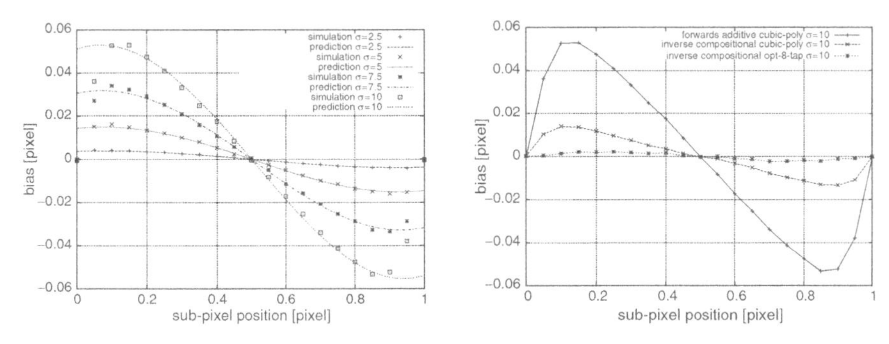

Fig. 5.14 Comparison of simulation result and theoretical prediction of the noise induced bias. *Left*: Forwards-additive DIC algorithm and cubic polynomial interpolation. *Right*: Comparison of forwards-additive and inverse-compositional algorithm for cubic-polynomial interpolation and inverse-compositional algorithm using an optimized eight-tap filter

the motion is proportional to the variance of the assumed Gaussian white noise distribution and inversely proportional to the sum of the squares in the local gradients in the intensity pattern, $\sum_i [\nabla F(x_i)]^2$.

Figure 5.14 shows a comparison of the theoretical noise-induced displacement bias according to the second term in Eq. (5.95) and the result of a numerical simulation using the image on the right in Fig. 5.11. For the numerical simulation, the interpolation bias was determined by applying the DIC algorithm to uncontaminated images. As can be seen in the left graph of the figure, the discontinuity at the integer pixel location is averaged over a distance proportional to the bias level. Aside from the integer pixel locations, the prediction is in excellent agreement with the simulation data for noise levels up to ten gray values using 8-bit images. For higher noise levels, the first-order approximations used will lead to increasing discrepancies. The data shown in the left graph of Fig. 5.14 was obtained using a forwards-additive implementation of the DIC algorithm. The right graph shows a comparison between the forwards-additive and the inverse-compositional algorithms for the same noise level of ten gray values. Even when using the same cubic polynomial interpolation method in both algorithms, the inverse compositional algorithm performs substantially better, and shows significantly less noise-induced bias. Furthermore, as can be seen in the graph for optimized eight-tap interpolation, all noise-induced bias can be virtually eliminated using appropriate interpolation filters.

Cubic interpolation was used in the digital image correlation algorithm. The simulation data is in excellent agreement with the theoretical predictions for the entire range of noise added. When no noise is added to the images, the graph shows the same interpolation bias as seen in the previous section. With the addition of noise, a significant amount of noise-induced bias is added, consistent with the derivation in the previous sections and the graph in Fig. 5.13. It is noted that the amount of bias can be substantially reduced using better interpolation filters. The numerical simulation confirms the predicted bias reduction for higher-order interpolation filters shown in Fig. 5.13.

5.7.2 Confidence Margins from the Covariance Matrix

Error bounds can also be obtained directly from the Hessian matrix used in the template matching algorithm. The inverse of the Hessian matrix can be used as an approximation for the covariance matrix. For the case of planar motion, the Hessian matrix is given by Eq. (5.60), and the covariance matrix can be found as

$$\mathbf{C} = \begin{bmatrix} \sum\left(\frac{\partial F}{\partial x}\right)^2 & \sum\frac{\partial F}{\partial x}\frac{\partial F}{\partial y} \\ \sum\frac{\partial F}{\partial x}\frac{\partial F}{\partial y} & \sum\left(\frac{\partial F}{\partial y}\right)^2 \end{bmatrix}^{-1} \tag{5.98}$$

Using the normalized residual of the cost function $1/N\sum|G-F|^2$, one can obtain confidence margins for the displacements (and higher-order terms, if used) estimated by digital image correlation. Under the assumption that independent Gaussian random noise is added to both images, we have $2\sigma^2 = 1/N\sum|G-F|^2$, and one obtains the same result as in Eq. (5.97)

$$Var(t) = \frac{\sum|G-F|^2}{NH_{11}} = \frac{2\sigma^2}{\sum(\partial F/\partial x)^2} \tag{5.99}$$

For two-dimensional applications, the covariance matrix describes a confidence ellipse, and the larger eigenvalue can be used to determine a scalar confidence bound for the displacements obtained by digital image correlation. Figure 5.15 shows a comparison between the theoretically predicted displacement noise according to

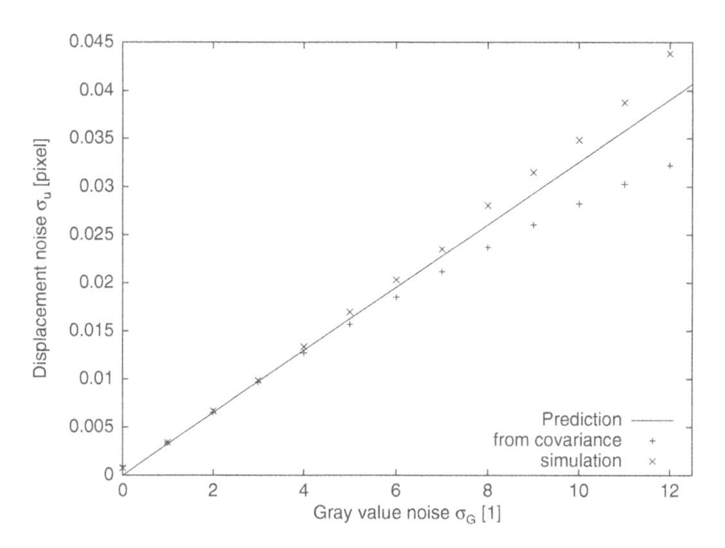

Fig. 5.15 Comparison between predicted and actual displacement noise as a function of gray value noise added to the images for a subset size of 15×15 pixels

Eq. (5.97), the noise predicted using the covariance matrix and the actual variance in digital image correlation for the right image from Fig. 5.13 and a subset size of 15×15 pixels. For noise less than 5 gray values in an 8-bit image, as is the case for the vast majority of practical applications, both predictions match the actual data almost perfectly. This justifies the use of the covariance matrix to obtain error bounds on displacement estimates obtained by digital image correlation in many applications. It is noted that for higher noise levels, this is no longer the case and predicted noise levels according to Eq. (5.99) are lower than the actual ones. This is due to the fact the derivatives of the gray values cannot be evaluated on the noise-free image data, since it is generally not known.

The approximate covariance matrix gives us a tool to not only determine if, but *how well*, the correspondence problem introduced at the beginning of the chapter can be solved. Earlier, it was shown that whether a solution to the correspondence problem exists depends on the invertibility of the Hessian matrix, i.e., on whether the approximate covariance matrix can be computed. With an estimate of the noise level in the images, the covariance matrix can be used to directly provide an estimate of the uncertainty in digital image correlation without actually applying the complete algorithm. It is noted that the terms in Eq. (5.98) can be evaluated very efficiently even for large subset sizes, since the summations can be implemented as a box-filter with a low operation count that is independent of the filter window. This permits a rapid analysis of images to automatically determine regions suitable for further analysis with the DIC algorithm, or the determination of a subset size required to reach a desired confidence interval for the displacements.

Chapter 6
In-Plane Measurements

Two-dimensional deformation and motion measurements were the foundation of the early applications in digital image correlation for solid mechanics measurements. In all cases, a nominally flat specimen (with or without a geometric discontinuity) was imaged while being subjected to nominally tensile loading. Throughout the loading process, it was assumed that the specimen deformed within the original planar specimen surface.

6.1 Constraints and Applicability

Three basic assumptions are commonly employed when using two-dimensional digital image correlation to estimate object motions. First, the specimen is assumed to be nominally planar. Second, the object plane is parallel to the sensor plane in the camera. Third, the specimen is loaded so that it is deformed within the original object plane.

6.1.1 Object Planarity Constraints

The object planarity assumption is applicable, at least approximately, for initially flat specimens. Such specimens may contain geometric discontinuities (e.g., cracks, notches, complex cut-outs) or gradients in material properties without affecting this assumption.

6.1.2 Object Deformation Constraints

In a general sense, it is assumed that a nominally planar object is subjected to a combination of in-plane tension, in-plane shear or in-plane biaxial loading so that the

M.A. Sutton et al., *Image Correlation for Shape, Motion and Deformation Measurements: Basic Concepts, Theory and Applications,* DOI: 10.1007/978-0-387-78747-3_6,

specimen deforms predominantly within the original planar surface. When cracked or notched specimens are loaded in this manner, Poisson's effect in the crack tip region (which results in small amounts of out-of-plane motion) is assumed to be small relative to the applied in-plane deformations. When planar specimens with material property gradients are similarly loaded, the assumption remains the same; in-plane deformations.

6.2 Uniaxial Tension of Planar Specimen

In this section, a uniaxial tension loading experiment of a nominally planar specimen is performed to demonstrate the appropriate and successful use of 2D-DIC for measurement of surface deformations.

6.2.1 Experimental Considerations

The material used in these experiments is a ductile polymer, Basell Hostacom CR 1171 G, used in transportation vehicle applications. As shown in Fig. 6.1, the material is provided in plate form with a thickness of 3.06 mm. Dog-bone tensile specimens with a straight gage region are machined from the as-provided plate with dimensions and orientation shown in Fig. 6.1.

All experiments are performed in an MTS 50 kip loading frame. Hydraulic fixtures with flat specimen platens are used to grip the specimens. All tensile loading is performed using displacement control with the crosshead moving at a constant rate of 0.040 mm/s throughout the experiment. For comparison to the 2D-DIC strain measurements, an MTS extensometer with a 25.4 mm gage is attached to the specimen. Extensometer and tensile load data are recorded using TESTAR software and digital control system. Figure 6.2 shows the specimen with the extensometer attachment.

6.2.2 Imaging Considerations

As shown in Fig. 6.2, one surface of the specimen is lightly coated with white spray paint to generate a black and white random pattern on the surface. All images are acquired using a 55 mm Nikon lens and a Q-Imaging camera with 8-bit intensity resolution and 1,392 by 1,024 square pixel camera resolution. Continuous specimen illumination was provided by a halogen light with attached infrared filter adjusted to ensure adequate contrast and reasonably uniform illumination with minimal heating of the polymer specimen. Figure 6.3 shows a photograph of the experimental optical

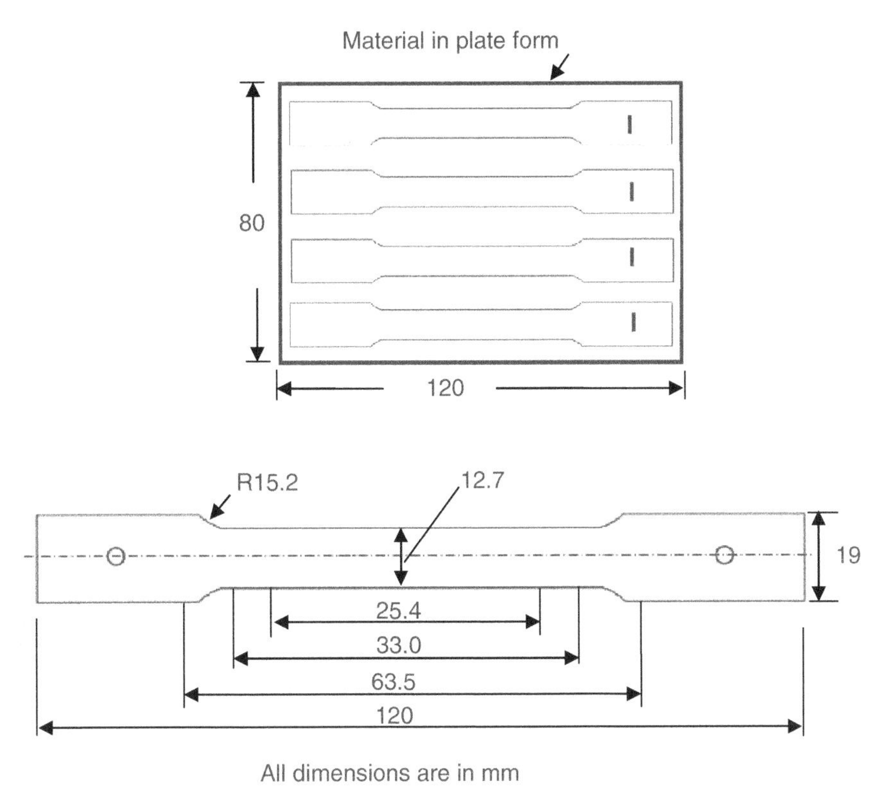

Fig. 6.1 Geometry and dimensions of Basell Hostacom CR 1171 G polymer tensile specimens and their orientation within sheet. All specimens are 3.06 mm thick

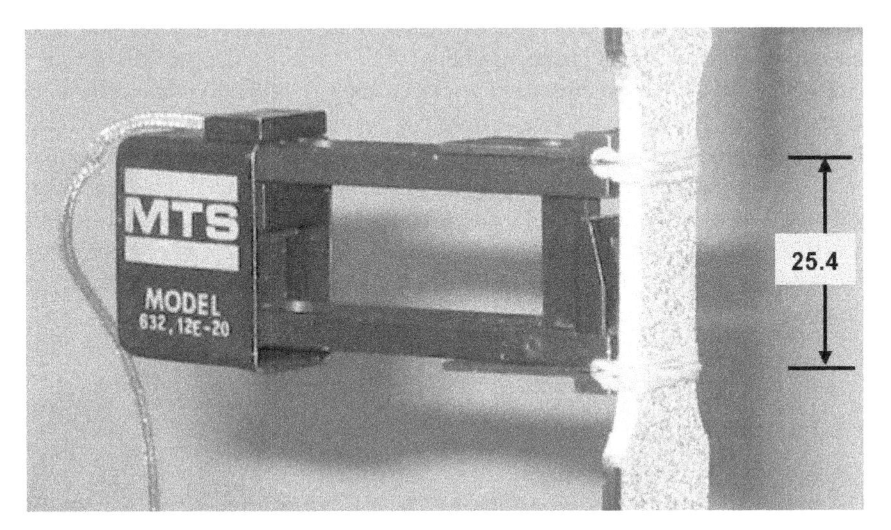

Fig. 6.2 Tensile specimen with extensometer attachment, providing an independent measure of the average strain over the specified gage length. All units in millimeters

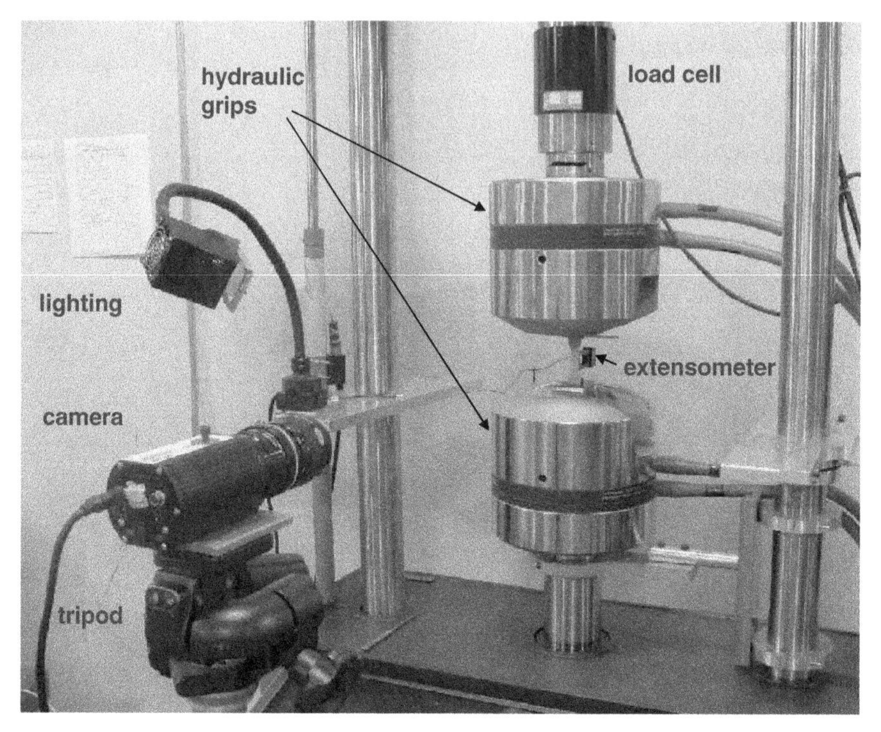

Fig. 6.3 Experimental setup for uniaxial loading of dog-bone specimen. Camera is $\approx$0.50 m from the specimen

setup. Figure 6.4 shows the reference image obtained by the optical system and used for 2D-DIC surface deformation measurements, along with the strain analysis region and the subset size used in image matching.

To maintain a constant crosshead displacement rate of 0.040 mm/s, which corresponds to an average strain rate of 0.0016 s^{-1}, images are acquired continuously throughout the experiment at 1/2 Hz using an exposure time of 100 μs. A total of 45 images are acquired for processing. Synchronization of the imaging, extensometer and load data is performed through direct input of MTS extensometer and load signals into the image acquisition system, VicSnap, ensuring a one-to-one correspondence between the (a) image, (b) instantaneous load and (c) instantaneous average strain.

6.2.3 Experimental Results

The MTS extensometer data is defined to be the engineering strain results. The 2D-DIC displacement data is converted into Lagrangian large strain values using Eq. (A.5) in Appendix A by assuming $\partial d_3/\partial x_1 = \partial d_3/\partial x_2 = 0$ and obtaining the

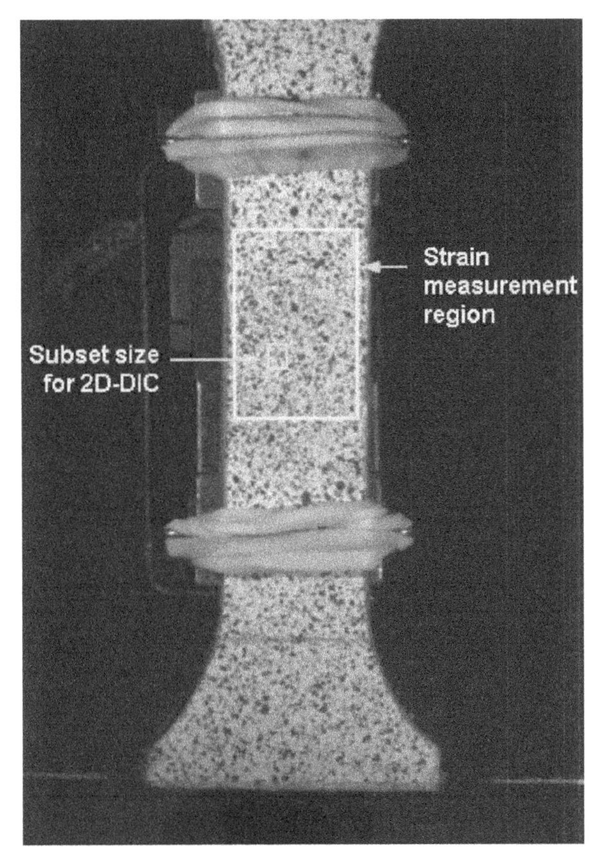

Fig. 6.4 Reference image for 2D-DIC measurements. Scale factors are $s_x = s_y = 18.4$ pixels/mm. Subset size of 25×25 pixels is shown on the specimen

local gradients at each point (x, y) by using a 15×15 array of displacement data points centered at (x, y). The Lagrangian strain tensor results, $E_{11}(x_i, y_i)$, are converted to engineering strain for direct comparison to the extensometer results using the expressions

$$\varepsilon^i_{engr} = \sqrt{1+2(E_{11})_i} - 1$$

$$\varepsilon^{avg}_{engr} = \frac{\sum_{i=1}^{N} \varepsilon^i_{engr}}{N} \tag{6.1}$$

Figure 6.5 shows the engineering stress-engineering strain curves obtained using both ε^{avg}_{engr} from 2D-DIC and the extensometer results. As shown in Fig. 6.5, the extensometer and strain gage results are indistinguishable up to, and beyond, maximum loading, eventually separating when the axial strain exceeds 3%.

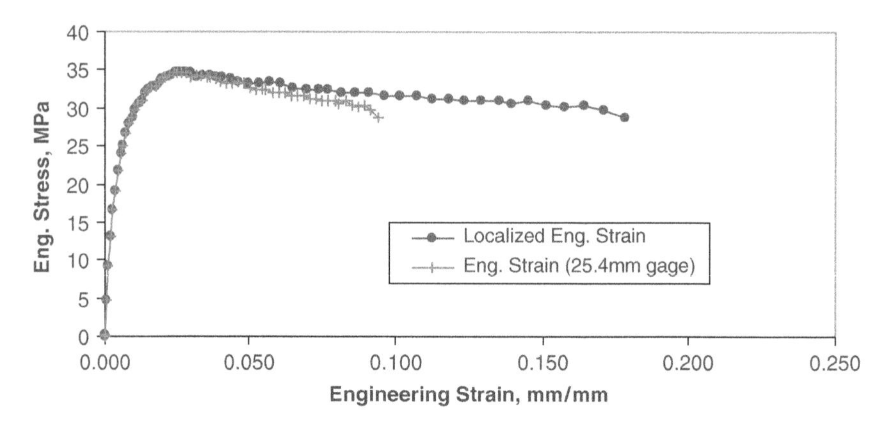

Fig. 6.5 Engineering stress versus strain obtained using 2D-DIC and extensometer for polymer specimen

To understand the cause for the deviation between the two measurement methods, Fig. 6.6 shows photographs of two fractured tensile specimens and an as-manufactured, unloaded dog-bone specimen. As shown in Fig. 6.6, the tensile specimens failed in the region where localized deformation (necking) occurs. Since 2D-DIC is averaging the strain data over the region containing the localized necking zone, the extensometer-based strain measurements in the gage region will be smaller than the 2D-DIC measurements.

Figure 6.7 shows both the stress–strain data and the measured 2D uniaxial tensile strain field, ε_{xx}, at three load levels: (1) elastic region for material response; (2) point near maximum load, nominally elastic; (3) region dominated by localization and large plastic deformation. Inspection of Fig. 6.7 shows that the average strain obtained using 2D-DIC measurements containing the strain localization zone are much larger than the nominal average strain obtained from the extensometer.

6.2.4 Discussion

As shown in Fig. 6.7, the 2D-DIC method is capable of accurately determining surface strains during uniaxial loading of planar specimens. An important advantage of the method is the full-field nature of the measurement data, providing quantitative evidence of important features such as strain localization during the loading process. Direct comparison of the full-field uniaxial strain measurements to extensometer data is achievable through averaging of the point-by-point strain measurements.

An additional advantage of the 2D-DIC method is that all three planar surface strain fields are obtained simultaneously. Figure 6.8 shows the two additional strain fields, ε_{yy} and ε_{xy}, obtained when acquiring the local uniaxial tensile strain data.

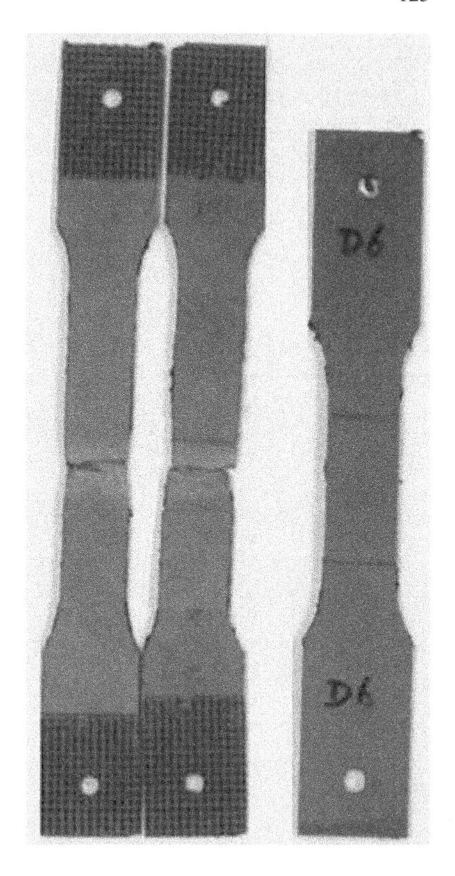

Fig. 6.6 Two fractured Hostacom Basell tensile specimens and an unloaded specimen for comparison

The transverse strain field, ε_{yy}, can be used to quantify Poisson's ratio in the elastic regime and the variation in the Poisson effect in the plastic deformation regime. For example, at load levels 1 and 2, the estimated Poisson's ratio within the circular region and defined by $-\varepsilon_{yy}^{avg}/\varepsilon_{xx}^{avg}$ is 0.33 and 0.29, respectively. When large plastic deformations occur in a localized region, the corresponding ratio is $\approx$0.10, though its relevance in characterizing material behavior is uncertain.

As will be shown in the following section, the effect of out-of-plane motion on in-plane strain measurements can be estimated by the ratio $(-\Delta z/z)$, where Δz is the out of plane displacement and z is the object distance from the lens pinhole. As shown in Fig. 6.3, the measured distance between the specimen and the camera lens is $\approx$0.50 m, a value that is a reasonable estimate for the object distance, z.

To estimate the out-of-plane displacement, a dial indicator is used to measure motion at a location near the specimen midpoint; the location is slightly offset from the maximum necking position. The maximum recorded displacement at steps 1, 2 and 3 are given in Table 6.1. As shown in Table 6.1, the effect of out-of-plane motion on in-plane strains is less than 2% of the axial tensile strain measured by

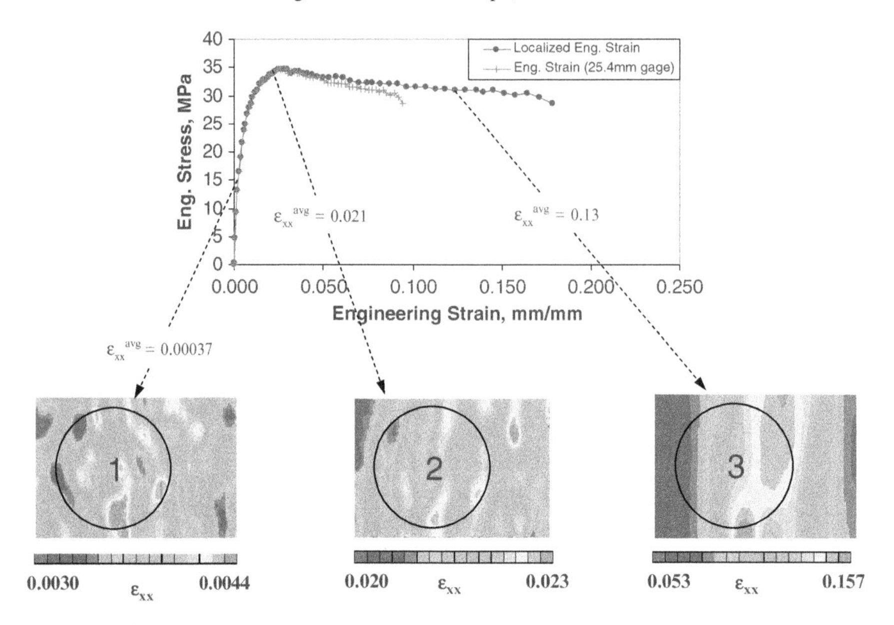

Fig. 6.7 Strain fields obtained by image correlation at three separate load levels during tensile experiment. Imaged region is 13.9 × 10 mm with scale factor of 18 pixel/mm. Loading rate is 0.04 mm/s. Strain recorded on graph is average of all strains within the 7.8 mm diameter circular region overlaid on each strain field

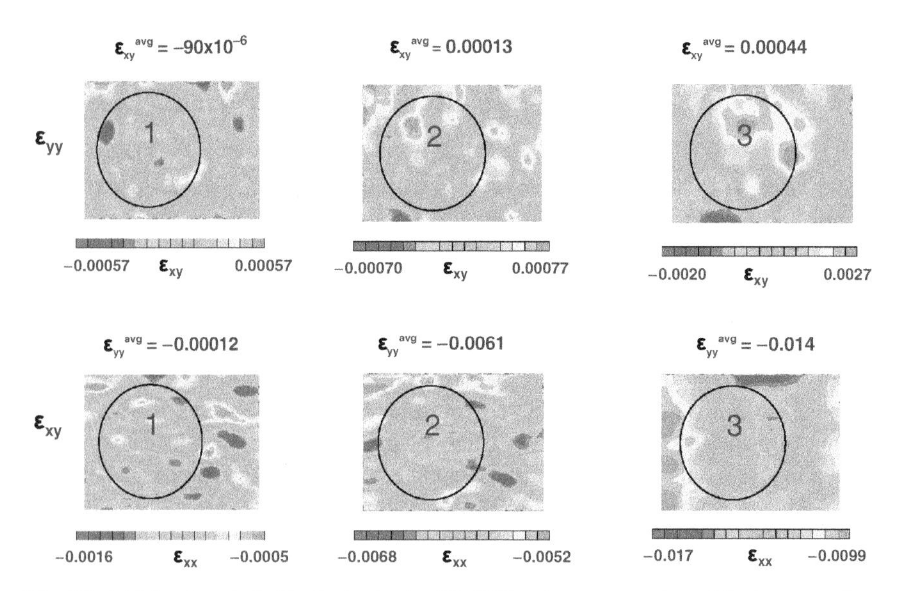

Fig. 6.8 Additional strain fields obtained using 2D-DIC and post-processing of displacement data. The strain ε_{yy} is the transverse strain field, representing Poisson effects. The shear strain field, ε_{xy}, should nominally be negligible for tensile loading of an isotropic material

Table 6.1 Effect of out-of-plane motion on 2D strain measurements

Out-of-plane motion and effects \ Measurement locations	①	②	③
Δz (m)	3×10^{-6}	20×10^{-6}	180×10^{-6}
$\Delta z/z$ (m/m)	6×10^{-6}	40×10^{-6}	360×10^{-6}
$[\Delta z/z]/\varepsilon_{xx}^{avg}$	0.017	0.002	0.003

2D-DIC at all three load values for the central point in the specimen, indicating that the effect of out-of-plane motion does not appreciably affect the uniaxial strain measurements.

Regarding the shear strain field, when $\varepsilon_{xx}^{avg} > 0.02$ the results in Fig. 6.8 show that the shear strain is appreciably smaller than the normal strains. In the elastic, small strain regime, the shear strain remains smaller than either of the normal strains, with variations in loading direction, specimen orientation and measured displacements contributing to the elastic shear strain estimates.

6.3 Out-of-Plane Motion

Implicit in the previous discussions is the fact that a specimen does incur small amounts of out-of-plane motion during the loading process. When using a single camera to obtain object deformations, relative out-of-plane motion of the object with respect to the imaging system will introduce image-plane displacement gradients. If these gradients are large enough, they will corrupt the in-plane displacement measurements and make it difficult or impossible to separate the true deformations for "pseudo image deformations" introduced by the rigid body out-of-plane motion.

6.3.1 Standard Lens Systems for Single Camera Measurements

For a simple lens system, Eqs. (2.17), (2.24) and (3.3) can be employed to define metric distance. Letting (x,y,z) be world coordinates (X_w, Y_w, Z_w) defined in the initial surface of a planar specimen. The sign convention for z is defined in Fig. 2.9, where z is positive when the object moves away from the lens. With this convention, the effect of out-of-plane motion can be described by replacing the original object distance, z, with $z + \Delta z$. In this case, the in-plane displacement field due to the out-of-plane motion, Δz, and the resulting strain field can be written as follows, where only the first order term in the expansion of $(1 + (\Delta z/z))^{-1}$ is retained.

$$
\begin{aligned}
u(\Delta z) &= x_s(z+\Delta z) - x_s(z) \cong \left(\tfrac{x \cdot f}{z}\right) \cdot S_x \cdot \left[-\tfrac{\Delta z}{z}\right] = x_i \cdot S_x \cdot \left[-\tfrac{\Delta z}{z}\right] \\
v(\Delta z) &= y_s(z+\Delta z) - y_s(z) \cong \left(\tfrac{y \cdot f}{z}\right) \cdot S_y \cdot \left[-\tfrac{\Delta z}{z}\right] = y_i \cdot S_x \cdot \left[-\tfrac{\Delta z}{z}\right]
\end{aligned}
\tag{6.2}
$$

$$
\begin{aligned}
\Rightarrow \varepsilon_{xx} &= \tfrac{\partial u}{\partial x_s} \cong -\left[\tfrac{\Delta z}{z}\right] \\
\varepsilon_{yy} &= \tfrac{\partial v}{\partial y_s} \cong -\left[\tfrac{\Delta z}{z}\right] \\
\varepsilon_{xy} &= \tfrac{1}{2}\left(\tfrac{\partial u}{\partial y_s} + \tfrac{\partial v}{\partial x_s}\right) \cong 0
\end{aligned}
$$

Consistent with the sign convention for z, Eq. (6.2) shows that out-of-plane motion away from the image plane decreases image magnification and introduces a negative normal strain in all directions [289].

6.3.2 Telecentric Lens System for Single Camera Measurements

For a more complex lens system, such as a combination of elements resulting in a object based telecentric lens, the imaging equations in Chapter 2 can be employed with some modifications.

Specifically, for a telecentric lens the effect of out-of-plane motion has been reduced by arranging elements so that light passing through the entrance pupil is nearly parallel to the object axis. In this case, Eq. (6.2) is oftentimes modified by replacing the physical object distance by an "effective" distance that is many times larger. Thus, for telecentric lenses, the corresponding displacement field has the following approximate form:

$$
\begin{aligned}
u(\Delta z) &\cong x_s \cdot \left[-\frac{\Delta z}{z_{\text{effective}}}\right] \\
v(\Delta z) &\cong y_s \cdot \left[-\frac{\Delta z}{z_{\text{effective}}}\right] \\
\Rightarrow \varepsilon_{xx} = \varepsilon_{yy} &\cong -\left[\frac{\Delta z}{z_{\text{effective}}}\right]; \varepsilon_{xy} \cong 0
\end{aligned}
\tag{6.3}
$$

By increasing the "effective" object distance, the recorded image plane displacement field due to of out-of-plane motion will be reduced. Here, the image distance γ will be constant in the experimental studies.

6.3.3 Out-of-Plane Translation Experiments

Figure 6.9 presents two photographic views of the camera arrangement employed to simultaneously acquire (a) stereo-vision images using the two outer cameras and

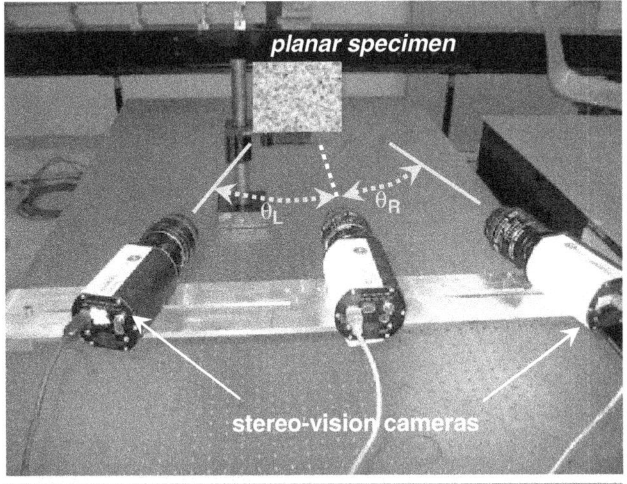

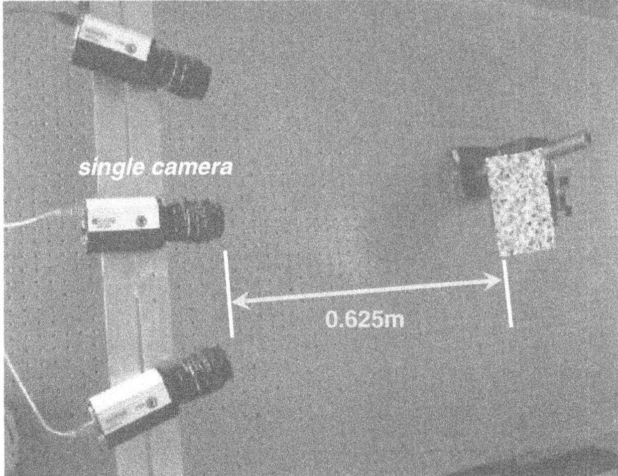

Fig. 6.9 Experimental setup for simultaneous (a) stereo-vision measurements using outside cameras and (b) two-dimensional measurements using center camera. Specimen translated out of plane and synchronized images acquired by all cameras

(b) single camera images using the center-mounted camera and standard Nikon lens oriented to be perpendicular to the planar object.

Figure 6.10 shows the single camera experimental arrangement used to acquire images with a telecentric lens. As in the previous experiment, the single camera is oriented perpendicular to the planar object. Table 6.2 summarizes the vision system parameters used to construct the three optical systems.

As shown in Table 6.2, the true magnification factors are $\simeq$0.045 and 0.50 for the two-dimensional cameras with standard lens and telecentric lens, respectively. It is noted that a separate experiment setup is required for the telecentric lens since it has a fixed focal length and a true magnification of 0.50 that made it incompatible with the system geometry shown in Fig. 6.10.

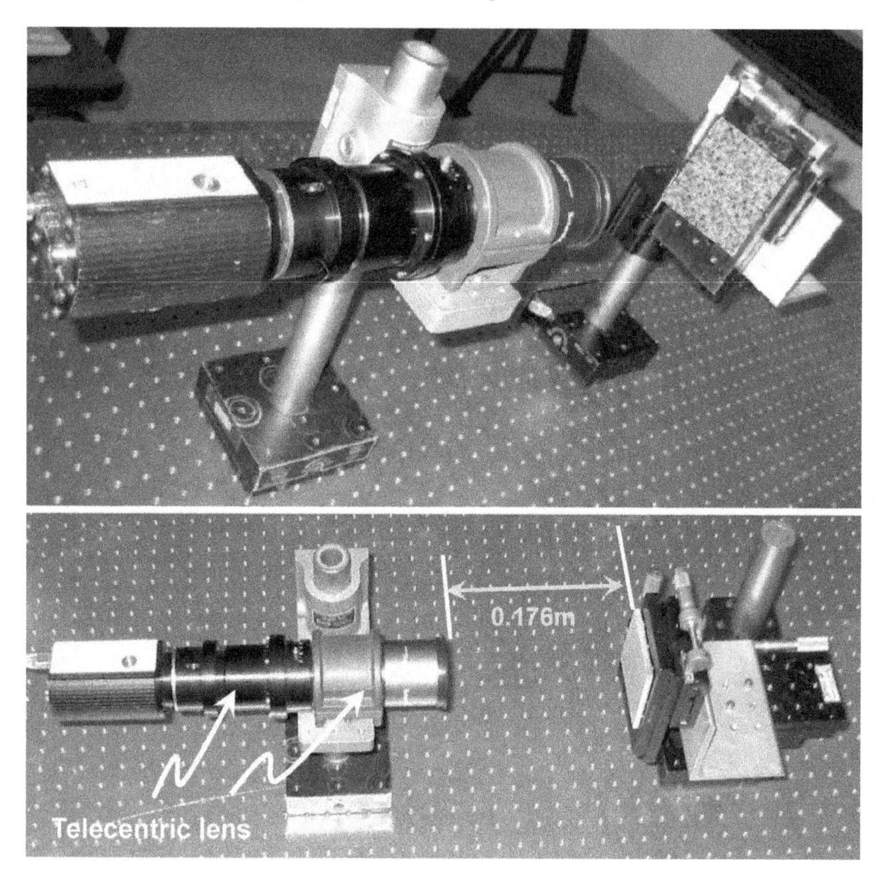

Fig. 6.10 Experimental setup for two-dimensional measurements using single camera with telecentric lens. Specimen translated out of plane and images acquired after each motion

Figure 6.11 shows the two random patterns used in this study. The pattern on the left is imaged by the three cameras shown in Fig. 6.9, with an in-plane scale factor of $\simeq$10 pixels/mm. The pattern on the right is imaged by the telecentric lens which has a fixed true magnification of 0.50 and a scale factor of $\simeq$108 pixels/mm. The 41$\times$41 subset size used in all analyses is selected to ensure adequate contrast throughout the region of interest.

6.3.4 Stereo-vision Calibration

As shown schematically in Figs. 4.3, 4.5, 4.7 and I.1, stereovision systems use multiple camera views to estimate all three components of displacement simultaneously. Hence, the measured three-dimensional displacement field should be such that the in-plane components of displacement are independent of the out-of-plane motion.

Table 6.2 Optical system components

	Stereo-vision system	Single Camera System Standard Lens	Single Camera System Telecentric Lens
Cameras	2 Q-Imaging QICAM fast 1394 (12 bit, 1392×1040)	Q-Imaging QICAM fast 1394 (12 bit, 1392×1040)	Q-Imaging QICAM fast 1394 (12 bit, 1392×1040)
Lenses	2 Nikon AF Nikkor (f = 28 mm, 1:2.8D)	Nikon AF Nikkor (f = 28 mm, 1:2.8D)	Schneider-Kveuznach Xenoplan 1:2 0.14/11
Lighting	1 Halogen Light	1 Halogen Light	1 Halogen Light
Table	Newport Optical Bench	Newport Optical Bench	Newport Optical Bench
Object	Aluminum plate with random pattern 1	Aluminum plate with random pattern 1	Aluminum plate with random pattern 2
Scale factor (at $W = 0$ mm)	9.5 pixels/mm ($M_T \simeq 0.0442$)	9.5 pixels/mm ($M_T \simeq 0.0442$)	107.5 pixels/mm ($M_T \simeq 0.500$)
Subset size	41×41 pixel square	41×41 pixel square	35×35 pixel square
Step size	5 pixels	5 pixels	5 pixels
Strain window size	9 points	9 points	7 points
Shape function	Affine	Affine	Affine
Data extraction	R = 200 pixels circular region at image center	R = 200 pixels circular region at image center	R = 200 pixels circular region at image center
Software	VIC-3D	VIC-2D	VIC-2D

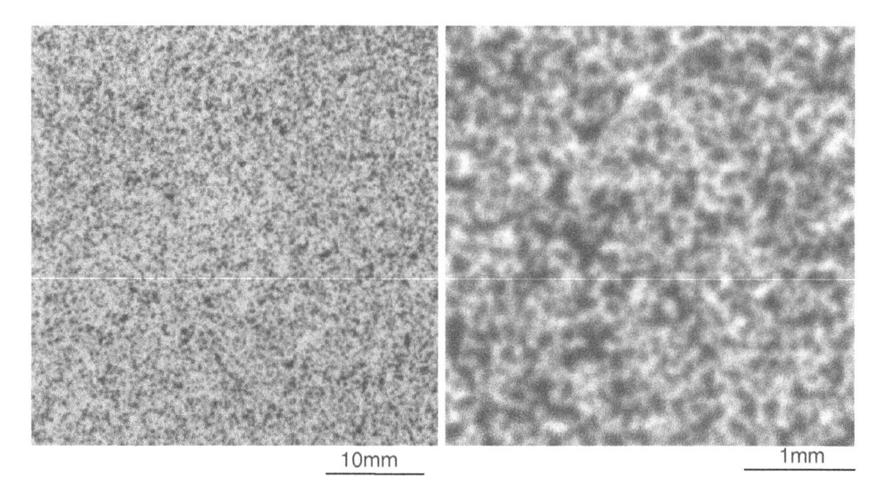

Fig. 6.11 Random pattern image for combined stereo-vision and single camera with standard lens (*left image*) and single camera with telecentric lens (*right image*)

Table 6.3 Calibration results for stereovision system

Parameter	Camera 1	Camera 2	Camera 1 → 2 transformation	
Center x (pixels)	703.2	703.0	Alpha (degrees)	−3.58
Center y (pixels)	457.0	128.9	Beta (degrees)	+35.39
fS_x (pixels)	6228.6	6243.8	Gamma (degrees)	−2.42
fS_y (pixels)	6228.6	6243.8	T_x (mm)	−421.13
Skew	0	0	T_y (mm)	+8.21
$\kappa_1^{normalized}$	−0.0787885	−0.129095	T_z (mm)	+139.7
κ_1	$-2.0309E-09$	$-3.3114E-09$	Baseline (mm)	443.8

Calibration of the stereovision system is performed using 52 images of a translated and rotated planar dot pattern with reasonably well known spacing. Table 6.3 shows the camera parameters obtained via the calibration process. Inspection of Table 6.3 shows that the lens center for camera 1 is near the geometric center of the sensor plane (696,520). The right camera has an estimated location for the lens center that is shifted by nearly 400 pixels, with a somewhat higher lens distortion correction coefficient. Though somewhat anomalous, other parameters such as the included pan angle (stereo angle) between the optical axes of the two cameras, $\beta \simeq 35°$, are in good agreement with protractor estimate of 36°. The relatively small values for the tilt angle, α, and the swing angle, γ, are also consistent with visual observations.

It is noted that the distortion factor, $\kappa_1^{normalized}$, represents the contribution of third order radial distortion. The "normalized" version is used to reduce round-off error during computations and is defined by the formula $\kappa_1^{normalized} = \kappa_1 (fS_x)^2$ so that corrections in the non-dimensional parameter $r \cdot (fS_x)^{-1}$ at the outer edge of the sensor plane are $O(10^{-4})$. When converted to pixel units, the corrections at the outer edge of the sensor plane are on the order 0.40 and 2.40 pixels for camera 1 and camera 2, respectively.

Though the distortion correction for camera 2 is large near the outer edge, the overall calibration has a standard deviation of residuals in the pixel positions for all images is of 0.026 and 0.028 pixels for cameras 1 and 2, respectively. The calibration residuals indicate that the calibration is adequate for stereo-vision measurements, though not as low as seen in previous experiments where residuals on the order of 0.012 pixels were obtained.

6.3.5 Experimental Results

Figure 6.12 shows the typical horizontal (U) and vertical (V) displacement fields obtained using (a) a 2D system with standard lens, (b) a 2D system with telecentric lens and (c) a stereo-vision system. Since only rigid body out-of-plane displacement is applied, the presence of gradients $\Delta U/\Delta x$ or $\Delta V/\Delta y$ in the measured sensor-plane displacement fields indicate the presence of strain measurement errors due to out-of-plane motion; if the gradients are high (low), then the sensitivity of strain measurements to out of plane motion is high (low).

Defining $W = -\Delta z$ so that the out-of-plane displacement towards the camera is positive, Fig. 6.13 presents a line plot for $U(x, y_0)$ for $W = 5$ mm, where the sensor position $y_s = 520$ pixels corresponds to the world coordinate system y_0.

The measured average ε_{xx} and ε_{yy} values[1] as a function of W for the (a) 2D camera with standard lens, (b) 2D camera with telecentric lens and (c) stereovision system are shown in Figs. 6.14 to 6.16, respectively.

Figure 6.17 shows the measured strains as a function of W for all three cases. It is important to note that the slope of the linear fit to the measurement data has units of (strain $\times$ 10^{-6})/mm and represents the amount of strain error per mm of out-of-plane motion.

6.3.6 Discussion

Using Eq. (6.2) and the slope of the best linear fit in Fig. 6.14 for the standard lens, the estimated object distance for the 2D system is $\simeq$0.66 m. This first-order estimate is in good agreement with the physically measured distance of $\simeq$0.62 m from the front of the lens to the object.[2]

[1] For each out of plane motion, the average values are obtained using strain estimates in a region with diameter of 400 pixels and centered at pixel (696, 520).

[2] Since the object distance used in Eq. (6.3) is measured from the pinhole location (i.e., effective lens "center"), the measured distance to the front of the lens is expected to be a lower bound.

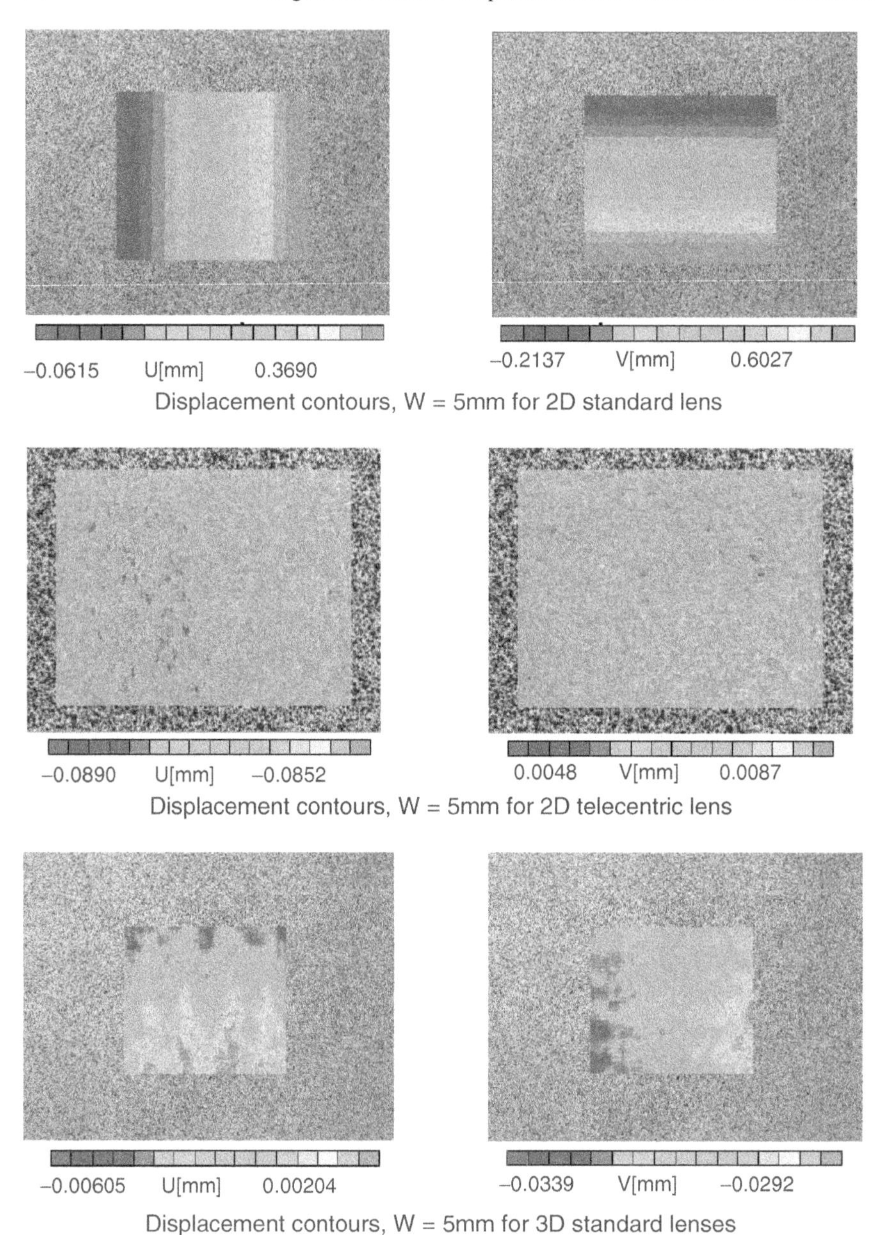

Fig. 6.12 Typical horizontal and vertical displacement fields obtained using VIC-2D for 2D images and VIC-3D for calibrated stereo pair for out-of-plane displacement towards the camera, $W = 5$ mm

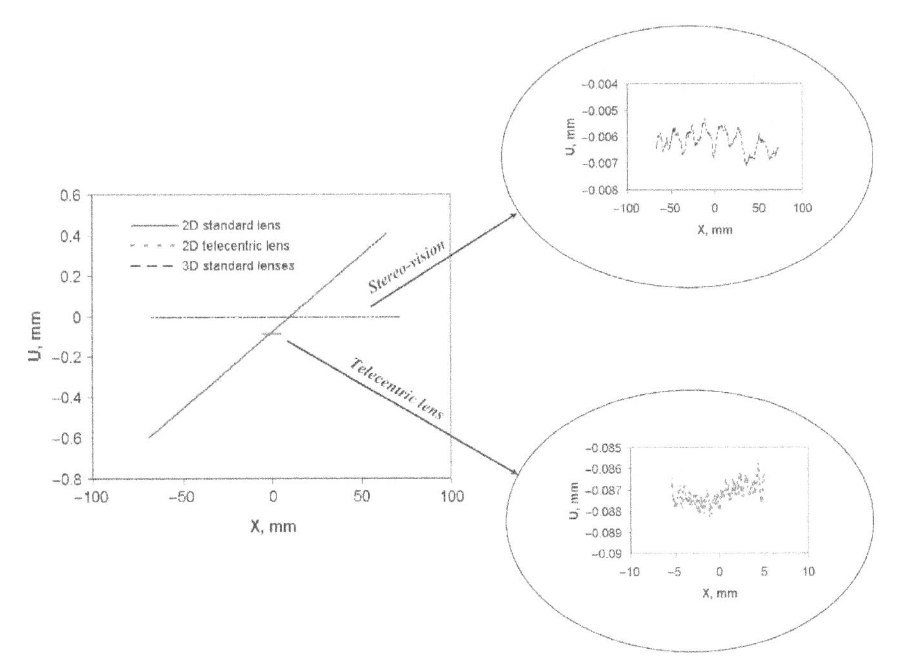

Fig. 6.13 Typical form of measured horizontal displacement, $U(x, y_0)$ along a y-line through the image center for $W = 5$ mm using all three optical systems. Variability in the measurements is clearly visible in expanded views

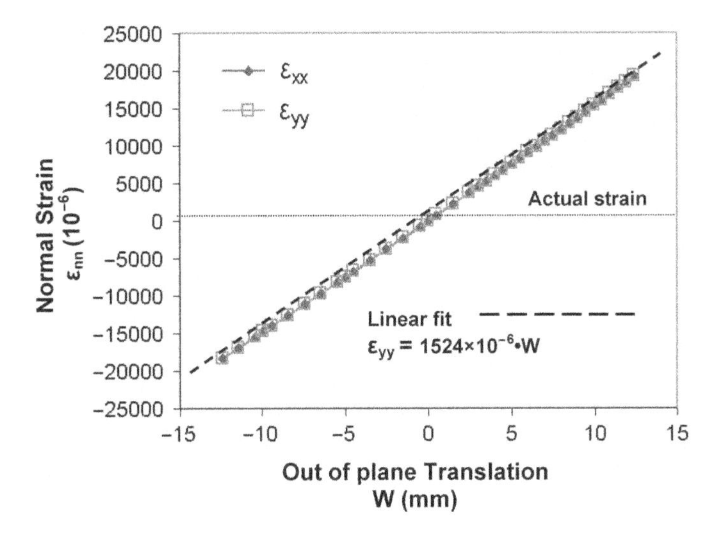

Fig. 6.14 Measured normal strains using single camera with standard lens and a range of out-of-plane motions. Linear best fit has slope of 1524 μs/mm

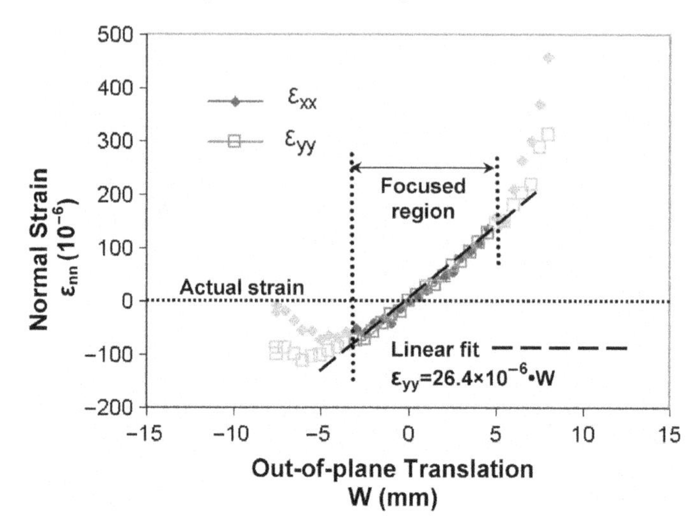

Fig. 6.15 Measured normal strains using single camera with telecentric lens and a range of out-of-plane motions. Linear best fit has slope of 26.4 μs/mm

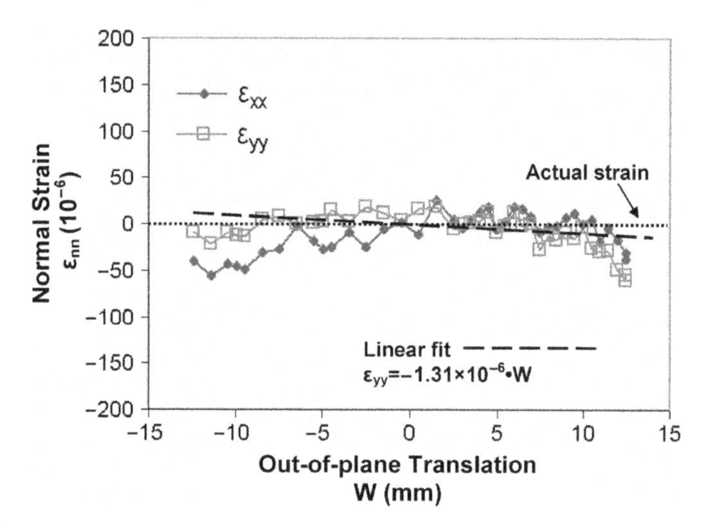

Fig. 6.16 Measured normal strains using stereo-vision system with two standard lenses and a range of out-of-plane motions. Linear best fit to ε_{yy} has slope approaching zero, indicating out-of-plane motion does not introduce errors into in-plane displacements

Following the same process for the telecentric lens, the data in Fig. 6.15 indicates that the effective object distance is $\simeq$37.3 m, nearly 60 times larger than measured when imaging with a standard lens. As shown in Fig. 6.15, image defocus

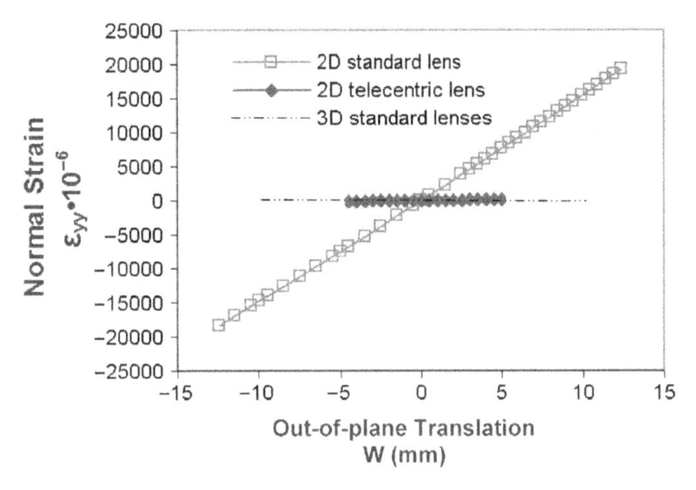

Fig. 6.17 Measured normal strains using all camera systems

occurred for $W \geq 5$ mm. In this regime, highly non-linear trends are measured with a rapid increase in measured strain. The rapid increase in strain is believed to be primarily due to defocus.[3]

As shown in the contour plot for stereovision in Fig. 6.12, the stereovision line plot in Fig. 6.13 and the average strain values in Figs. 6.16 and 6.17, a typical stereovision system does not have discernible displacement gradient in either in-plane direction due to out-of-plane motion. Since a stereo-vision system simultaneously determines the three-dimensional position before and after experiencing out-of-plane translation, the data clearly shows that the in-plane motions measured by a stereo vision system are not corrupted by the out-of-plane displacement component.

Figure 6.17 shows that, in comparison to a 2D system using an f28 lens for imaging, both the 2D system using a telecentric lens for imaging and the stereovision system with standard lenses are insensitive to out-of-plane motions within the range of focus for each system; the maximum strain error at the edge of focus for the telecentric lens $\simeq$135 $\mu\varepsilon$.

6.3.7 Remarks

A telecentric lens acts very much like a telescope, while retaining high magnification and reasonably good light transmission. A well-designed telecentric lens also reduces out-of-plane sensitivity to a manageable level so that out-of-plane motion will not contribute significantly to in-plane strain measurement error. Disadvantages

---

[3] The measured 2D displacement data obtained from the standard lens and telecentric lens and used to construct Figs. 6.14 and 6.15 is not corrected for spatial distortion.

of high quality commercial telecentric lenses are (1) physical size and weight, (2) cost and (3) fixed magnification. The Schneider Xenoplan lens used in these studies is 0.3 m long with a retail cost on the order of $2,500.

The stereovision calibration parameters for camera 2 in Table 6.3 indicate a rather large offset of the center location with a corresponding increase in distortion correction. The large number of parameters obtained during the highly non-linear calibration process can result in physically unrealistic optimal parameter estimates for a lens-camera combination. Even so, the complete set of optimal parameters provides a solution set that will accurately estimate 3D positions for points within the calibration volume. Of course, if one prefers, constrained optimization could be performed requiring that the center (C_x, C_y) correspond with the image center. The remaining parameters would then be optimized during the calibration process to obtain a corresponding optimal calibration set that minimizes the error metric.

The error analyses presented in Sections 5.6 and 5.7 indicate that the presence of Gaussian noise increases image-based measurement errors. Assuming noise $\Gamma_N = 1\%$, results suggest interpolation bias and measurement standard deviation due to Gaussian noise of approximately the same magnitude (0.004 pixels). These theoretical estimates are slightly smaller than shown in Fig. 6.13, where 2D measurements indicate a range of noise on the order of 0.02 pixels or a standard deviation $\simeq$0.005 pixels.[4] The experimental measurement variability is consistent with values reported in the literature for similar experimental configurations.

6.4 Development and Application of Far-Field Microscope for Microscale Displacement and Strain Measurements

Optical microscope systems offer investigators the opportunity to make high resolution measurements in highly localized areas of interest. By combining high resolution images of nominally planar specimens with 2D digital image correlation concepts, the localized variations in material response can be quantified near flaws, sharp material property gradients or local geometric features [241, 290].

6.4.1 Problem Description: Measurement of Crack Closure Load During Fatigue Crack Growth

With the pioneering work of Elber in the early 1970s, the concept of plasticity induced crack closure and its direct effect on the driving force for fatigue crack growth

[4] The experimental evidence includes the effect of (a) quantization of each 8-bit intensity value, (b) vibrations, (c) thermal fluctuations and other experimental factors not considered in the theoretical estimates.

(FCG) has been a hallmark of fatigue modeling. To quantify the individual contribution of specific FCG parameters (e.g, plasticity and environment), investigators have developed specific experimental designs to highlight (or minimize) an effect.

6.4.2 Fatigue Specimen Geometry, Material and Surface Preparation

The geometry of the 2 mm thick extended compact tension specimen and the location of the pattern are shown in Fig. 6.18.

To provide additional information for modeling studies, both a crack mouth opening displacement device (CMOD gage) and also a back face strain gage are incorporated into the experiment. A region $\simeq$4 mm wide along the crack line and ahead of the crack is treated to obtain a high quality random pattern when viewed by the optical system. Images of this region obtained during the fatigue process are used to quantify crack tip displacements.

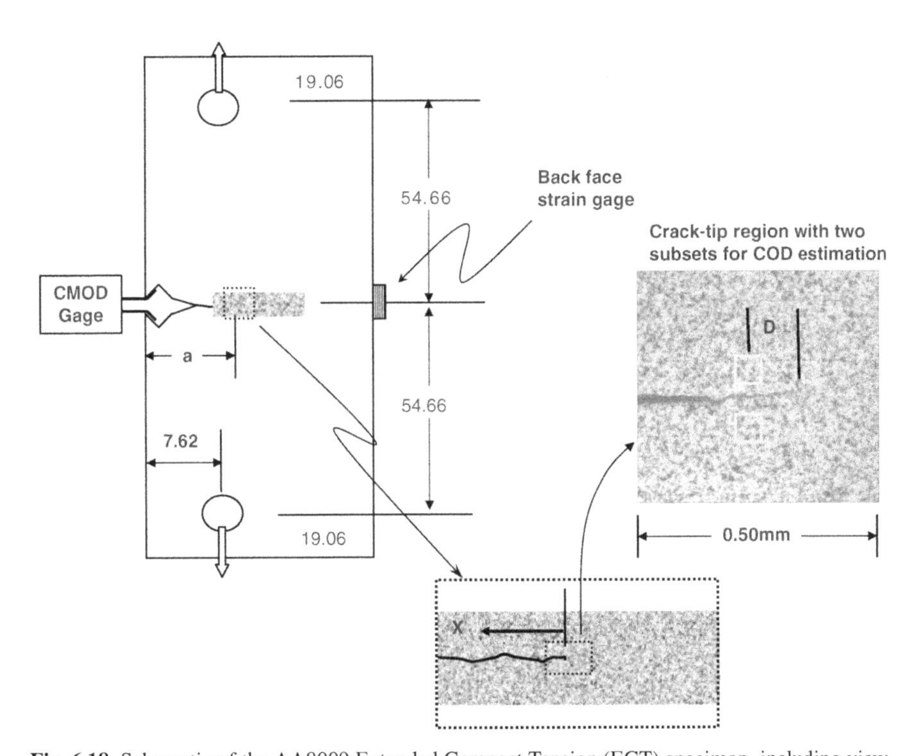

Fig. 6.18 Schematic of the AA8009 Extended Compact Tension (ECT) specimen, including view of the random pattern applied to the specimen surface. Thickness is 2 mm. All dimensions are in millimeters

Table 6.4 Material properties for AA8009

Yield stress, σ_Y	428 MPa
Ultimate stress, σ_{ULT}	482 MPa
Modulus of elasticity, E	82 GPa
Poisson's ratio	0.30
Grain size	$\simeq 0.50\ \mu m$
Dispersoid size	$\simeq 0.060\ \mu m$

The material selected for the fatigue study is AA8009. AA8009 is a specialty advanced powder metallurgy, heat-resistant aluminum alloy (Al, Fe, V, Si) originally considered for use by NASA in aerospace application. Material properties and limited micro-structural information for AA8009 are given in Table 6.4. The fine-grain specialty alloy is selected to minimize roughness-induced closure in the wake region of the crack. Under these conditions, crack tip measurements that define the onset of crack closure should provide a quantitative measure of the "inherent" FCG response of a flawed material undergoing nominally tensile loading.

6.4.3 Validation Specimen and Preparation

The geometry of the uniformly loaded validation specimen is shown in Fig. 6.19. The 2.3 mm thick specimen is manufactured from AL2024-T351. Near the horizontal centerline, the region is cleaned and polished. After surface preparation, a random pattern is applied to the specimen in a small region. After random pattern application, the specimen is again cleaned to remove any patterning material from the surrounding area. After cleaning is completed, miniature single-axis strain gages are bonded to the specimen on either side of the random pattern. The bonded gages are attached to a Micro-Measurements strain gage readout unit to verify their functionality.

6.4.4 Pattern Application

Two methods are employed to apply speckle patterns with appropriate density on the validation and fatigue specimens for image correlation. The first method used E-beam lithography to fabricate a mask with the desired random pattern [255, 290]. Using the mask, contact lithography is performed to transfer the random pattern onto the specimen surface. To make a high contrast pattern, the transferred pattern is manufactured using tantalum. Figure 6.20 shows two typical lithographic pattern applied to the polished aluminum surface, where the average spot size is either 4 or 20 μm, depending upon the size pattern required.

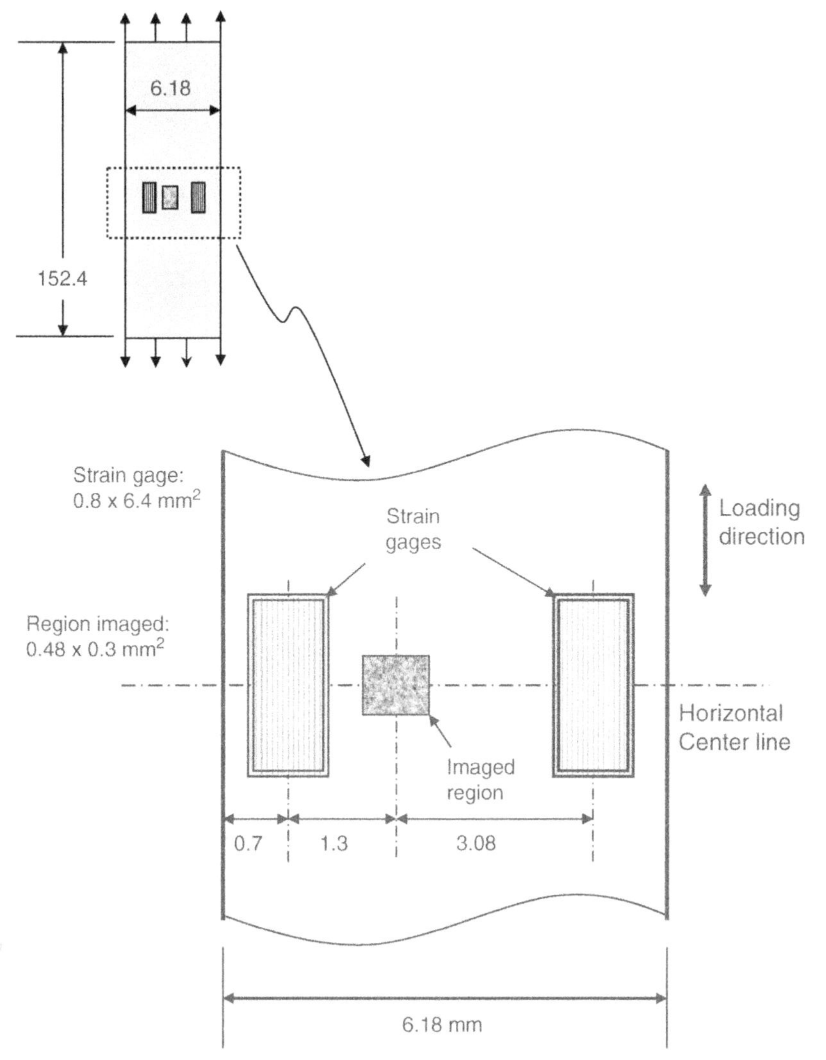

Fig. 6.19 Schematic defining local relationship of strain gages and micro-scale image region used to validate experimental 2D vision-based strain measurements. Material is AL2024-T3All. Dimensions in millimeters

The second method, which is simpler but less controllable, applies filtered toner powder to the polished surface [277, 285, 286, 290]. Attachment of the powder to the specimen surface is performed by briefly heating the surface to $\simeq$200°F so that the pattern partially melts and adheres to the surface. Figures 6.21 and 6.23 show a toner powder pattern, where the average diameter of the powder is $\simeq$11 μm prior to heating.

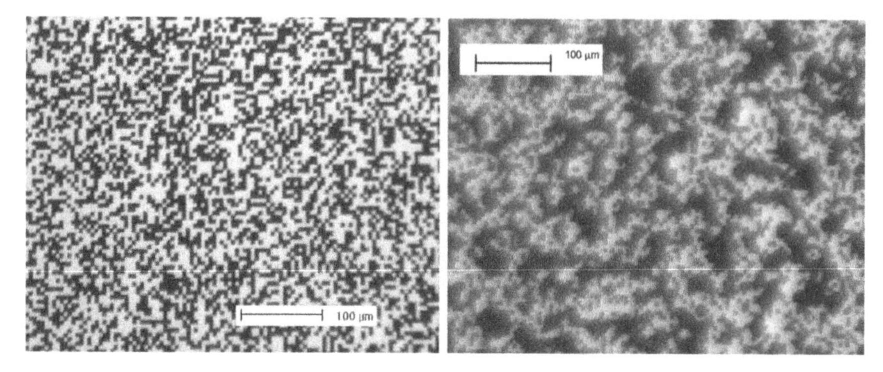

Fig. 6.20 Typical photolithography patterns with 4 μm (*left image*) and 20 μm average speckle size on polished aluminum specimens

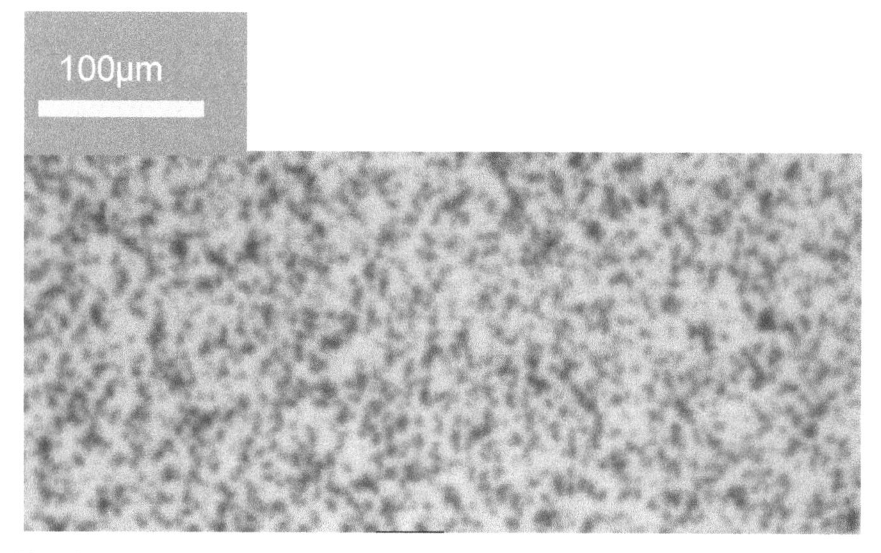

Fig. 6.21 Typical filtered toner powder pattern with 11 μm average powder size on polished aluminum specimens. Brief heating to 100°C bonded partially melted powder to specimen surface

6.4.5 Optical Setup for Imaging

Figure 6.22 shows the setup used to record high magnification images of the specimen as it is undergoing cyclic tensile loading.

A: The Questar OM-1 Far Field Microscope (Q-FFM) is used to acquire high magnification images of the specimen. The Q-FFM is a parabolic-mirror based lens with outer diameter of 89 mm, 30-1 range in field of view (0.5–15 mm), working distances from 0.5–1.5 m and a corresponding range in values for $\mathcal{N}$ from approximately 9 to 17, respectively, and an estimated diffraction limit of 2.7×10^{-6} m.

B: Specimen lighting is provided by a fiber optic illuminator.

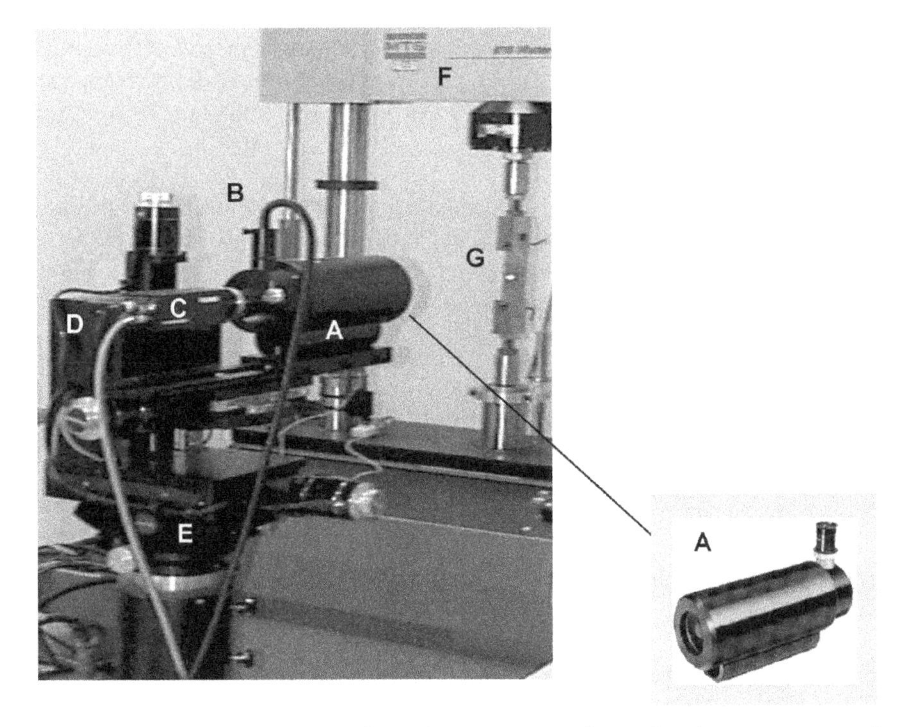

Fig. 6.22 Arrangement of optical far-field microscope system for crack region measurements. *A*: Questar OM-1 Far Field Microscope; *B*: Fiber optic illuminator; *C*: Digital CCD Camera; *D*: Three-axis translation stage; *E*: Sturdy Questar adjustable rolling microscope mount; *F*: MTS hydraulic tensile test frame; *G*: Fracture specimen for fatigue crack growth studies

C: The images are digitized by a Pulnix 9701 progressive scan 768(H)×484(V) pixel array with non-square pixels 11.6 μm(H)×13.6 μm(V) and 8-bit direct digital output. The camera has a full-frame shutter capable of operating between 1/60th and 1/16,000th of a second.

The camera is mounted to integrated translation stages for independent translation in three orthogonal directions (D). The translation stages are mounted to a sturdy, movable camera stage (E). Also attached to the stage is an integral control unit for digital control of the translation process (not shown) The ECT specimen (G) to be viewed is mounted in the pin-hole grips of the MTS 20 kip test system (F).

6.4.6 Validation Experiment

The validation tensile experiment is performed in a 20 kip, MTS servo-hydraulic uniaxial loading system. The specimen shown in Fig. 6.18 is attached to the MTS loading frame using hydraulic grips. The quasi-static experiment is performed in displacement control, with load monitored to ensure nominally elastic response during the loading and unloading processes. Acquisition of the load cell, strain gage and image signals is synchronized so that the image-based strain values can be compared to the strain gage measurements.

Since the goal of the validation experiment is to confirm that the relative displacement between image subsets is being determined accurately, the image based strain estimates are determined by selecting three pairs of 91×91 subsets. Each pair of subsets is separated by 250 pixels along the loading direction. At each load level, P, the image based strain is obtained using the equation

$$\varepsilon_{\text{axial}}^{\text{avg}}(P) = \frac{1}{3}\left[\sum_{i=1}^{3}\left(\frac{u_i^{\text{top}}(P) - u_i^{\text{bottom}}(P)}{250}\right)\right] \tag{6.4}$$

Both image based data and strain gage data were obtained every 45 N up to 1780 N. A least squares best fit straight line is applied to the $\varepsilon - P$ data for both strain gage and image-based data.

6.4.7 Fatigue Experiment

The fatigue experiment is performed in the 20 kip, MTS servo-hydraulic uniaxial loading system on the ECT shown in Fig. 6.18 at a loading frequency of 11 Hz under mode I conditions. For this study, the fatigue process is performed under K-control with an R-ratio of 0.05 and a constant[5] $\Delta K = 4.4$ MPa $\text{m}^{1/2}$.

The entire experiment-specimen loading, image acquisition, image storage, camera movement to maintain focus, camera movement to ensure growing crack tip remains in field of view – can be performed in either manual mode or in an automated mode for the fatigue experiments. The process is as follows:

- Load cycle, load and displacement data from MTS system sent to common PC.
- Digital images output from Pulnix camera stored via video board in same PC with time stamp.
- X–Y–Z translation stage control boards also contained in common PC.

The fatigue process proceeds as follows. The test is initiated by cycling at 11 Hz until a preset number of fatigue cycles is completed (e.g. 1,000 cycles). Then, the frequency is reduced to 0.01 Hz and 100 images are acquired by synchronizing the image acquisition with the loading process. At the end of the imaging cycle, the load is held constant briefly as the specimen is translated out-of-plane to verify that the specimen is clearly focused. Once focus is confirmed, images are analyzed to locate the crack tip and horizontal translation is performed to ensure the current

[5] In this case, an estimate for the cyclic Irwin plastic zone size, $r_{p\ \text{cyclic}} \simeq 40$ μm and the maximum Irwin plastic zone size during fatigue, $r_{p\ \max} \simeq 100$ μm. These values are 20X smaller than the specimen thickness, confirming that the plastic zone is fully contained.

crack tip is located $\simeq$250 pixels from the right edge of the image (see crack tip Inset in Fig. 6.18). This process is continued until the user specified cycle limit has been reached.[6]

6.4.8 Post-processing to Determine COD

Once the experiment is completed, post-processing is performed using two-dimensional digital image correlation to determine the crack opening displacement (COD) as a function of applied loading at various locations behind the crack tip. Figure 6.23 shows graphically the process that is used.

First, the image corresponding to the minimum load during the fatigue process (P_0) is selected and the crack tip is located in this image; it is considered the "reference" image for estimating COD with the current crack tip location. Next, distances of interest behind the crack tip D_1 and D_2 are selected. For example, at location D_1,

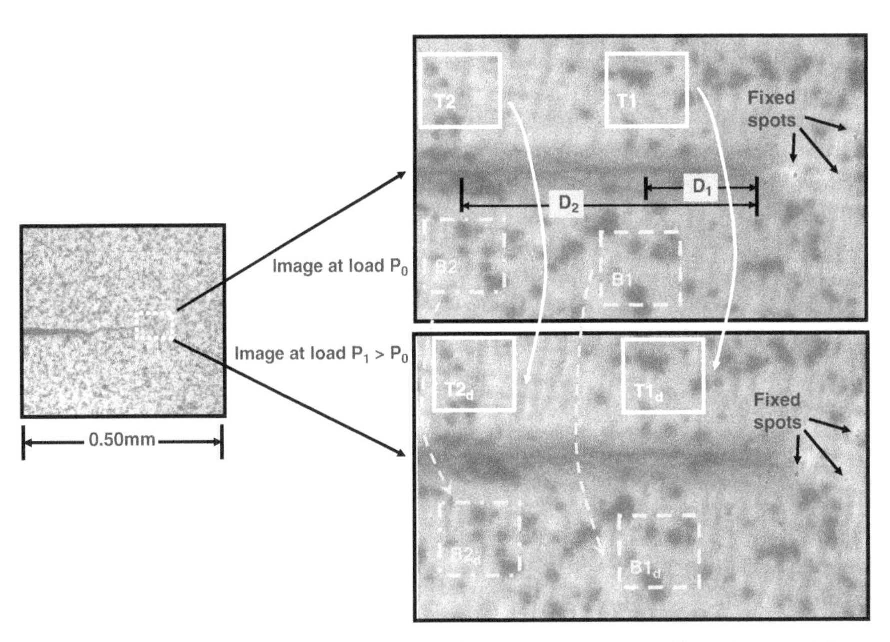

Fig. 6.23 Small local subsets and local COD displacement measurements. Errors increased due to anisotropy of subset intensity pattern and local fixed spots due to dust or water on lens or glass in front of sensor plane that tend to bias image-based displacement measurements

[6] All fatigue experiments and imaging reported in [241] were performed in the laboratory of Dr. Robert Piascik in the former Mechanics and Materials Branch, Building 1205, NASA Langley Research Center.

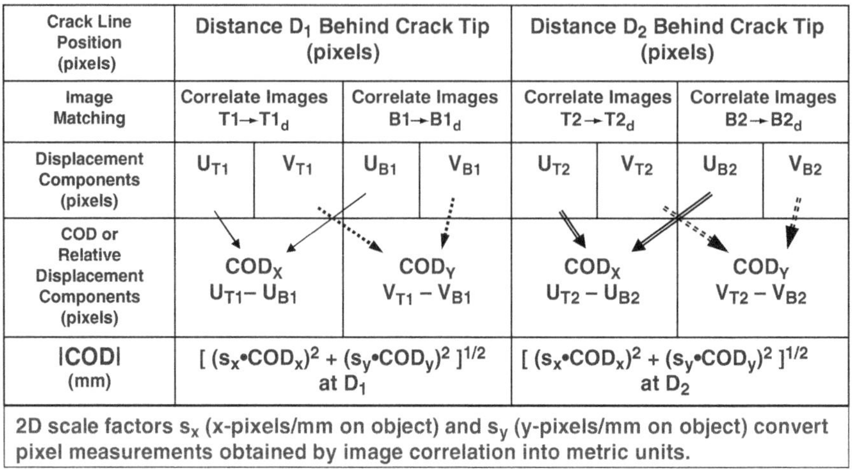

Crack Line Position (pixels)	Distance D_1 Behind Crack Tip (pixels)				Distance D_2 Behind Crack Tip (pixels)			
Image Matching	Correlate Images $T1 \rightarrow T1_d$		Correlate Images $B1 \rightarrow B1_d$		Correlate Images $T2 \rightarrow T2_d$		Correlate Images $B2 \rightarrow B2_d$	
Displacement Components (pixels)	U_{T1}	V_{T1}	U_{B1}	V_{B1}	U_{T2}	V_{T2}	U_{B2}	V_{B2}
COD or Relative Displacement Components (pixels)	COD_X $U_{T1} - U_{B1}$		COD_Y $V_{T1} - V_{B1}$		COD_X $U_{T2} - U_{B2}$		COD_Y $V_{T2} - V_{B2}$	
|COD| (mm)	$[(s_x \bullet COD_x)^2 + (s_y \bullet COD_y)^2]^{1/2}$ at D_1				$[(s_x \bullet COD_x)^2 + (s_y \bullet COD_y)^2]^{1/2}$ at D_2			
2D scale factors s_x (x-pixels/mm on object) and s_y (y-pixels/mm on object) convert pixel measurements obtained by image correlation into metric units.								

Fig. 6.24 Procedure for determining COD at positions, D_1, and D_2 along crack line

subsets above the crack line (T1) and below the crack line (B1) are identified that contain sufficient contrast for image matching. This process is repeated at location D_2 to obtain subsets T2 and B2.

Once the subset pairs are selected in this "reference" image, each subset is located in the next image with load $P_0 + \Delta P$ by performing image correlation to match T1 to $T1_d$, B1 to $B1_d$, T2 to $T2_d$ and B2 to $B2_d$. The 2D image matching process outputs the in-plane displacement vector for each subset. Processing of the displacement vectors (U,V) for each subset to obtain COD (mm) is outlined in Fig. 6.24. The procedure outlined in Fig. 6.24 is repeated to compare subset pairs in all 100 images acquired during the imaging cycle, determining COD as a function of (a) distance behind the crack tip, D_i, (b) tensile loading, P, and (c) crack length, a.

6.4.9 Experimental Results

For the validation experiment outlined in Section 6.4.6, Fig. 6.25 presents a direct comparison between (a) the strain data obtained from both gages and (b) the average strain data obtained from the 2D image matching process for the range of applied tensile loads.

The data in Fig. 6.25 confirms that the relative displacement measurements obtained during micro-scale image matching are accurate and in agreement with independent measurements.

By fitting the raw relative displacement data versus load with a least squares best fit straight line, results show that each displacement has a standard deviation of 0.05 pixels. With a scale factor of 1,667 pixels/mm, the displacement error is $\simeq$30 nm in metric distance.

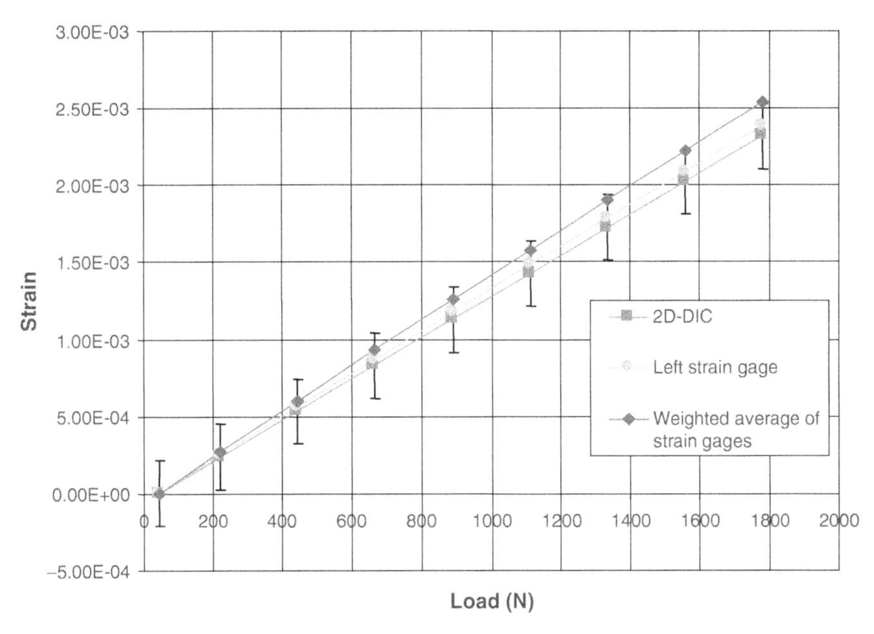

Fig. 6.25 Comparison of strain gage measurements and 2D digital image correlation results for tensile loading of 2024-T3 aluminum specimen, primarily in the linear elastic region

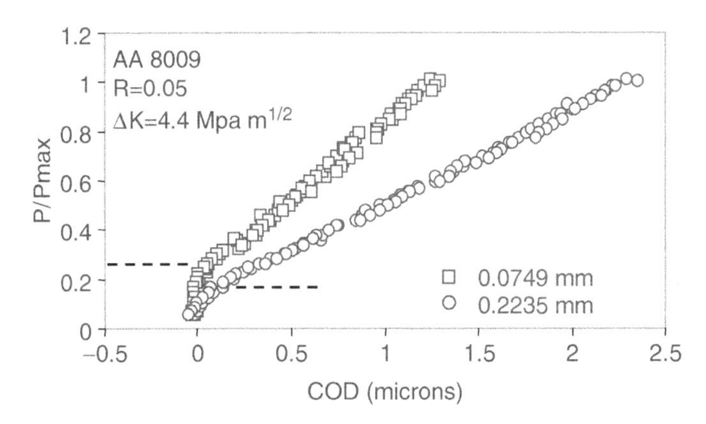

Fig. 6.26 Measured crack opening displacement at two locations behind current crack tip on extended compact tension specimen as a function of non-dimensional applied loading parameter for AA8009 steel alloy. Dashed lines denote transition between closed crack with contacting surfaces and fully open crack

For the fatigue experiment outlined in Section 6.4.7, Fig. 6.26 shows the type of experimental COD versus P data that is obtained for points located $D_1 = 75\ \mu\text{m}$ and $D_2 = 224\ \mu\text{m}$ behind the current crack tip.

As shown in Fig. 6.26, during the early stages COD has a minimal increase as the load rises. At a well-defined location, the surface COD data shows a rapid increase with additional load. The demarcation in slope is well-known to correspond to the local loss of contact as the crack opens; the corresponding force is oftentimes designated the **closure load**, P_{CL}.

The data in Fig. 6.26 shows that P_{CL}/P_{MAX} is smaller for points farther from the crack tip, indicating that the crack opens in a "peeling" manner, initially losing contact for points removed from the tip and then peeling open near the tip as the load increases.

6.4.10 Discussion

Microscope optics, including catadioptric components such as the Q-FFM, have severe limits on depth of field. If it is assumed that the diameter of the circle of confusion corresponds to the diffraction limit, then Eq. (2.13) provides an estimate for the image focus zone. Using the physical data for the Q-FFM optics, Table 6.5 shows the DOF for two Q-FFM configurations. The top line in Table 6.5 corresponds to the highest magnification and smallest FOV (0.5×0.5 mm^2) with a DOF $\simeq 0.05$ mm (0.0012 in.). Within this zone, the specimen will have an image that has clarity consistent with the diffraction limit. For larger Δz motions, image sharpness will degrade as the edges of the pattern blur. As shown for the telecentric lens in Section 6.3, subset-based image correlation will still be possible until considerable defocus occurs and pattern matching is no longer achievable.

With this small DOF, it is important that specimen preparation be adequate for imaging in the microscope and includes

- Planarity of the overall specimen
- Minimization of local depressions/height variations in imaged region
- High quality polishing to remove scratches
- Minimization of the physical height of the applied pattern

An advantage of FFM systems is that, within this small DOF, the large object distance reduces pseudo-strain errors. For example, if it is assumed that image correlation is possible for $\Delta z = 0.1$ mm (0.0025 in.), Eq. (6.4) gives a small estimated strain error of 1.8×10^{-4} that would be embedded in the measured displacement field.

Since the image analysis process employed to estimate COD uses images such as shown in Fig. 6.23, two points require further elaboration.

Table 6.5 Depth of field estimates for Questar FFM

Object distance	Lens $\mathcal{N}$ (f/D_{LENS})	Focal length, f (m)	Field of view (10^{-3} m)	d_{COF} (10^{-6} m)	Depth of field (10^{-3} m)
0.56 m	8.7	0.53	0.50	2.5	0.05
1.56 m	17.0	1.02	4.70	2.5	0.24

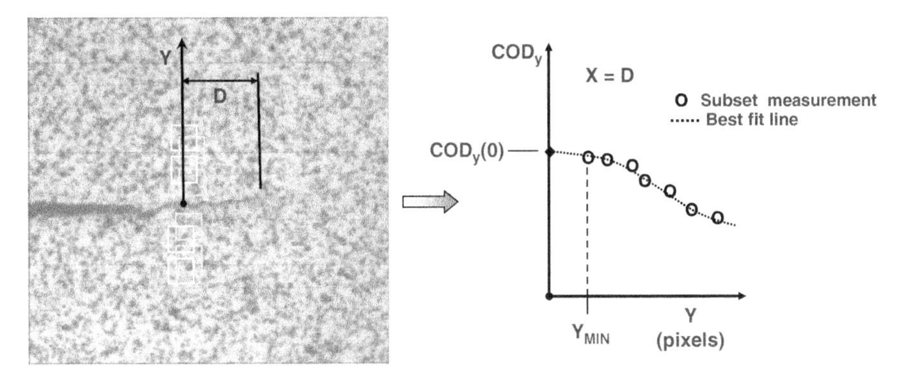

Fig. 6.27 Extrapolation of subset-based crack opening displacement measurement to crack line. Subsets lie along vertical line at distance D behind current crack tip

- Small subset intensity pattern
- Fixed points in image

Since it is preferred to obtain displacement measurements close to the crack line (COD is theoretically defined to be the relative displacement of points **on opposite crack faces**), this necessitates the use of smaller subsets that are close to the crack line. Since it is difficult to obtain a high quality pattern everywhere along the crack line, this requirement results in selection of subsets that have:

(a) Few pixels (e.g., 13×13) with low contrast (see Section 5.7.1, Eqs. (5.95) and (5.97) for definitions of pattern contrast in terms of spatial gradients in the intensity pattern) and
(b) Anisotropy in the subset contrast

Anisotropy in the subset intensity pattern will cause interpolation-bias [252, 253] in the measured motions. Since the recorded intensity pattern will have low levels of noise, small subsets with low contrast will introduce higher variability in the measured motions. In both cases, the simplest way to decrease such errors is to increase subset size to improve both isotropy and contrast of the subset pattern. As a result, there is always a trade-off when estimating COD using the procedure in Fig. 6.23.

Given the difficulties noted above, another approach has been used successfully and is shown schematically in Fig. 6.27. In this approach, larger subsets along a vertical line can be selected to reduce subset-based displacement errors. Extrapolation to the crack line using a least squares fit to the measurements is performed to estimate components or magnitude of COD at the crack line.

6.4.10.1 Image Artifacts

The spots highlighted in the images shown in Fig. 6.23 are *image artifacts*. They do not correspond to points on the object and hence do not move with the object during deformation. Such features are important to identify in any image-based measurement system, and are discussed in Section 10.1 as part of the Practical Considerations review.

It is worth noting that image distortions are not discussed. In fact, such distortion corrections can be very important when points away from the image center are considered. In such cases, calibration using methods in Section 4.1 or distortion removal using methods in Section 3.2 is recommended. To confirm that the effects of distortion on the COD measurements are small for this application, a series of in-plane translations are performed using the FFM system and the measured displacement field evaluated. Results showed that the small motions incurred by points near the crack line during an imaging cycle will not require distortion correction.[7]

6.5 Inverse Methods: Material Property Measurements Using Full-Field Deformations and the Virtual Fields Method

An area of growing interest in the mechanics and materials community is the use of full-field deformations with an appropriate computational model to extract local material properties. The approach has seen remarkable growth in the past few years, with articles describing applications in fracture mechanics, friction stir welds and other heterogeneous materials. In this section, the authors describe the use of full-field 2D-DIC measurements to extract linear elastic properties from nominally homogeneous polymer foam. The model approach used to couple the experimental measurements with the mechanical response of the foam is the Virtual Fields Method (VFM). Given the relative newness of the approach, a brief summary of the concepts as applied to a relatively simple example is provided in the following section.

6.5.1 Virtual Fields Method

The Virtual Fields Method[8] provides a relatively simple, robust modelling approach to quantify material properties when full-field deformation measurements are available for the region of interest. Such information can be provided by 2D-DIC for nominally planar objects undergoing planar deformation or 3D-DIC for more complex loading situations.

---

[7] The manufacturer of the FFM reported distortions of 1 in 15,000.

[8] The theoretical developments described in this section are adapted from developments by Prof. Fabrice Pierron and Dr. B. Guo, Arts et Métiers ParisTech, France.

6.5.2 Derivation of VFM Equations for Cantilever Beam Undergoing Planar Deformations

The VFM is based on the principle of virtual work [94,95]. For static loading conditions without body forces, the principle can be written as follows:

$$
\begin{aligned}
&-\int_{v} \underline{\underline{\sigma}} : \underline{\underline{\varepsilon}}^{*} dV + \int_{\partial V} \overrightarrow{T} \cdot \overrightarrow{u}^{*} dS = 0 \\
&\underline{\underline{\sigma}} : \underline{\underline{\varepsilon}}^{*} = \Sigma_i \Sigma_j (\sigma_{ij} \varepsilon_{ij}^{*}) \\
&\overrightarrow{T} \cdot \overrightarrow{u}^{*} = \Sigma_i (T_i u_i^{*})
\end{aligned}
\tag{6.5}
$$

where $\underline{\underline{\sigma}}$ is the stress tensor across the volume of the solid (denoted V), $\overrightarrow{T}$ is the traction vector applied over the external surface of the solid (denoted ∂V), $\overrightarrow{u}^{*}$ is a continuous vector test function known as the virtual displacement field and $\underline{\underline{\varepsilon}}^{*}$ is the virtual strain tensor derived from $\overrightarrow{u}^{*}$.

This equation is also known as the weak formulation of equilibrium and is the basis of the finite element method. Equation (6.5) is valid for any kinematically admissible virtual field. By kinematically admissible, one means that the virtual displacement field must be continuous and differentiable (C^0) and must be zero on the boundary where displacements are prescribed. It should be noted that the virtual strain field can be discontinuous. These quantities do not relate to anything physical in the material and hence can be considered as mathematical test functions used to write global equilibrium relationships. Given the general nature of the approach, VFM is a robust tool for defining integral equilibrium equations.

To implement the VFM method, the stress components in Eq. (6.5) are replaced using the relevant constitutive equations for the material system being studied. In our example, we assume an isotropic, linear elastic material response with plane stress conditions in the $X-Y$ plane (see Fig. 6.28). The stress–strain relationships

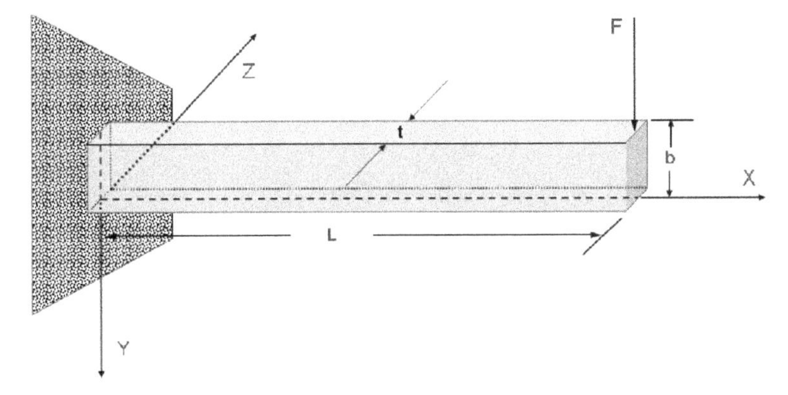

Fig. 6.28 Cantilever beam with thickness t, width b and length L

in the $x-y$ plane can be written:

$$\begin{pmatrix} \sigma_{xx} \\ \sigma_{yy} \\ \sigma_{xy} \end{pmatrix} = \begin{bmatrix} \frac{E}{1-\nu^2} & \frac{\nu E}{1-\nu^2} & 0 \\ \frac{\nu E}{1-\nu^2} & \frac{E}{1-\nu^2} & 0 \\ 0 & 0 & \frac{E}{2(1+\nu)} \end{bmatrix} \begin{pmatrix} \varepsilon_{xx} \\ \varepsilon_{yy} \\ 2\varepsilon_{xy} \end{pmatrix} \tag{6.6}$$

Assuming homogeneous material properties and combining Eqs. (6.5) and (6.6), we have:

$$\begin{aligned} &-\tfrac{E}{1-\nu^2}\left(\int_V \varepsilon_{xx}\varepsilon^*_{xx}dV + \int_V \varepsilon_{yy}\varepsilon^*_{yy}dV\right) - \tfrac{\nu E}{1-\nu^2}\left(\int_V \varepsilon^*_{xx}\varepsilon_{yy}dV + \int_V \varepsilon_{xx}\varepsilon^*_{yy}dV\right) \\ &-\tfrac{2E}{1+\nu}\int_V \varepsilon_{xy}\varepsilon^*_{xy}dV + \int_{\partial V} \vec{T}\cdot\vec{u}^*dS = 0 \end{aligned} \tag{6.7}$$

Equation (6.7) is the primary result from the principle of virtual work for the special case being studied. Since Eq. (6.7) is valid for each kinematically admissible virtual displacement field and associated virtual strain field, selection of at least two independent virtual displacement fields should result in a system of equations that can be solved for the unknown material properties, E and ν.

6.5.3 Virtual Fields for Cantilever Beam Specimen

Consider a simple cantilever beam of thickness, t, undergoing in-plane deformation as shown in Fig. 6.28. To use Eq. (6.7) and model the beam structure, it is assumed that (a) the applied loading results in planar deformations, (b) plane stress conditions exist and (c) the material is isotropic and linear elastic.

For this case, a kinematically admissible virtual displacement field and its associate virtual strain field are as follows:

$$\begin{cases} u^* = 0 \\ v^* = -x \end{cases} \Rightarrow \begin{cases} \varepsilon^*_{xx} = \frac{\partial u^*}{\partial x} = 0 \\ \varepsilon^*_{yy} = \frac{\partial v^*}{\partial y} = 0 \\ \varepsilon^*_{xy} = \frac{1}{2}\left[\frac{\partial u^*}{\partial y} + \frac{\partial v^*}{\partial x}\right] = -\frac{1}{2} \end{cases} \tag{6.8}$$

Combining Eqs. (6.7) and (6.8), one has the following:

$$\left(\frac{E}{1+\nu}\right)\cdot\left[\int_V \varepsilon_{xy}dV\right] + FL = 0 \tag{6.9}$$

It is noted that the virtual work of the external forces at $x=0$ is zero because both components of the virtual displacement field are zero. Since the virtual displacement at $x=L$ is in the direction of the applied force, F, the inner product is FL.

Since the problem is 2D, the volume integral can be transformed into a surface integral over the rectangular region $0 \leq x \leq L$ and $0 \leq y \leq b$, hereafter designated S^P. Moreover, since it is assumed that the full field displacement measurements can be converted into surface strain data over N discrete points of surface S^P (e.g. using 2D-DIC to obtain the displacement components on the front surface), we have:

$$\frac{E}{1+\nu} = -\frac{F}{bt\,\overline{\varepsilon_{xy}}} \tag{6.10}$$

where

$$\overline{\varepsilon_{xy}} = \frac{1}{n}\sum_{i=1}^{n} \varepsilon_{xy}^{i}$$

$n =$ number of data points on the surface

To solve for both unknowns, a second kinematically admissible virtual displacement field and its associate virtual strain field are necessary. A virtual Bernoulli bending field is suitable here so that we have[9]:

$$\begin{cases} u^* = xy \\ v^* = -\frac{x^2}{2} \end{cases} \Rightarrow \begin{cases} \varepsilon_{xx}^* = y \\ \varepsilon_{yy}^* = 0 \\ \varepsilon_{xy}^* = 0 \end{cases} \tag{6.11}$$

Combining Eqs. (6.7) and (6.11), one has the following:

$$\frac{E}{1-\nu^2}\int_V y\,\varepsilon_{xx}dV + \frac{\nu E}{1-\nu^2}\int_V y\,\varepsilon_{yy}dV = \frac{FL^2}{2} \tag{6.12}$$

Following the same procedure as used to develop Eq. (6.10), the volume integral can be transformed into a surface integral over S^P. Considering surface strain data over N discrete points on a surface S^P (e.g. using 2D-DIC to obtain the displacement components on the front surface), one can write:

$$\frac{E}{1-\nu^2}\overline{y\varepsilon_{xx}} + \frac{\nu E}{1-\nu^2}\overline{y\varepsilon_{yy}} = \frac{FL}{2tb} \tag{6.13}$$

where

$$\overline{y\varepsilon_{xx}} = \frac{1}{n}\sum_{i=1}^{n} y^i \varepsilon_{xx}^{i}$$

$$\overline{y\varepsilon_{yy}} = \frac{1}{n}\sum_{i=1}^{n} y^i \varepsilon_{yy}^{i}$$

---

[9] Equation (6.10) can be verified by noting that the shear force is transmitted through each section between the clamp and the loading point and integration of the shear force along x gives the same expression.

Equations (6.10) and (6.13) can be combined to give

$$\begin{bmatrix} \overline{\varepsilon_{xy}} & -\overline{\varepsilon_{xy}} \\ \overline{y\varepsilon_{xx}} & \overline{y\varepsilon_{yy}} \end{bmatrix} \begin{pmatrix} \frac{E}{1-\nu^2} \\ \frac{\nu E}{1-\nu^2} \end{pmatrix} = \begin{pmatrix} -\frac{F}{bt} \\ \frac{FL}{2bt} \end{pmatrix} \tag{6.14}$$

where

$$\frac{E}{1+\nu} = \frac{E}{1-\nu^2} - \frac{\nu E}{1-\nu^2}$$

The approach described above requires that the investigator choose appropriate virtual fields for the problem of interest, with solutions to the resulting equations (e.g. Eq. (6.14)) providing estimates for the material parameters. Baseline computational studies have shown conclusively that material parameter estimates are a strong function of the choice of the virtual fields, with some choices providing better accuracy than others.

As a consequence of the observed dependency, two additional approaches have been used to improve the stability and accuracy of the VFM method. First, a procedure was developed to derive optimal virtual fields for parameter estimation [21]. Second, and more recently, a finite element (FE) based procedure was developed to use piecewise virtual fields expanded over an FE set [22] for material parameter estimation. Since the latter method has been shown to be robust and efficient, the FE procedure also has been implemented in this work for comparison to the results obtained using Eq. (6.14).

6.5.4 Experimental Studies

A low density polyurethane foam[10] is used in these studies. A PCO Sensicam camera with 12-bit intensity resolution and 1.3 mega-pixel spatial resolution is used in these studies to acquire digital images. Figures 6.29 and 6.30 show the experimental setup and a close-up of the region of interest on the foam specimen. Static loading is applied using calibrated weights ranging from 0.015 N (1 g) to 0.46 N (30 g). End clamping of the relatively soft foam specimen is performed via flat platens in a screw-driven test fixture. Repeated experiments confirmed that the clamping arrangement is not critical since the results obtained from the combined 2D-DIC and VFM approach is insensitive to the actual load distribution in the clamp region.

Images are acquired after each 0.15 N load increment. Two-dimensional digital image correlation software designated Correli [114, 115] is used to perform the image analysis.[11] Subset size is 16 $\times$ 16 pixels^2 and the spacing between subsets (step size) is 8 pixels, resulting in a 154 $\times$ 110 array of displacement data throughout the analysis region.

[10] Foam used is type 85735K16 from McMaster-Carr, IL, (USA), with a reported density of $\approx$ 30 kg m^{-3}.

[11] The image analysis was performed by Prof. F. Pierron et al., Arts et Métiers ParisTech, France.

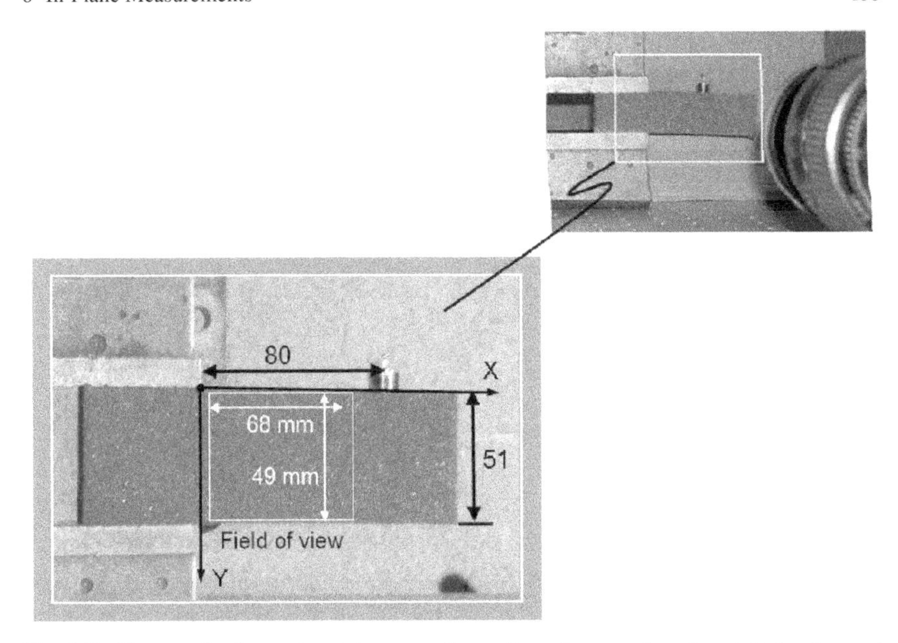

Fig. 6.29 Photograph of experimental setup with cantilever beam undergoing static loading and also a close-up of analysis region. Camera lens is located ≈0.50 m from specimen and image scale factors $s_x = s_y = 18.2$ pixels/mm on the object. Force applied at mid-thickness using calibrated weights. All units are in millimeters

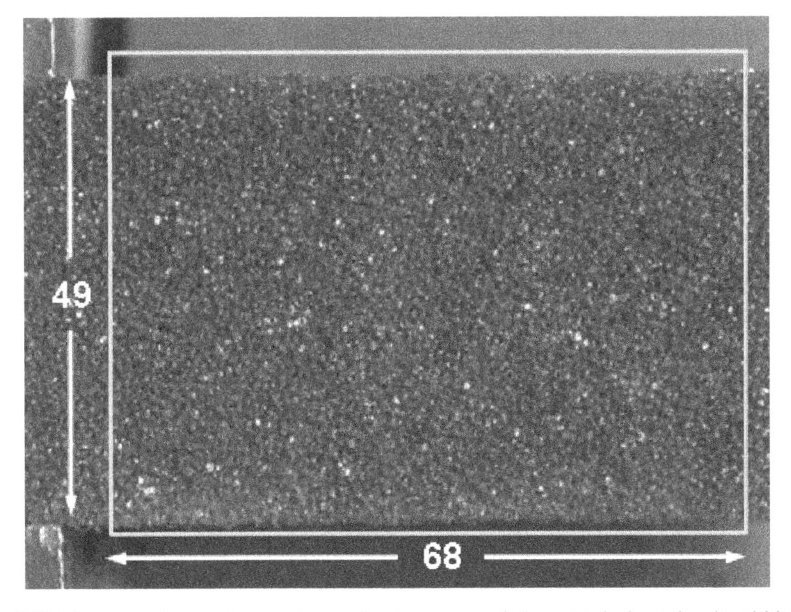

Fig. 6.30 Foam specimen reference image for image correlation. Analysis region is within the rectangular region. All dimensions in millimeters

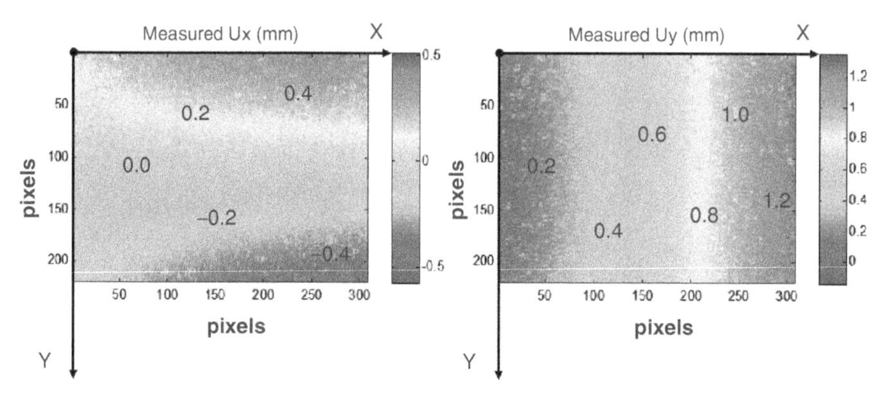

Fig. 6.31 Horizontal and vertical displacement fields in cantilever beam specimen. All dimensions in millimeters

Figure 6.31 shows the raw displacement fields for a load of 0.29 N. Since the vertical axis (Y) is downward (see Figs. 6.28 and 6.29), the measured vertical displacement due to mechanical loading will be positive. Furthermore, since the X-axis is along the beam specimen length, the computed shear strain will be positive (e.g. $\partial U_y / \partial X \geq 0$).

Inspection of the results indicates the relatively small subset size and modest intensity contrast results in relatively high levels of variability in the measured displacements. However, the general trends are clearly discernible and a combination of displacement smoothing and optimization for material parameters estimation will be shown to be robust, even in this case.

6.5.5 Smoothing by 2D Finite Element Methods

Finite element based smoothing methods have been a staple of the mechanics community for decades (e.g. see [285]). Since the focus in previous discussions has been on local smoothing, in this section a Finite Element (FE) smoothing method is employed in order to reduce displacement measurement "noise". The FE mesh incorporates bilinear triangular elements, with the element mesh size ranging from 17 to 25 pixels in each direction, corresponding to inclusion of 12–20 measurement points in each element, respectively. Figure 6.32 shows an example of reconstructed displacement maps together with the related mesh.

Figure 6.33 shows the in-plane strain fields ε_{xx}, ε_{yy} and ε_{xy} obtained using the smoothed FE fields obtained with a mesh size of 25 pixels (approximately 20 data points per element).

Inspection of Fig. 6.33 shows that one can visualize (a) the linear bending ε_{xx} strain field, being approximately zero near the centerline, (b) the small transverse strain field, ε_{yy}, and (c) a shear strain field that is maximum at the center and near zero values at the horizontal edges, as expected. Though the strain values are rather

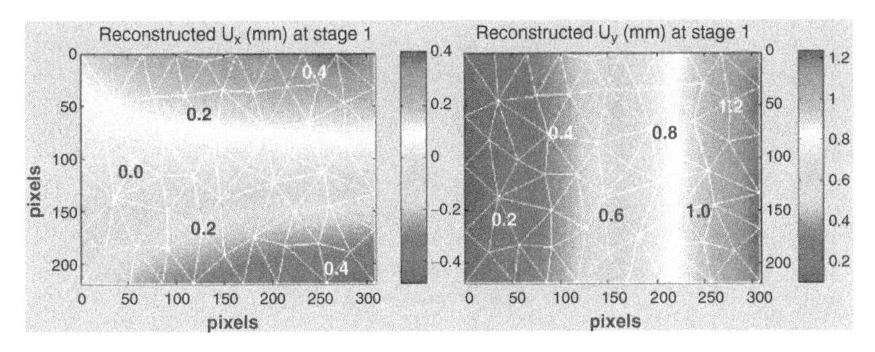

Fig. 6.32 Horizontal and vertical displacement fields after smoothing of experimental data. FE mesh size is 20 and applied loading, $F = 0.29$ N

small and the noise is relatively high, it will be shown that the integral formulation of the VFM will have a smoothing effect, resulting in consistent estimates for the material parameters.

6.5.6 Material Property Results

Using Eq. (6.14) with smoothed strain values at 154×110 positions corresponding to the subset centers used for image correlation, Table 6.6 provides a summary of the results obtained. Here, the effect of mesh size on the estimated material properties is included for completeness.

Using the FE-based approach to obtain "optimal" virtual fields, Table 6.6 also reports the estimated values for E and ν for different smoothing mesh sizes. As shown in Table 6.6, Young's modulus is consistently estimated for both types of virtual fields and all mesh sizes. For the linear elastic case under consideration, this is not surprising since the strain field results are most sensitive to variations in E.

Results for Poisson's ratio show far more variability, especially when using the investigator-selected virtual fields. However, when using the optimized virtual fields, variability in ν decreased significantly, confirming that the virtual fields used for parameter estimation can have a significant impact on the results.

Finally, it is noted that the parameter identification process using the virtual fields method is direct in this case, with computation time of a few seconds on a standard PC.

6.5.7 Discussion

The use of 2D and 3D-DIC offers investigators the opportunity to combine large, full-field data sets with computational and theoretical models to predict a wide range of physically important variables, including

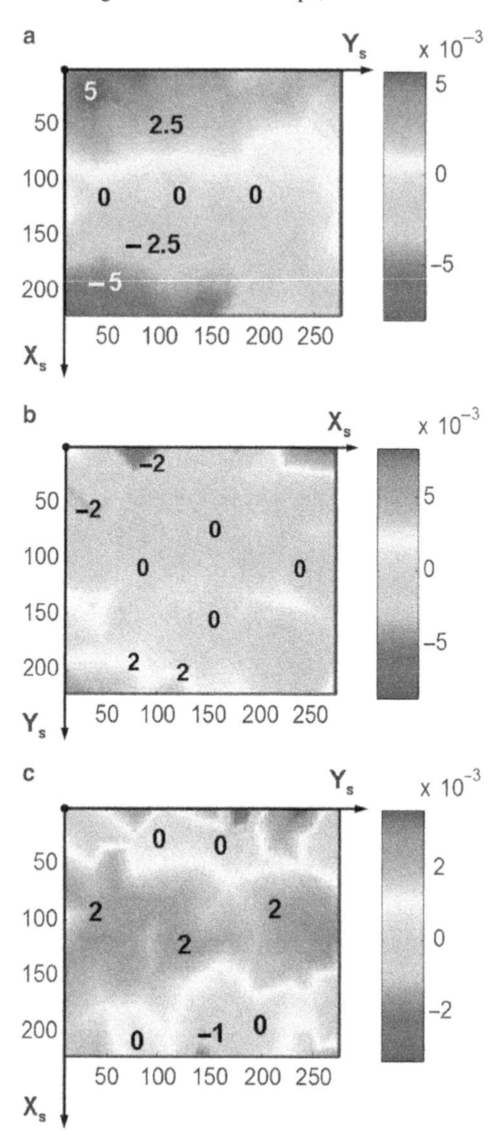

Fig. 6.33 Strain fields **a** ε_{xx}, **b** ε_{yy}, and **c** ε_{xy} obtained from smoothed displacement field results. FE mesh size is 25 pixels and applied loading, $F = 0.29$ N

- Time- and/or load-dependent displacement boundary conditions
- Local stress–strain response in heterogeneous engineering materials
- Pressure distribution on a specimen boundary with known material properties
- Dynamic material response under high rate loading (e.g. Hopkinson bar studies) and
- Dynamic traction conditions for specimen with known properties undergoing time-varying loading

Table 6.6 Material property estimates for E and ν as a function of FE and smoothing mesh size

	E, Young's Modulus (kPa)		ν (Poisson's ratio)	
Mesh size (pixels)	Equation (6.14)	FE Based	Equation (6.14)	FE based
17	400	398	0.501	0.335
19	417	413	0.535	0.326
21	424	415	0.496	0.346
23	429	408	0.217	0.353
25	415	417	0.320	0.372
Mean	417	404	0.414	0.333
Coef. var. (%)	2.7	3.9	33.4	10.7

6.6 Accurate 2D Deformation Measurements in a Scanning Electron Microscope: Basic Concepts and Application

As noted in Section 6.4.10, the spatial resolution of an optical imaging device is controlled by its diffraction limit. For the Q-FFM system, the diffraction limit is $\simeq 3\ \mu m$. Scanning Electron Microscope (SEM) systems offer investigators the ability to make measurements at extremely high spatial resolution.

6.6.1 Imaging in an SEM

In contrast to the proven performance of optical methods with modern scientific grade cameras recording images directly in digital form, SEM systems acquire images by raster-scanning an electron beam across a specimen surface, in a manner that is qualitatively similar to that used to record images using an optical analog camera. Figure 6.34 presents a schematic of an SEM system [52–54, 284]. After e-beam generation in a thermal emission gun (TEG) or field emission gun (FEG), the beam passes through a series of electromagnets and is focused onto the specimen surface. The e-beam rastering process and conversion to a digitized image is shown in Fig. 6.35.

In an SEM system, e-beam steering during the scanning process is performed using electromagnetic "lenses". Since a typical modern SEM system is open loop on e-beam position (i.e., the e-beam position is not a control parameter), positional errors will occur during the scanning process due to environmental and system variables. Variables may include combinations of the following factors; electromagnetic field variable fluctuations, time shifts between scan lines, beam positioning variations, scan distortions, environmental factors (e.g., thermal fluctuations, mechanical vibrations, air currents).

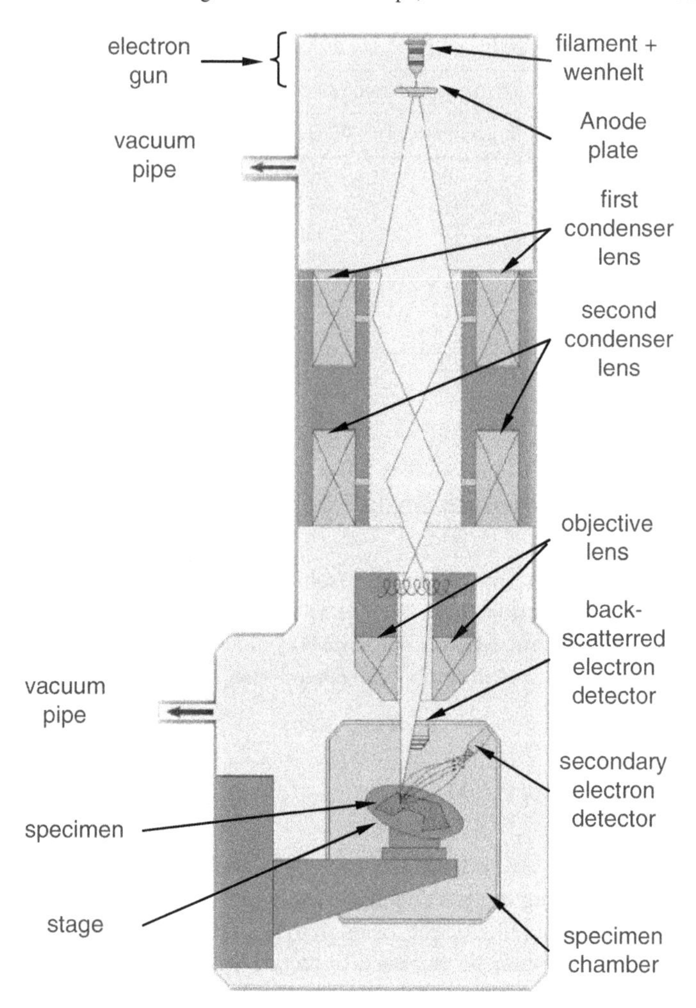

Fig. 6.34 Schematic of a typical Scanning Electron Microscope and imaging process

6.6.2 Pattern Development and Application

High magnification imaging requires an appropriately small random pattern for image correlation. Two approaches are developed for use at low and high magnification; lithographic pattern application for low magnification (1–20 μm pattern size) and pattern rearrangement for high magnification (50–250 nm). Details are outlined in a recent article [255].

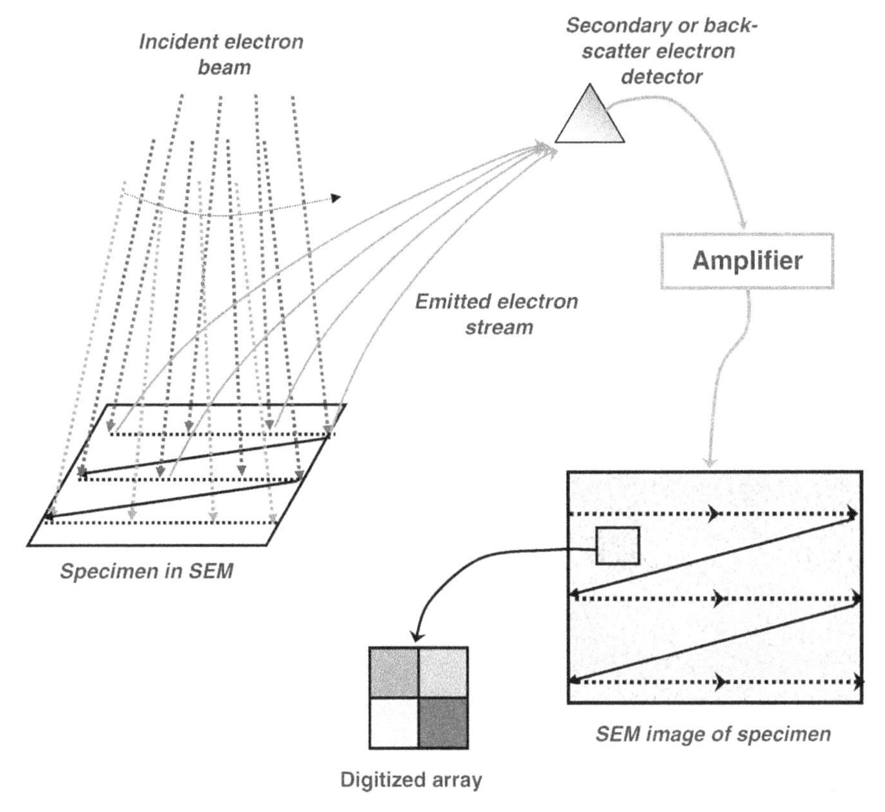

Fig. 6.35 SEM scanning process schematic and process used to obtain digital array representing final image. E-beam rastering is shown to proceed by scanning horizontal rows. Beam dwell time at each "pixel" location along row is preset. Delay time between scanning rows is required to reset the e-beam. Either secondary or back-scatter electron detector is selected to record the SEM image

6.6.2.1 Lithographic Pattern

The procedure to develop a micro-scale random pattern [255] using lithographic methods is shown schematically in Fig. 6.36a. Typically, the mask is an aluminum-on-glass construction. The random pattern can be developed using a focused ion beam process. Figure 6.37a shows a typical SEM image obtained using tantalum on an aluminum alloy. The one shown is constructed to have an average speckle size of 20 μm and is appropriate for low magnification in an SEM.

6.6.2.2 Rearranged Nano-coating

The procedure is briefly outlined in Fig. 6.36b. First, the specimen is coated with a nano-scale layer that has a significantly higher atomic number than the base material (e.g, tantalum or gold on aluminum). The method outlined in Fig. 6.36b has

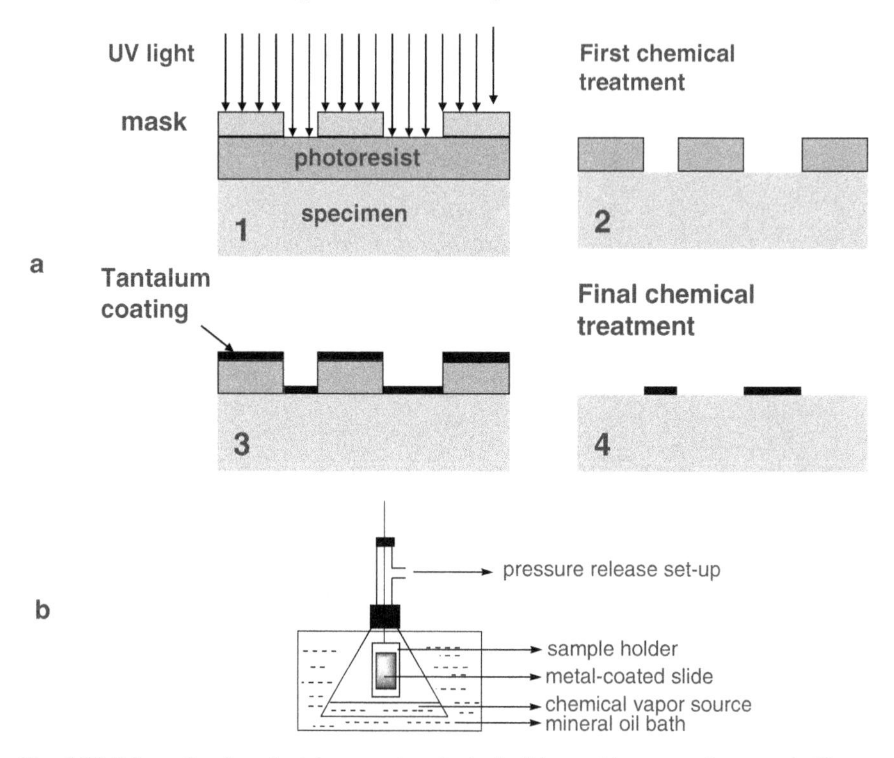

Fig. 6.36 Schematic of methodology used to **1** obtain lithographic pattern in range 1–20 μm, **2** convert thin gold film to nano-scale random pattern in range 50–500 nm, **3** coat specimen with tantalum, and **4** remove photoresist to obtain random pattern on specimen

been used successfully for layers ranging from 3 to 25 nm [255]. Next, the coated specimen is placed in a container and exposed to heated vapor for a few hours. The presence of a rearranged pattern can be determined by directing a laser beam onto the surface and viewing the presence or lack of laser scattering. Direct observation of the pattern in an SEM provides quantitative information regarding the size and distribution of the pattern on the specimen surface. Figure 6.37c shows an image of the pattern obtained rearranging a gold nano-layer on aluminum.

6.6.3 Digital Image Correlation to Quantify SEM Imaging System Errors

Direct application of image correlation can be used to identify local shifts in image position due solely to inaccuracies in the SEM imaging process. The process is as follows:

- Pattern a specimen for the magnification of interest using one of procedures in Section 6.6.2

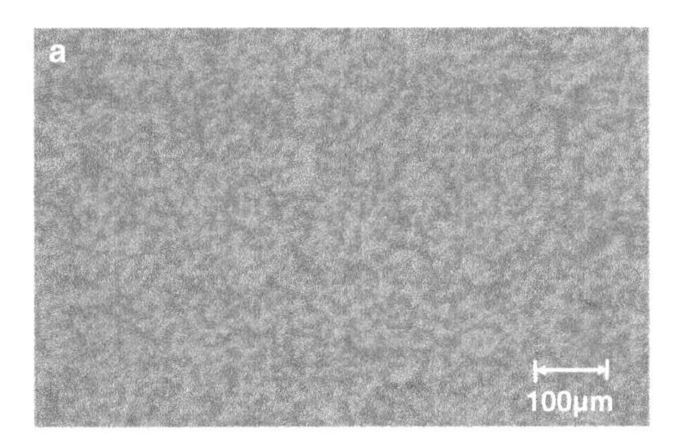

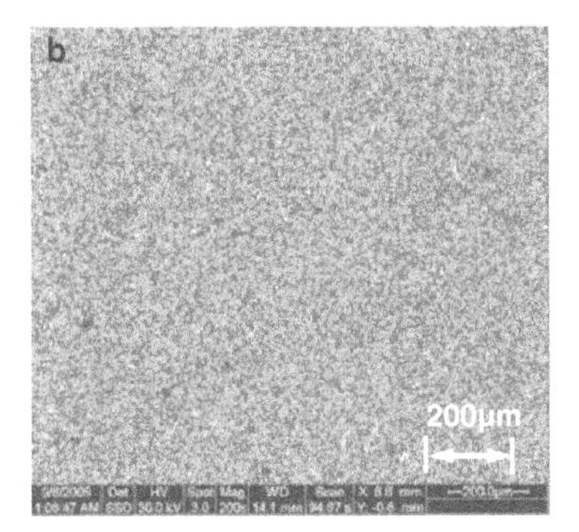

Fig. 6.37 **a** SEM secondary electron images of lithographic tantalum on 8009 aluminum, **b** SEM backscatter images of lithographic gold on Al2024 at 200X, and **c** SEM backscatter images of re-arranged nano-layer of gold on Al2024 at 5000X

- Place patterned specimen in SEM chamber such that adequate conductivity is ensured
- Select all SEM scanning parameters (e.g., dwell time, spot size, scan voltage, scan time, images integrated)
- Acquire and store images

6.6.3.1 Image Acquisition and Analysis

Table 6.7 presents both the subset-matching parameters and the SEM imaging parameters used to acquire a series of consecutive single images using three different SEM systems; (a) FEI Nano Nova, (b) Zeiss Gemini and (c) Hitachi 4800 S. The high magnification images are acquired in each system using the same specimen and mounting procedure. The specimen is manufactured from Al2024-T351 and has a rearranged gold nano-layer. Specimen size is $\simeq 50.8 \times 10 \times 1$ mm.

Several single images of the specimen are acquired using each SEM system. In a manner similar to previous work by the authors [284], digital image correlation is performed by comparing a "reference" image to all other images. Figure 6.38 shows typical displacement patterns obtained using each of the SEM systems.

Table 6.7 SEM scan parameters and field of view

Brand	Model	Detector	Accelerating voltage (kV)	Spot size	Magnification factor($\times 10^3$)	Working distance (mm)	Dwell time (μs)	Image size (square pixels)	Total acquisition time (s)	Field of view (μm)	nm/pixels
FEI	Nova Nano 630	SE	10	2	10	6.3	45	1024 by 884	40.7	29.8 by 25.7	29.1
Zeiss	Gemini	SE	15	9A	5	7.0	25.7	1024 by 760	20.2	22.7 by 17.0	22.2
Hitachi	4800 S	SE	15	$\simeq 3$	5	3.8	52.3	1280 by 896	60	25.6 by 17.9	20.0

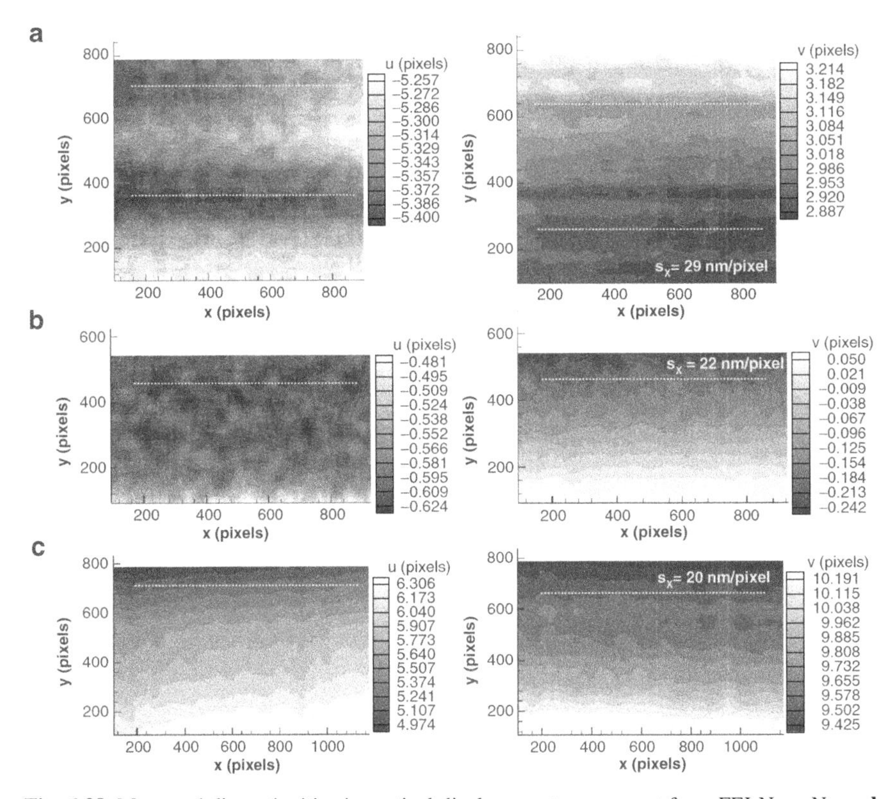

Fig. 6.38 Measured discontinuities in vertical displacement component for **a** FEI Nova Nano, **b** Zeiss Gemini and **c** Hitachi 4800 S

Table 6.8 presents a summary of (a) the intensity pattern matching metrics results and (b) measured step changes in the displacement components.

6.6.3.2 Discussion of Results

As shown in Fig. 6.38 and Table 6.8, all SEM systems exhibit unanticipated image shifts during the scanning process. The sudden shifts in image position have been measured and shown to occur (a) perpendicular to each scan line in Fig. 6.35 (i.e., in the vertical direction, $V(x,y)$), (b) parallel to the scan direction (i.e., in the horizontal direction, $U(x,y)$) or (c) in both directions simultaneously. Figure 6.39 presents conceptually how unwanted shifts in image position can be introduced during the scanning process. Though the source of the shifts is not known precisely, issues such as (1) specimen charging, (2) electromagnetic field fluctuations and (3) environmental factors (e.g., thermo-mechanical variations) have been reported as having an effect on image stability.

Table 6.8 Digital image correlation results from untranslated SEM image comparisons

Brand	Model	Detector	Subset size/spacing (pixels)	Intensity (bits)	Correlation metric/ shape function	Min/max in U-step change (pixels)	Min/max in U-step change (nm)	Min/max in V-step change (pixels)	Min/max in V-step change (nm)
FEI	Nova Nano 630	SE	43/5	8	NCC/Linear	0.051/0.222	1.50/6.46	0.093/0.238	2.72/6.94
Zeiss	Gemini	SE	43/5	8	NCC/Linear	0.043/0.044	0.95/0.97	0.082/0.87	1.82/1.94
Hitachi	4800 S	BSE/SE	43/5	8	NCC/Linear	0.272/0.369	5.44/7.38	0.168/0.228	3.36/4.54

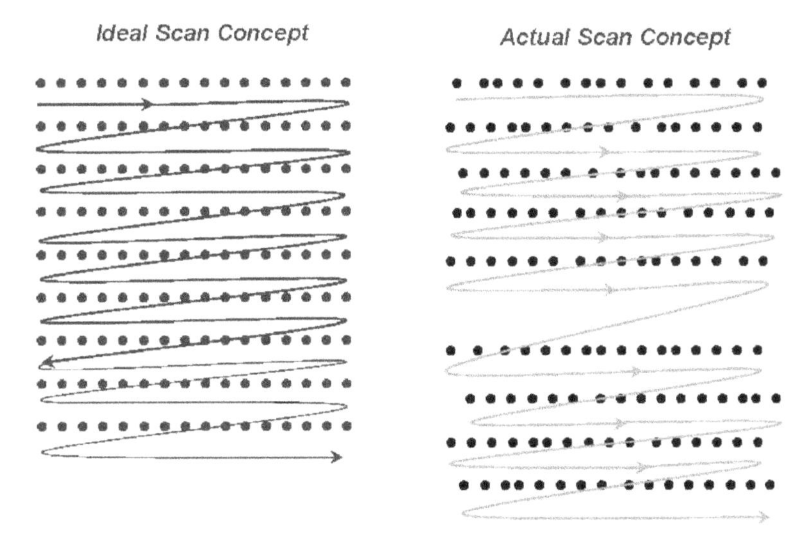

Fig. 6.39 Idealization of scanning process. Local dwell position and beam repositioning errors during line scan are shown

The magnitude of the shift in the image plane can be relatively large, exceeding $\frac{1}{2}$ of a pixel in some modern SEM systems. As shown in Table 6.8, the system exhibiting the smallest spatial shift is the Zeiss Gemini using the SE detector, with less than 1 nm displacement shift, which corresponds to $\simeq$0.04 pixels at the relatively high magnification employed.

Dozens of experiments have shown that the location(s) of the positional shift are essentially random in both directions. As such, image integration[12] has been shown to be an effective method for minimizing the effect of such shifts when performing digital image matching to extract underlying object deformations. Of course, image averaging will tend to blur the integrated image by the amount of the random shifts. Since image correlation requires adequate contrast to maintain accuracy during the image matching process, large shifts (e.g. shifts > 1 pixel) have been shown experimentally to cause difficulty during the correlation process.

6.6.4 Digital Image Correlation for Elastic Deformation Measurements

As noted in Section 6.6.3, the SEM imaging process introduces small, randomly located, shifts in image position. In addition to these random variations, the scanning process shown schematically in Fig. 6.35 also has two additional image distortions

[12] Integration of eight images has been shown to be sufficient to eliminate shifts when comparing consecutive images.

that require consideration for optimal accuracy. These are oftentimes considered as being uncorrelated functions and hence can be described independently as follows:

- Non-random, time-dependent image distortions (drift-induced)
- Non-random, time-independent spatial distortions

The time-varying distortion may be defined at each pixel location (x_s, y_s) by a time dependent velocity function, $v(x_s, y_s; t)$ [282]. This is shown schematically in Fig. 6.40 for a typical sensor location, where the velocity vector for a fixed sensor location is estimated by using digital image correlation to obtain the displacement of the current subset centered at this location at several times, t_i. Using finite difference or other appropriate method to estimate $v(x_s, y_s; t)$, the drift displacement correction at each pixel can be obtained by integration of v over time.

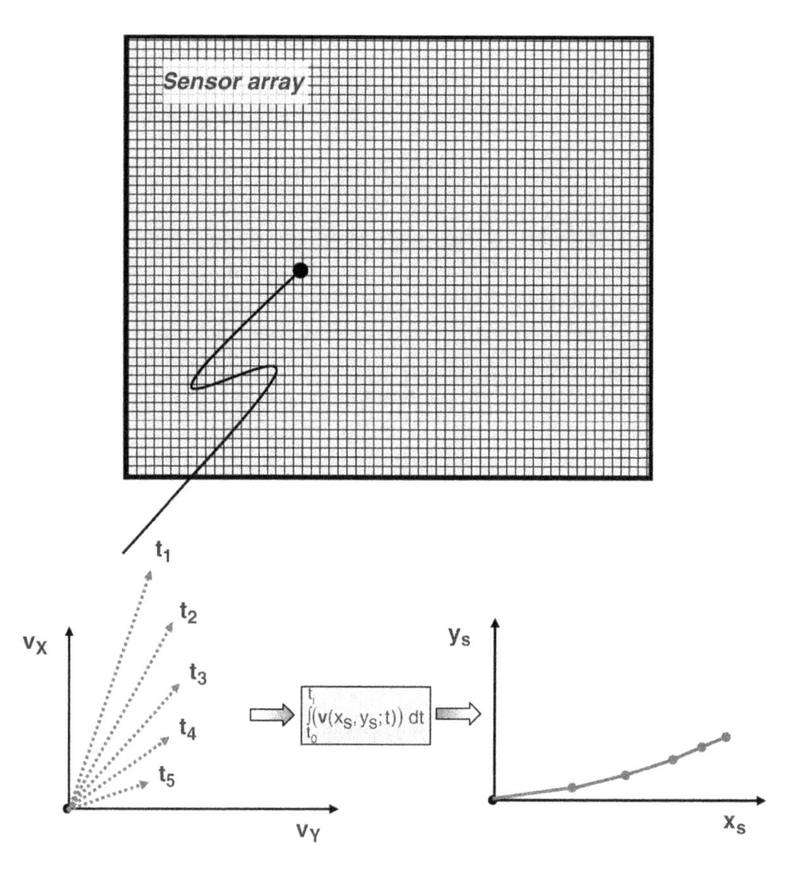

Fig. 6.40 Schematic of time-varying distortion at a pixel location. Velocity vector is estimated knowing SEM scan parameters (dwell time, beam repositioning time) and sensor location. Time-dependent drift displacement is estimated by integration of velocity vector

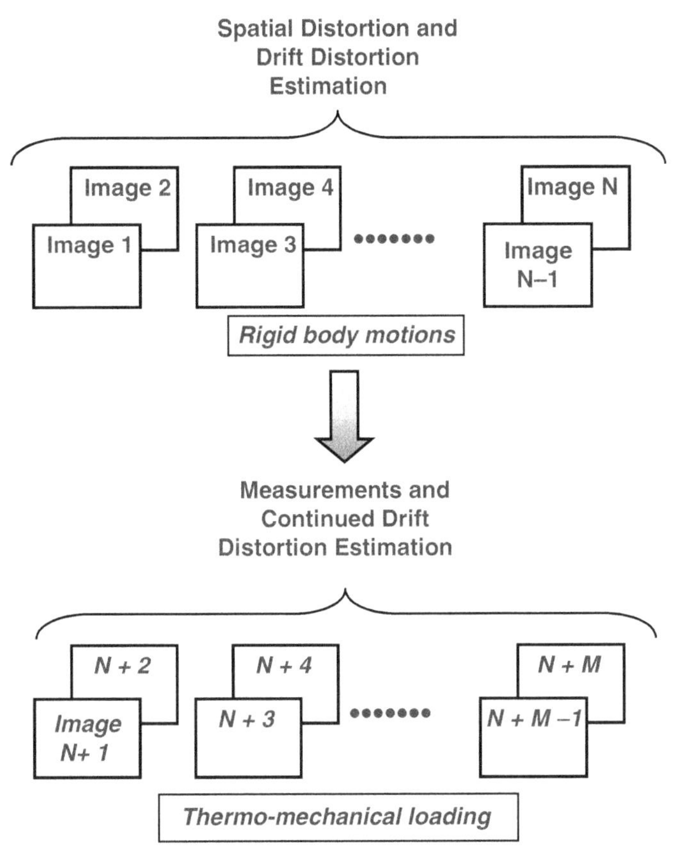

Fig. 6.41 Overview of the SEM measurement and distortion removal procedures. Note that image averaging is assumed to eliminate random shifts in image position. Spatial distortions can be estimated using either a priori or parametric distortion functions

Spatial distortion can be estimated using a series of rigid body translations. The procedure is described in detail in previous publications and is shown schematically in Fig. 6.41. It is noted that the image scale factor (pixels/nm) is typically estimated during the translation process.

6.6.5 Tensile Experiment

The flat aluminum specimen with rearranged nano-coating is the same as used for distortion removal. An FEI Quanta-200 SEM is used to view a 26 × 22 μm region on the surface. Image integration over 16 images with 2 × 2 binning (equivalent array size is 512 × 442) is used to eliminate random image shifts, giving a scale factor ≃0.5 nm per "pixel".

The flat specimen is mounted in a miniature tensile stage and placed in the SEM chamber [282, 283]. After evacuating the chamber, the procedure described in Fig. 6.41 is used to obtain the drift and spatial distortion correction fields.

6.6.6 Experimental Results

Figures 6.42 and 6.43 show the two vector components for the drift field and the spatial distortion field, respectively.

After correcting all images obtained during the tensile loading stage, the average elastic stress–strain response is obtained. Typical stress- strain results obtained using computer vision are shown graphically in Fig. 6.44.

Table 6.9 shows both the corrected image-based strain values and standard deviations in the corrected strains within the field of view during a repeated stress–strain experiment. For the axial strain ε_{yy}, Table 6.10 shows (a) the uncorrected axial strain, (b) drift corrections, (c) spatial distortion corrections and (d) corrected axial strain.

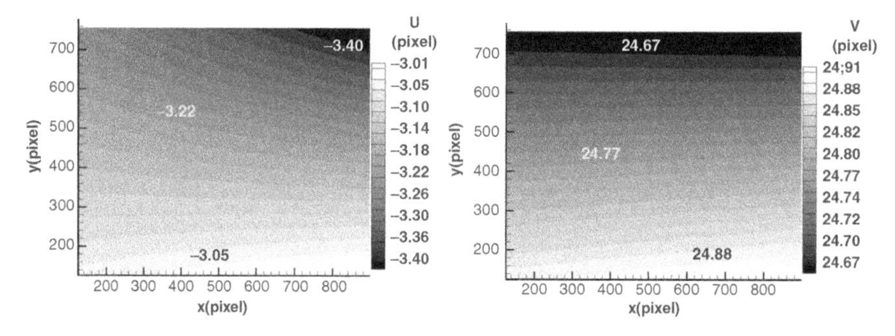

Fig. 6.42 Measured horizontal and vertical time-dependent drift fields for 22 × 25 μm field of view in FEI Quanta 200 SEM in high vacuum mode. Estimated scale factor is 40 pixels/μm. Time is ≃1800 s after initiating scan process in SEM

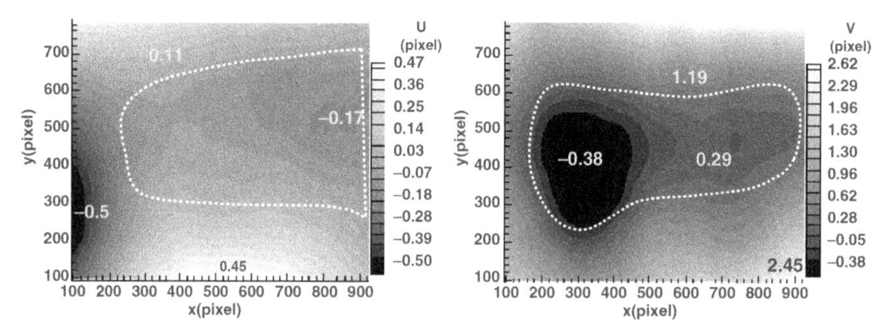

Fig. 6.43 Measured horizontal and vertical spatial distortion fields for 22 × 25 μm field of view in FEI Quanta 200 SEM in high vacuum mode. Estimated scale factor is 40 pixels/μm. Dotted line encloses domain where spatial gradients are somewhat lower, resulting in smaller spatial distortion in this region

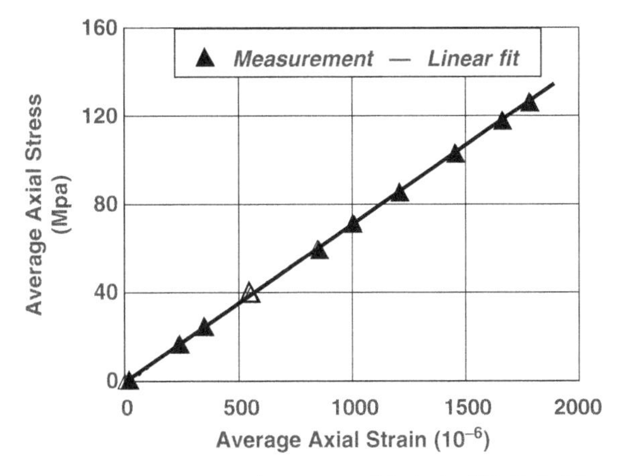

Fig. 6.44 Average stress versus measured average axial strain. 2D digital Image processing of SEM images used to obtain strains throughout loading process

Table 6.9 Measured average axial stress, corrected axial strain and corrected transverse strain

σ_{yy} (MPa)	Strain gage ε_{yy} ($\times 10^{-6}$)	Image-based ε_{yy} ($\times 10^{-6}$)	Image-based $\Sigma_{\varepsilon_{yy}}$ ($\times 10^{-6}$)	Image-based ε_{xx} ($\times 10^{-6}$)	Image-based $\Sigma_{\varepsilon_{xx}}$ ($\times 10^{-6}$)	Image-based $-\varepsilon_{xx} / \varepsilon_{yy}$
24.9	347+/−3	350	62	−83	54	0.24
48.2	673+/−2	677	51	−249	41	0.37
68.5	959+/−3	956	63	−387	53	0.40
92.0	1,286+/−3	1,290	71	−446	114	0.34
112.7	1,577+/−2	1,573	78	−457	105	0.29
132.0	1,846+/−2	1,850	81	−631	87	0.34

6.6.7 Discussion

As shown in Fig. 6.42, the drift distortion field is relatively small after 30 min and the gradients in both the U and V displacement fields are fairly uniform, with the V-field gradients suggesting a drift strain error in ε_{yy} on the order of 400 µs.

In contrast, as shown in Fig. 6.43, the spatial distortion fields are quite complex, with large gradients occurring in the outer regions and relatively small gradients occurring in the central region outlined by the white dotted lines. As an object point moves from one portion of the field to another, spatial distortions occur.

As shown in Fig. 6.44, the stress versus corrected elastic strain data follows the expected linear response of the aluminum specimen. The slope of the best fit straight line is 71.5 MPa, which is in good agreement with literature values for the modulus of elasticity in aluminum alloys, providing quantitative evidence that the corrected strain data is usable in the elastic deformation regime.

Though not shown, the strain variations are generally random throughout the field. Thus, the standard deviations in strain components shown in Table 6.9 are appropriate statistical metrics and provide quantitative estimates for variability in the strain field. It is also noted that the random character of the variability provides quantitative evidence that the spatial distortion and drift distortion fields calculated during the experiment are adequate to correct the images.

6.6.8 Remarks

As shown in Table 6.10, the distortion corrections are less than 0.0012 for the entire experiment. These corrections are absolutely essential to accurately estimate the relatively small elastic strains in the specimen.

However, if the experiment is focused on measuring large plastic strains (e.g., $\varepsilon_{\mathrm{distortion}}/\varepsilon_{\mathrm{mech}} < 0.10$), then the magnitude of both spatial and drift distortions shown here are relatively small and probably can be neglected. In this case, making measurements in an SEM is straightforward, requiring only an estimate for the scale factor and application of a random pattern adequate for the magnification of interest.

Regarding the relatively sharp changes in displacement shown in Fig. 6.38, the small size of the shifts (on the order of 5 nm) suggests that the effect will not be observable for relatively low magnification in an SEM when using digital image correlation to acquire the data. Letting Λ be the scale factor and ζ being the shift in nm, if random variability in image correlation errors are Π pixels, then when the inequality given by $\Lambda \leq 0.5\,\Pi\,\zeta^{-1}$ is met, the shifts will most likely not be observable in the measurements. For example, suppose we have the following conditions

- Image is recorded with a 1K $\times$ 1K pixel array
- Random image correlation error of 0.02 pixels is observed in measurements
- Shifts of 10 nm are known to be present

In this case, $\zeta = 10$ nm, $\Pi = 0.02$ pixels and $\Lambda \leq 1$ pixel/μm. For a 1K pixel array, the smallest allowable field of view that can be used without observing the effects of local shifts is 1 mm. This field of view corresponds to a magnification of $\simeq$250X on an FEI Quanta 200. Hence, using the assumed conditions, an SEM magnification smaller than 250X will result in images that are unlikely to have observable shifts in the measured displacements.

Table 6.10 Measured corrections and axial strains

Uncorrected ε_{yy} ($\times 10^{-6}$)	Drift correction ε_{yy} ($\times 10^{-6}$)	Spatial correction ε_{yy} ($\times 10^{-6}$)	Corrected ε_{yy} ($\times 10^{-6}$)	Strain gage ε_{yy} ($\times 10^{-6}$)
5	103	242	350	347
24	191	462	677	673
9	321	626	956	959
27	448	815	1,290	1,286
86	484	1,003	1,573	1,577
110	637	1,103	1,850	1,846

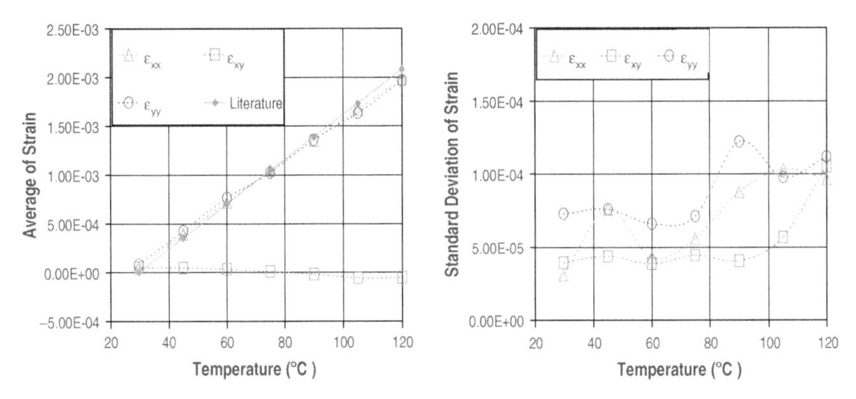

Fig. 6.45 Measured average thermal strain components and corresponding standard deviation for Al2024-T351 at 10,000X. Field of view is 25 by 22 μm. All images obtained using backscatter electron detector in FEI Quanta 200

SEM imaging offers unique opportunities to make important measurements at reduced length scales. Recent work has shown that the concepts outlined above can be extended to a variety of applications including

- Thermal strains during specimen heating
- Effects of other environmental factors on material behavior
- Three-dimensional shape and deformation

Figure 6.45 shows recent results [158] using 2D digital image correlation and SEM imaging to make non-contacting thermal strain measurements at high magnification, confirming again that the method is capable of a wide range of applicability.

With regard to use of the method in an Environmental SEM at modest humidity and low pressures, the authors have performed limited experiments. The key issue confronting the effort is reduced contrast and increased noise in the image due to effects such as electron beam interactions with the environment and increased specimen charging effects. Though the preliminary experiments are marginally successful, additional studies are needed to optimize image quality in similar situations.

Chapter 7
Stereo-vision System Applications

7.1 Stereovision System Design Considerations

There are several initial considerations for a typical stereo-vision system, such as shown schematically in Fig. 4.3 or Fig. I.1. These include (a) anticipated object motion with corresponding (a-1) field of view on object, (a-2) system depth of field and (a-3) spatial resolution of camera; (b) surface lighting; (c) camera exposure time and (d) surface diffusivity.

Suppose the region of interest on the object is expected to move several millimeters to the right and then several millimeters towards one of the cameras during the loading process. If the stereo-vision system(s) are stationary, then the field of view for each camera must be arranged so that the final position of the object remains within the image for all cameras. Thus, both the depth of field and field of view for each camera-lens combination must be constructed to obtain a "focused imaging volume" that is appropriate for the situation. Assuming a fixed lens size, then the following are noted and shown schematically in Fig. 7.1.

- Reduction (increase) in aperture size
  - Increases (decreases) $\mathcal{N}$ and correspondingly increases (decreases) the depth of field within which a focused image is obtained
  - Requires increase (decrease) in object illumination to maintain high contrast in image for accurate pattern matching
- Increase (decrease) in focal length of lens
  - Increases (decreases) $\mathcal{N}$ and correspondingly increases (decreases) the depth of field within which a focused image is obtained
  - Decreases (increases) angular field of view

M.A. Sutton et al., *Image Correlation for Shape, Motion and Deformation Measurements: Basic Concepts, Theory and Applications*, DOI: 10.1007/978-0-387-78747-3_7,

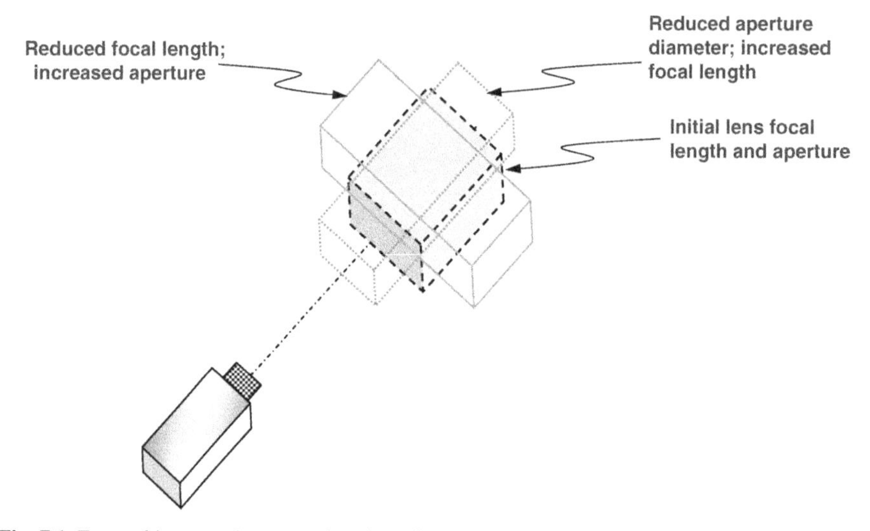

Fig. 7.1 Focused image volume as a function of lens focal length and image aperture

7.2 Quasi-Static Experiment: Four-Camera Stereo-Vision System for Large Deformation Measurements in Specimen Subjected to Out-of-Plane Bending

In this example, a total of four cameras arranged as two 2-camera stereo-vision system are employed to measure the three-dimensional (3D) displacements on both the front and back surfaces of a thin, ductile, single-edge-notched polymer specimen subjected to four-point, out-of-plane bending loading.

7.2.1 Problem Description: Compression and Out-of-Plane Bending of Flawed Polymer Sheets

Several new polymers have been developed specifically for use in transportation applications where the response of flaws in relatively thin sheets under quasi-static out-of-plane bending is required.

Since design modifications result in changes in the anticipated flaw loading conditions, validated computational models are needed to reduce time and cost. In recent years, modified cohesive zone models (CZMs) for crack growth have been proposed. In an effort to develop validated CZMs for use in transportation applications [310, 351], one approach is to measure several key quantities during loading of a specimen and then vary parameters in the model while matching the measured data. Proposed measurements include the following:

(a) Uniaxial material properties for ductile materials
(b) Loading and applied displacement

(c) Crack extension and crack path
(d) Crack front shape
(e) Surface deformations on both sides of a specimen during stable tearing

One approach that can be used to meet the requirements in (a)–(e) uses digital images to track crack extension and digital image correlation to measure full-field surface deformation during loading. A complicating factor in this experiment, which requires an edge-cracked specimen to be subjected to a combination of axial compression and out-of-plane bending, is that the flaw is expected to grow asymmetrically, requiring simultaneous imaging of both sides of the specimen during the loading process.

7.2.2 Geometry of Specimen

Prior to performing the bending experiments, both sides of the specimen were lightly coated with white paint and over-sprayed with black paint to obtain a random black-and-white speckle pattern having a spatial variation in intensity that is appropriate for displacement measurements. In order to improve the bonding between the coating and the specimen, a Rust-Oleum™ Specialty High Heat white enamel spray paint and Rust-Oleum Stops Rust Flat Black enamel spray paint were adopted.

Figure 7.2 presents both a schematic of the specimen geometry and a photograph of the as-patterned specimen. Figure 7.3 shows a close-up of the compression-bending specimen attached to the loading hinges.

7.2.3 Experimental Considerations and Arrangement of Two Stereovision Systems

The out-of-plane bending experiments employ a Tinius Olsen 5000 screw-driven test system. The bending experiments were performed using compression loading. An initial perturbation in specimen shape was introduced by the hinge offset so that out of plane bending of the specimen was also applied.

In order to obtain simultaneous stereo images of both sides of the polypropylene specimen during the compression-bending loading and stable tearing crack growth process, a dual stereo-vision system was developed. Figure 7.4 shows the Tinius-Olsen loading frame, specimen-hinge combination and the two stereovision systems.

Each stereovision system consists of the following components:

- Two matched Q-Imaging QICAM digital cameras
  - CCD resolution of 1,392 × 1,040 pixels
  - Square sensor pixel with dimensions 4.65 × 4.65 μm
  - 12 bits for intensity resolution

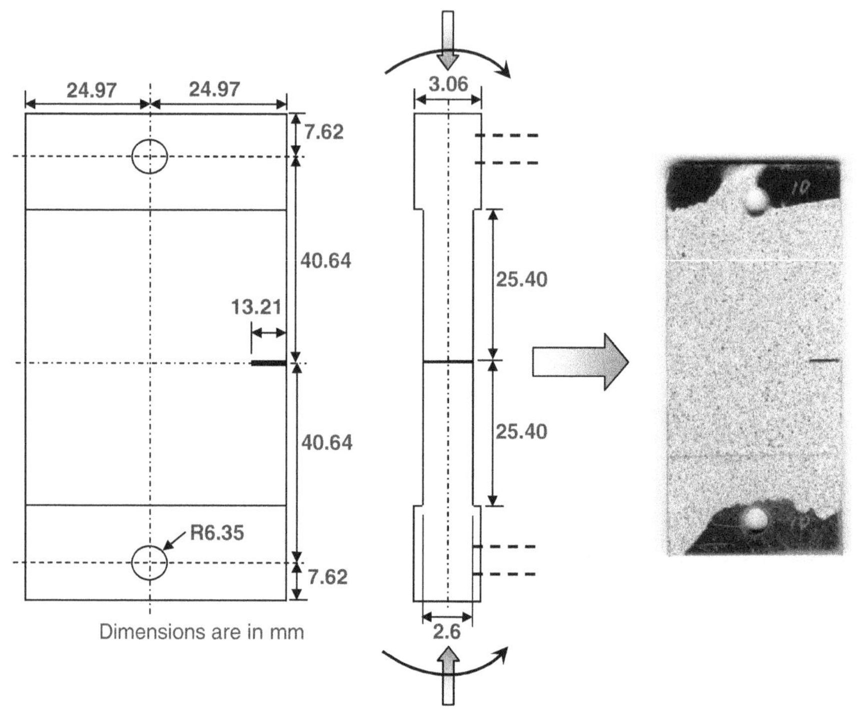

Fig. 7.2 Geometry for single edge cracked, out-of-plane bending specimen. Specimen on right is the black specimen thinly coated with white paint prior to being speckled with black paint to obtain random pattern for 3D image correlation

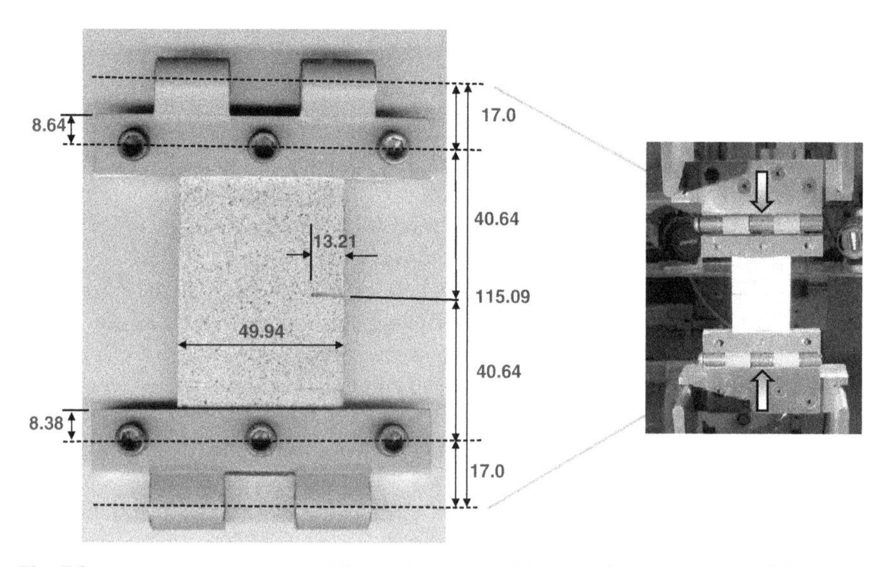

Fig. 7.3 As assembled specimen with loading hinges. Close-up of specimen shows hinge grips attached to matching hinges in loading frame. Hinge offset used to convert axial loading into a combination of out-of-plane bending and compression. All dimensions are in millimeters

Fig. 7.4 Experimental setup with two stereovision systems. The vision systems are synchronized and used to measure surface deformation on the front and back surfaces of the single edge cracked specimen during loading of the specimen. 0–1: Stereo-vision system Set I; 2–3: Stereo-vision system Set II; 4–5: Translation stages; 6: Tinius Olsen 5000 displacement controlled loading frame; 7: Fiber-optic lighting sources; 8: Assembled specimen and loading hinges

- Two matched Nikon AF Nikkor 28 mm f/2.8D ($\mathcal{N} = 2.8$) lenses.
- Object distance $\simeq$0.500 m for the experiments.
- Typical scale factors are $s_x = s_y \cong 10$–14 pixels/mm.
- The angle between the optical axis of each camera and the initial normal to the specimen's surface $\simeq 15^\circ$. This corresponds to an included pan angle between cameras of $\simeq 30^\circ$.
- Rigid cross member to minimize relative motion between cameras.
- Translation stages mounted between cross member and camera to simplify process for focus.

7.2.4 Calibration of the Camera Systems

Prior to performing the bending experiment, each stereovision system is calibrated using a target with uniformly spaced markers. Typical image pairs obtained during calibration with a dot grid pattern are shown in Fig. 7.5. During calibration, the target is tilted and rotated into different orientations while images are acquired. The calibration process is performed separately for the Set I and Set II stereovision systems. Tables 7.1 and 7.2 show the resulting camera parameters.

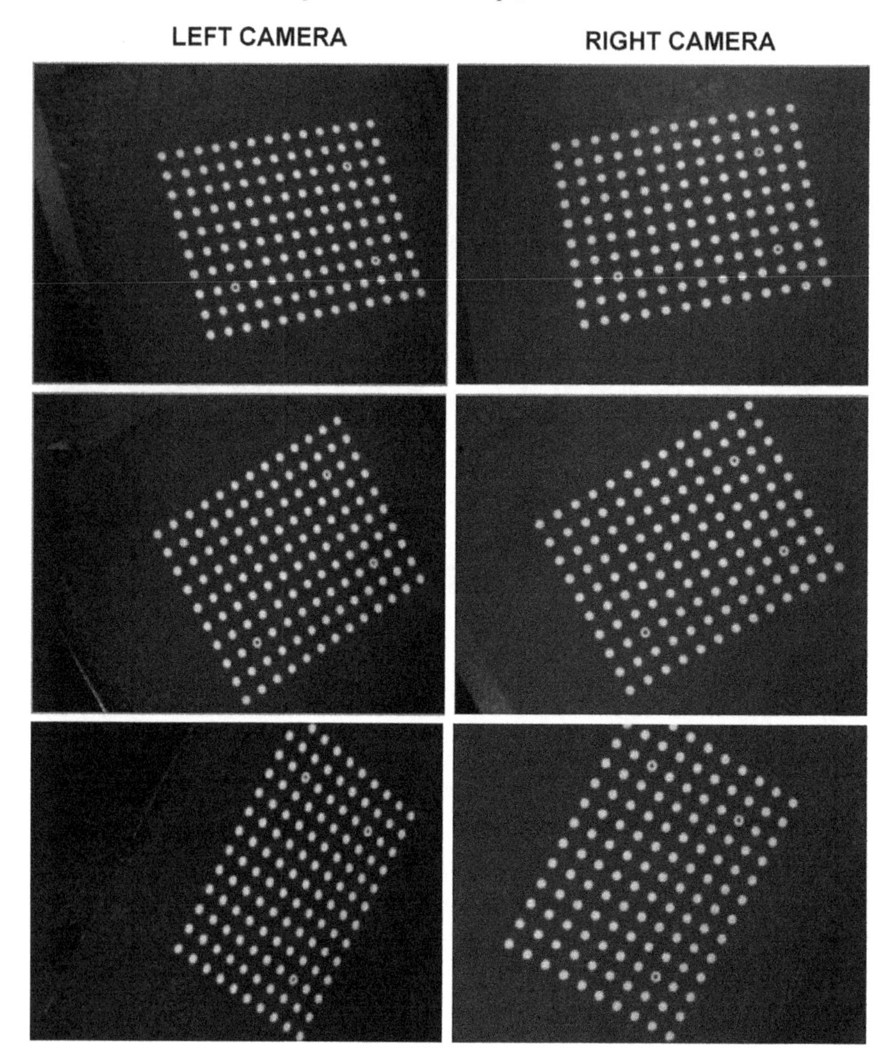

Fig. 7.5 Typical image pairs obtained during calibration process for Set I and Set II stereovision systems. Each pair of cameras is synchronized and 50 pairs of images acquired with grid in different orientation and position. Spacing of grid points is nominally 4 mm

The bundle adjustment calibration process determines (a) all of the intrinsic camera parameters (image center, scale parameters, image skew, distortion parameters) and the extrinsic parameters (pinhole location, orientation of each camera relative to target) and (b) the orientation and position of the target during the calibration motion sequence. Since the target orientations are not needed for measurements, they are typically discarded at the end of the calibration process. Table 7.1 shows the intrinsic and extrinsic camera parameters for each camera in both Set I and Set II stereo systems.

Table 7.1 Calibration results for stereovision system

	Set I			
Parameter	Camera 0	Camera 1	Camera 0 → 1 transformation	
Center x (pixels)	675.069	742.615	Alpha (degrees)	27.3566
Center y (pixels)	482.614	239.65	Beta (degrees)	−1.0405
fS_x (pixels)	6,303.01	6,306.68	Gamma (degrees)	−0.12459
fS_y (pixels)	6,303.01	6,306.68	T_x (mm)	−0.3707
Skew	0	0	T_y (mm)	247.703
$\kappa_1^{normalized}$	−0.13656	−0.10983	T_z (mm)	57.2036
κ_1	−3.4373E-09	−2.7646E-09	Baseline (mm)	254.223
Residual (pixels)	0.03290	0.04495	Residual after optimization	0.04041

Table 7.2 Calibration results for stereovision system

	Set II			
Parameter	Camera 2	Camera 3	Camera 2 → 3 transformation	
Center x (pixels)	713.227	806.308	Alpha (degrees)	26.7755
Center y (pixels)	447.755	365.898	Beta (degrees)	−0.7272
fS_x (pixels)	6,310.52	6,361.88	Gamma (degrees)	1.61751
fS_y (pixels)	6,310.52	6,361.88	T_x (mm)	−6.01635
Skew	0	0	T_y (mm)	251.74
$\kappa_1^{normalized}$	−0.09244	−0.107355	T_z (mm)	52.7531
κ_1	−2.3213E-09	−2.6958E-09	Baseline (mm)	257.279
Residual (pixels)	0.03955	0.067351	Residual after optimization	0.06862

Assuming that the image plane matching error is ± 0.01 pixels when locating the common image point in both views of each two-camera stereo systems used in the experiments, for the pan angle $\beta \simeq 30^\circ$ between the optical axes of the two cameras and image scale factors $S_x = S_y = 12$ pixels/mm, the estimated 3D positional errors are 0.005 mm for the in-plane translation and 0.020 mm for the out-of-plane displacement.

7.2.5 Post-processing and Special Considerations

The software VIC-3D is used to analyze all images acquired by both calibrated systems and perform image matching to quantify 3D displacement fields using both Set I and Set II stereovision systems. All analyses were performed with 25×25 subsets with a step size of 5 between subset centers. All displacements on both front and back surfaces are expressed in the same common global coordinate system (see Section 7.2.5.1 discussion), with all displacement components (U,V,W) defined in the same system. Here, the common global coordinate system is shifted so that the origin is at the initial crack tip location.

Displacement derivatives are obtained in the common coordinate system. The partial derivatives of the displacement field are computed from a quadratic polynomial approximation to the computed displacement field in a local neighborhood using a 5×5 array of measurements. The strain tensor is defined using the in-plane components of the Lagrangian strain tensor in Eq. (A.5). It is important to note that the strain tensor components defined by Eq. (A.5) are invariant with respect to arbitrary rigid body motion.

In addition to measuring the surface deformation fields, all images are used to locate the position of the initial and growing crack tips. Pattern features near the growing crack tip are used to improve the estimated crack tip position throughout the entire history of crack growth.

Due to the combination of bending flexibility, ductility of the thin sheet polymer specimen and the retardation of crack growth due to compressive loading on one side of the fracture surface induced by the out-of-plane bending load, it was expected that out-of-plane displacement would be relatively large during the loading process. It was later determined that the maximum out-of-plane displacement experienced by the specimen prior to failure was on the order of 40 mm.

To ensure that the specimen remained in focus for both stereo vision systems shown in Fig. 7.4, the setup was modified slightly.

- Reduction in aperture and an F22 ($\mathcal{N} = 22$) setting on lenses are used to maximize depth of field. The depth of field is $\simeq$50 mm for all cameras.
- Increased object illumination to compensate for reduced aperture and maintain adequate contrast throughout the experiment.
- As shown in Fig. 7.6, the specimen is placed (a) near the back of the depth of field for the camera(s) viewing the tensile surface and (b) near the front of depth of field for stereo system viewing compressive surface.

Since all four cameras must acquire images simultaneously during the experiment, additional modifications were made to the software and hardware to synchronize the process for acquisition of images throughout the experiment.

7.2.5.1 Common Coordinate System

To use the 3D position and displacement measurements from both stereo-vision systems for simulations, it is necessary that all data from the stereo-cameras 0 and 1 (Set I) and stereo cameras 2 and 3 (Set II) be transferred to a common coordinate system. To obtain a common coordinate system for Set I and Set II, a specially designed common coordinate system plate (CCSP) was manufactured.

As shown in Fig. 7.7, the 1.93 mm thick aluminum CCSP has six through-thickness holes (diameter <1 mm) machined in a well-defined arrangement. The CCSP is then painted with speckle patterns on both sides, with the location of all six holes clearly discernible on both sides of the plate after applying the speckle pattern.

The procedures to define the common coordinate systems are as follows. First, the stereovision systems Set I and II are put in place and oriented to image the

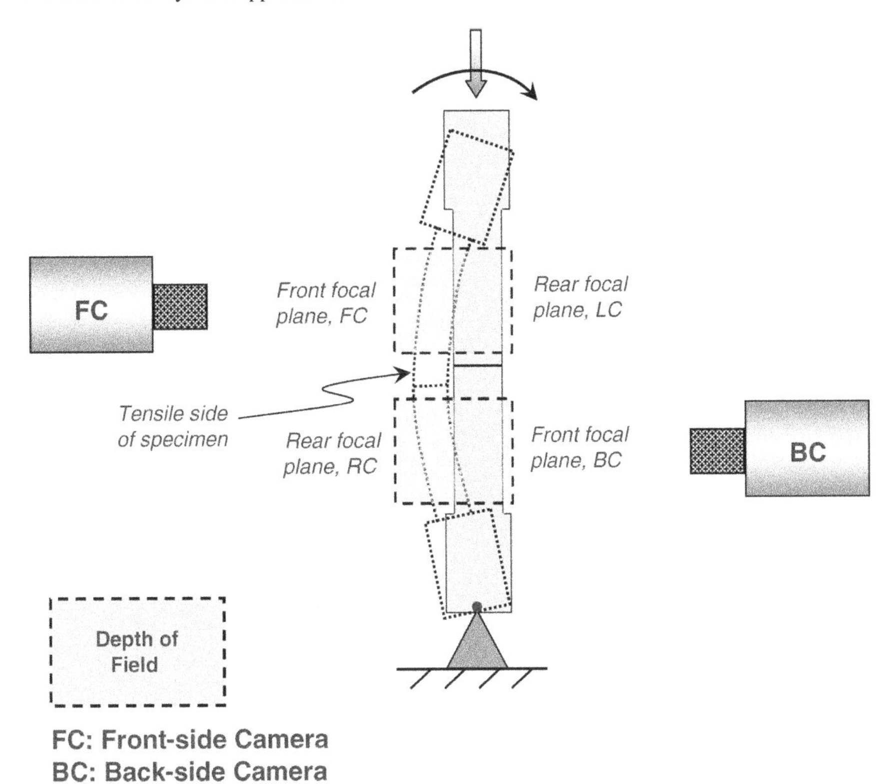

Fig. 7.6 Depth of field and out-of-plane motion direction. Front focal plane of FC is on tensile side. Front focal plane of BC is on compressive side

bending specimen. Second, each system is calibrated independently. When the calibrations are complete, the world coordinate system for Set I (Set II) is defined by the first stereovision grid image recorded by Set I (Set II). Third, the CCSP is placed in the approximate location of the actual bending specimen and images of both side of the speckled plate are obtained by the Set I and Set II stereo-vision systems. Fourth, the calibration results are used to determine the six extrinsic parameters relating camera 1 to the camera 0 coordinate system in Set I and camera 3 to camera 2 coordinate system in Set II. Fifth, with the stereovision Set I viewing the front surface of the specimen, image correlation is performed between camera 0 and camera 1 to obtain a full-field of 3D positions relative to the camera 0 coordinate system. Sixth, a plane is fit to the 3D data and the 3D positions are determined within the plane of all six holes. Seventh, use the 3D locations of holes A' and B' to define a unit vector along the X_f direction, $\mathbf{e}^{\mathbf{f}}_{\mathbf{x}}$, with the origin of the system (X_f, Y_f, Z_f) defined to be at A'. Eighth, the best-fit plane's normal, $\mathbf{e}^{\mathbf{f}}_{\mathbf{z}}$, and $\mathbf{e}^{\mathbf{f}}_{\mathbf{x}}$ are used to define the Y_f unit vector,[1] $\mathbf{e}^{\mathbf{f}}_{\mathbf{y}}$. Ninth, the unit vectors $(\mathbf{e}^{\mathbf{f}}_{\mathbf{x}}, \mathbf{e}^{\mathbf{f}}_{\mathbf{y}}, \mathbf{e}^{\mathbf{f}}_{\mathbf{z}})$ are used to define the

[1] The unit vector along the Y_f direction can be checked for consistency by using the best fit plane locations of holes A' and 6 shown in Fig. 7.7 for comparison.

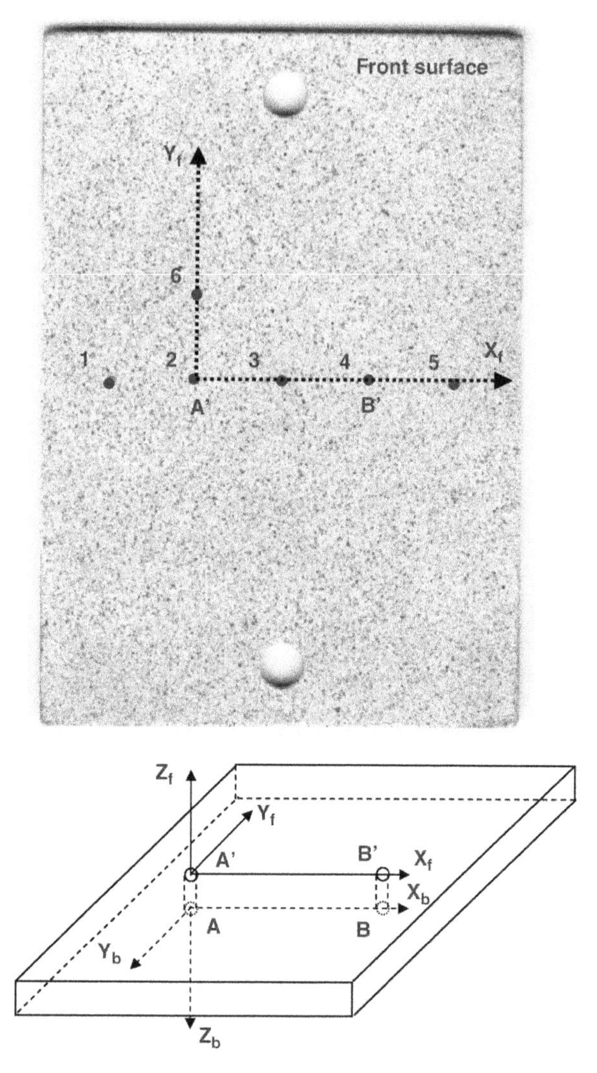

Fig. 7.7 Post-calibration CCS plate used to define common coordinate system for both stereovision systems. The planar specimen with known thickness includes five through-thickness small holes to define a common X-direction. One small hole is used to define the Y_f-direction within the specimen's front surface and perpendicular to the X-direction. The Y_b direction is defined to be opposite to Y_f. The Z-directions are defined by the cross-product of corresponding in-surface coordinate directions

transformation between the camera 0 coordinate system and the front surface coordinate system (e.g., the orientation and the relative location of A' and camera 0 origin). Finally, this transformation is applied to convert all Set I 3D position data for the front surface into the front surface coordinate system, (X_f, Y_f, Z_f).

The procedure outlined above is repeated for the Set II stereovision camera 2–3 system viewing the back surface. With all 3D position data obtained in the camera 2 coordinate system, the location of the origin A and the unit vectors $(\mathbf{e}_\mathbf{x}^\mathbf{b}, \mathbf{e}_\mathbf{y}^\mathbf{b}, \mathbf{e}_\mathbf{z}^\mathbf{b})$ are used to define the transformation between the camera 2 coordinate system and the back surface coordinate system (e.g., the orientation and the relative location of point A and the camera 2 origin).

Assuming the front surface system is to be used as the "reference", the (X_b, Y_b, Z_b) system is transformed to the (X_f, Y_f, Z_f) system by (a) applying a rotation matrix that rotates about the X_b axis by 180° and (b) translating in the Z_f-direction by the plate thickness, $(0, 0, -t_o)$. Figure 7.8 presents a schematic of the transformation process used to define the common coordinate system.

Assuming that a point is defined in the camera 2 pinhole system by $\mathbf{r_2}$, the transformations described in Fig. 7.8 are used to convert these coordinates into a common coordinate system. In equation form, the transformations can be written as follows:

$$\mathbf{r_b} = [\mathbf{R}]_{2-\mathbf{b}} \cdot \{\mathbf{r_2}\} - \{\mathbf{t}\}_{2-\mathbf{b}}$$

$$\mathbf{r_{f/b}} = [\mathbf{R}]_{\mathbf{b}-\mathbf{f}} \cdot \{\mathbf{r_b}\} - \{\mathbf{t}\}_{\mathbf{b}-\mathbf{f}}$$

where the final position $\mathbf{r_f}$ gives the common system coordinates for the point defined by $\mathbf{r_2}$ in the camera 2 system.

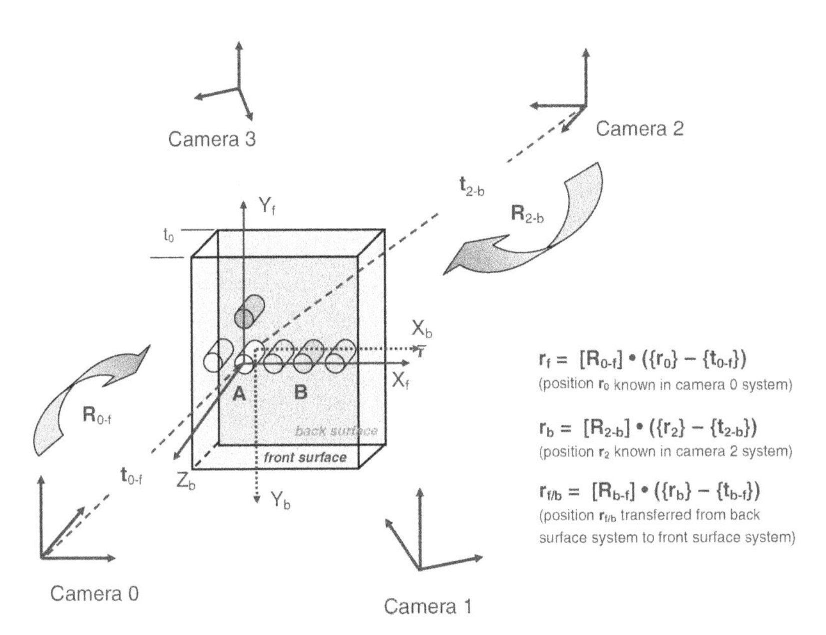

Fig. 7.8 Common coordinate system plate and schematic of coordinate systems associated with stereovision systems. Transformations for 3D position vectors, **r**, also shown

7.2.6 Experiments

Once calibration has been completed and images of the CCSP plate have been obtained, the polymer specimen (speckled on both sides) is placed in the loading system. Prior to applying load, synchronized reference images are acquired by the calibrated Set I and Set II stereovision systems.

The compression-bending process is performed under displacement control bending on 2.6 mm thick, pre-notched PP specimens. The recorded data during the experiment includes synchronized measurement of (a) applied compressive load, (b) applied far-field displacement and (c) images of from both sides of the specimen during loading at pre-specified times. The loading rate for a typical specimen is 0.104 mm/s (0.25 in./min) and images are typically acquired every 5 s throughout the loading process.

7.2.7 Experimental Results

Figure 7.9 shows different views of the specimen after undergoing stable tearing and then removal from the Tinius Olsen test system.

For this case, the maximum actuator displacement approached 67 mm. As can be seen from Fig. 7.10, crack growth only occurred on the front side of the specimen where tensile stresses are present; there is no crack growth on the back side where compression occurs. It is noted that crack growth continued with increasing actuator displacement until a displacement of about 50 mm. Additional actuator displacement beyond 50 mm did not result in visible additional crack growth on the front surface. Furthermore, the shape of the final specimen suggested that deformations were primarily localized along the specimen width and in the vicinity of the crack.

To obtain the crack path on the tension side, each crack extension increment from the previous crack tip to the current crack tip is defined as the distance between the two crack tip locations in the initial undeformed configuration, with the total crack extension defined to be the sum of all increments. To identify the material location of each growing crack tip, correspondence was simplified by comparing features in the reference and current location using software Flick32.[2] The estimated error in crack tip location is ± 2 pixels. The crack length between any two selected crack tips in the initial configuration is calculated with the help of software VIC-3D to obtain metric distance between matching sensor locations in the two camera views. Figure 7.9 shows the crack path on the front surface. Generally straight crack extension was observed throughout the experiment.

As shown in Fig. 7.11, the load reached a maximum just prior to crack extension, decreasing by 40% for $\Delta a = 1$ mm. From this point onwards, the load decreased linearly with crack extension until the no further crack growth was observed and the experiment was terminated.

[2] Flick32 and VIC-3D provided by Correlated Solutions, Incorporated. Web site located at www.correlatedsolutions.com.

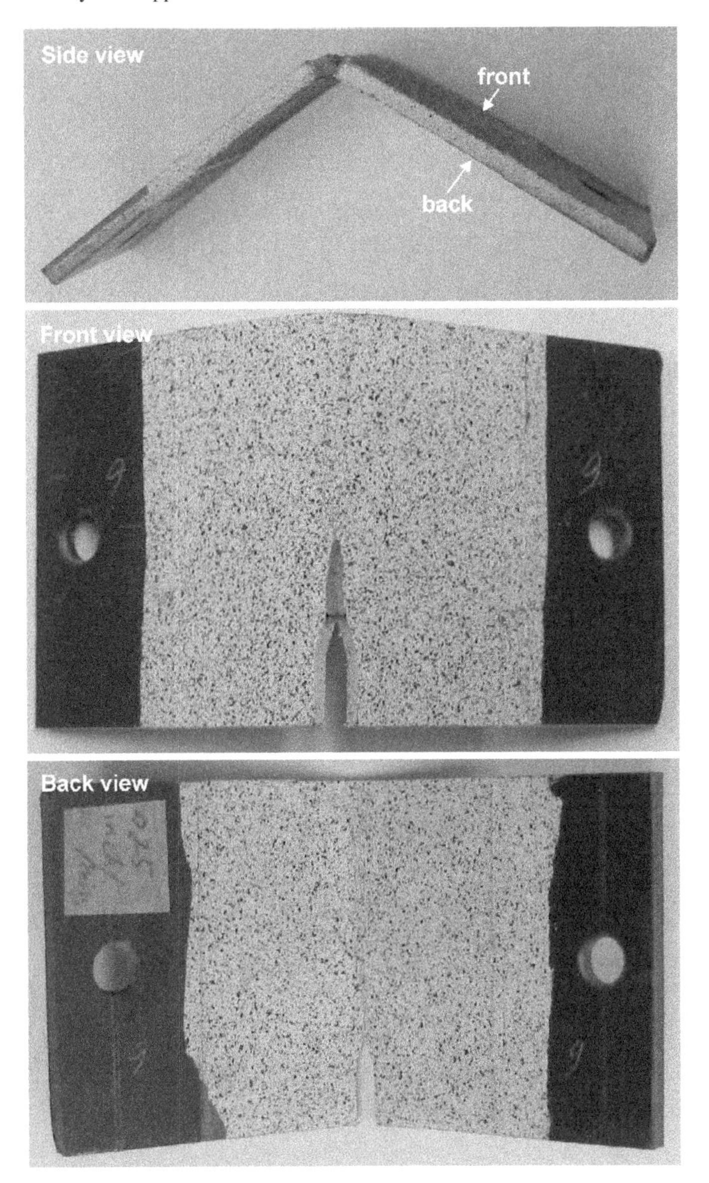

Fig. 7.9 Deformed specimen after load removal

Figures 7.12 and 7.13 show full-field contours for both W and ε_{yy} on the front and back surfaces, respectively, for an actuator displacement of 35.45 mm, load of 62.2 N and front crack extension of 5.56 mm. It should be pointed out that the common coordinate system for the front and back surfaces has been transferred to the notch tip of the front surface, with the origin at the notch tip and the X-axis along the notch and crack growth direction. Results clearly show that strains are localized in the central region along the crack line as bending resulted in a “hinge”

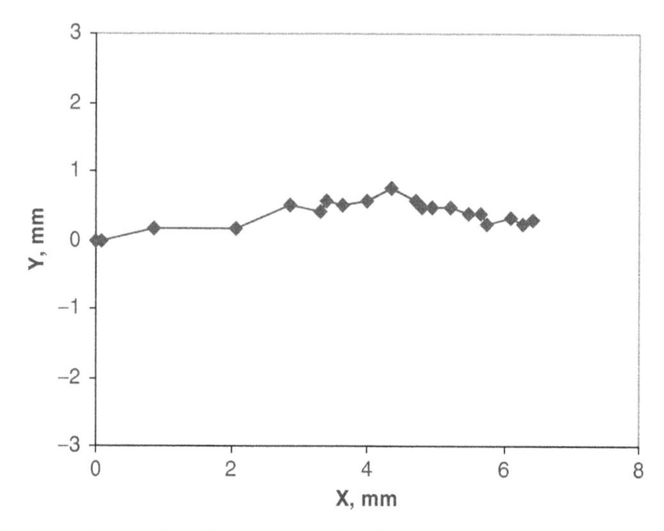

Fig. 7.10 Stable tearing crack path on front surface in material system defined in the reference, undeformed configuration

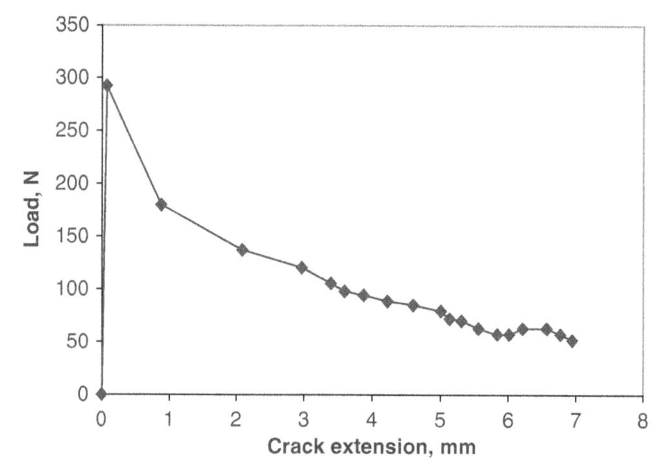

Fig. 7.11 Axial compressive load as a function of front surface crack extension

formation in this region, with the front surface nominally in tension and the back surface demonstrating compression due to bending.

7.2.8 Discussion and Practical Considerations

The major complexity in this application revolved around the need to ensure a common coordinate system for both stereo-vision systems. This led to development of

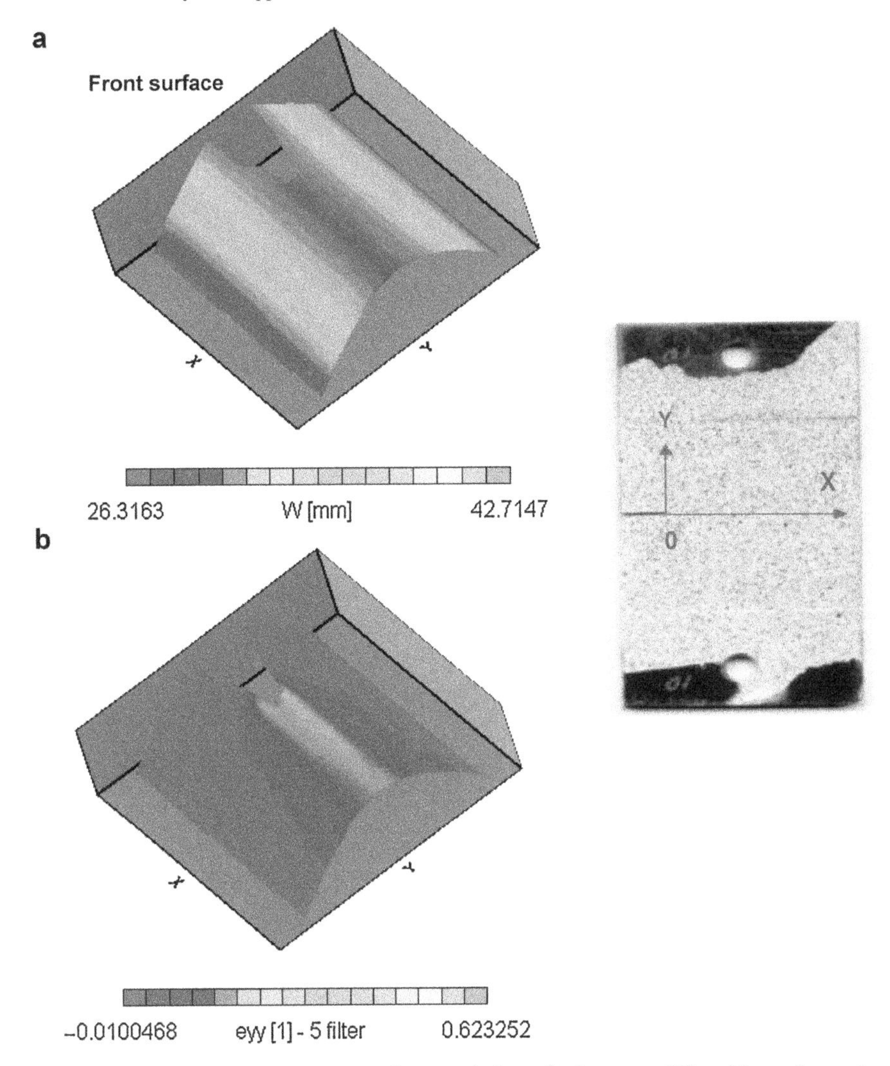

Fig. 7.12 Front-surface measurements of **a** out-of plane displacement (W) and **b** crack opening strain ε_{yy}, for compressive load of 62.2 N and front surface crack extension of 5.56 mm. Coordinate system located at original crack tip

a specialized specimen for calibration of both systems, the CCPS, so that common features were visible by both systems. Using images of the CCPS, and applying appropriate rigid body transformations, the coordinate systems for Set I and Set II were aligned.

A secondary complexity in this application dealt with the large out-of-plane motions that occurred during the experiment. In order to maintain focus in all four cameras as the specimen displaced towards (away) from a stereo-vision system nearly 50 mm. To meet this requirement, several modifications to the system were implemented.

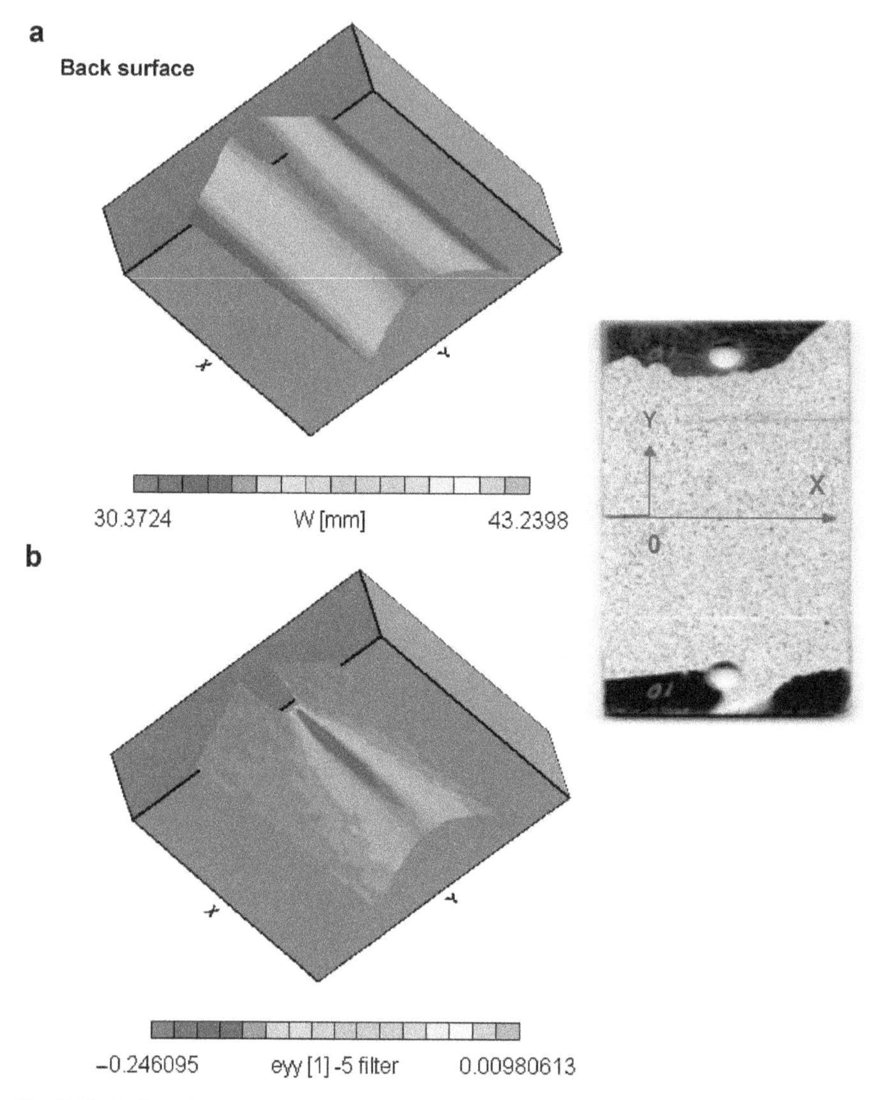

Fig. 7.13 Back-surface measurements of **a** out-of plane displacement (W) and **b** crack opening strain ε_{yy}, for compressive load of 62.2 N and front surface crack extension of 5.56 mm. Coordinate system located at original crack tip

- Reduced aperture to achieve $\mathcal{N} = 22$ on all lenses
- Increased lighting for both stereovision systems
- Arrangement of each stereovision system so that the specimen was moving within the depth of focus (see Fig. 7.6)

A less obvious complication was the gradient in rotation of the specimen due to bending of the specimen. This is shown schematically in Fig. 7.6 and physically

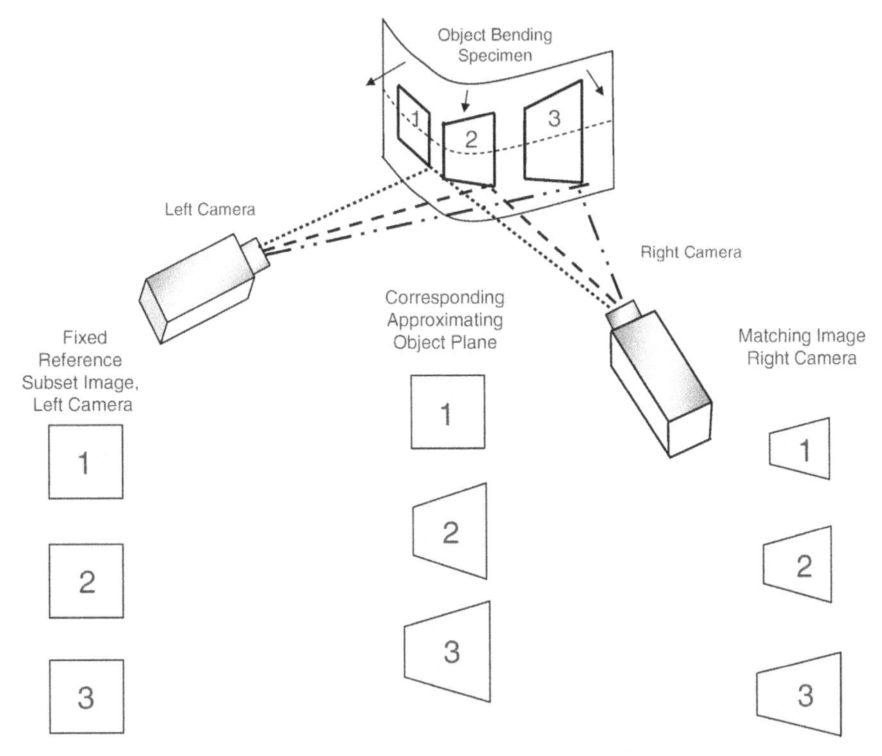

Fig. 7.14 Shape and size of matching subsets as function of relative angle between surface normal and optical axis

in Fig. 7.9. The effect of the camera viewing angle relative to the object normal is shown in Fig. 7.14. For example, a square subset in the left camera viewing region 1 on the object surface will be imaged as a compressed, trapezoidal shaped subset in the right camera. If the image in camera 2 is too small or too distorted, then it will be difficult to match subsets accurately. For this reason, the included angle between the two cameras was reduced to ensure that subset matching between views was achievable for the bending specimen.

7.3 Dynamic Experiment: High Speed Stereovision System for Large Deformation Measurements of Specimen Subjected to Combined Tension–Torsion Impact Loading

In this example, a two-camera high speed stereo-vision system is employed to measure the transient three-dimensional (3D) displacements of a thin, ductile, single-edge-cracked AL6061-T6 specimen subjected to transient mixed mode I/III loading.

7.3.1 Problem Description: Impact Loading of Single Edge Cracked Specimen Subjected to Mixed Mode I/III Impact Loading

In recent years, the ability to determine when a flaw will propagate, and in what direction the flaw will extend, has received increasing interest, especially for those cases where the applied loading results is a combination of mixed mode conditions in the near crack tip region [11, 12, 234, 291, 296, 298, 344]. Since the onset and direction of crack growth is believed to be governed by the fields existing along the crack front, an approach based on a combination of experimental surface measurements and computational modeling of the three-dimensional specimen geometry with estimated far-field loading (displacement) conditions will allow investigators to improve their understanding of conditions in the crack tip region during crack extension.

With the development of modern high speed cameras capable of synchronization, as well as the availability of software capable of analyzing stereo images and acquiring accurate 3D motion measurements, it is now possible to extract full-field, time-varying displacement data for dynamic events [306]. The enclosed example describes details regarding the measurement of deformations during mixed mode I/III impact loading of thin sheet materials.

7.3.2 Specimen Geometry and Preparation

Figure 7.15 shows the in-plane geometry of the 2 mm thick mixed-mode I/III specimens used in the impact experiments. Each specimen was machined in the LT orientation (i.e. notch is aligned with the transverse direction, perpendicular to the sheet's rolling direction) by laser cutting, with a notch length of 28.6 mm. Fatigue pre-cracking under nominally mode I conditions was performed using an 810 MTS test machine to develop an initially sharp flaw for subsequent dynamic tearing experiments. The total fatigue pre-crack length for all specimens was $\simeq$32.5 mm.

Prior to performing impact loading experiments, all specimens were lightly coated with an adherent material to obtain a random black-and-white speckle pattern having a spatial variation in intensity that is appropriate for displacement measurement in this application.

7.3.3 Experimental Considerations and Arrangement of Two Stereovision Systems

As shown in Figs. 7.15 and 7.16, the mixed mode I/III loading fixtures and loading frame are mounted within a Dynatup Drop Tower [298]. The 2,669 N load is

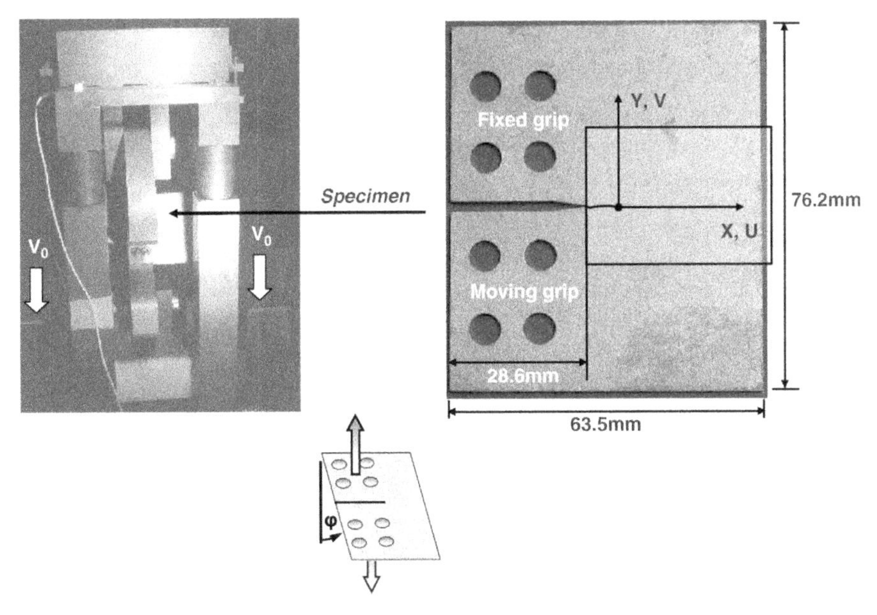

Fig. 7.15 Mixed mode I/III load frame for use in drop tower. Drop weight of 272 kg impacts lower striker bar with $V_0 = 1$ m/s. Specimen tilted out-of-plane relative to load, introducing a combination of tension and torsional moment. Angle $\varphi = 0°$ corresponds to mode I conditions, with increasing φ denoting initially mixed tension torsion conditions

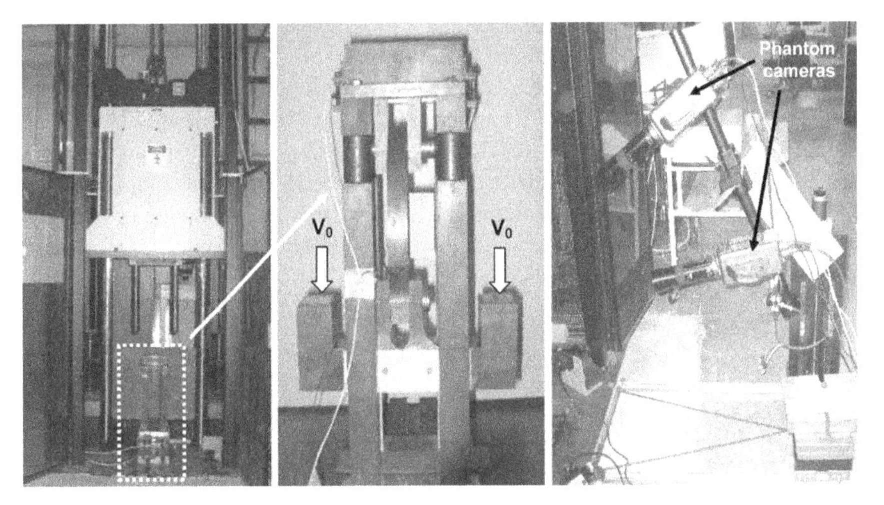

Fig. 7.16 Photographs of the drop tower with mode I/III fixture (left image), close-up of mode I/III loading fixture (center image) and stereovision arrangement with Phantom® V7.1 cameras (right image)

dropped and impacts the lower striker bar at a velocity $V_0 = 1$ m/s. The lower striker bar is connected to the lower grip through a clevis and high strength pin. The specimen is rigidly attached to the lower grip using four bolts and a cover plate.

To view the specimen surface as impact is occurring, a high speed stereovision system is employed. Images are acquired using the following system components:

- Two synchronized Vision Research Phantom V7.1 cameras with extended memory to allow continuous recording of 8-bit, full resolution images (800×600) at maximum rate for 1.3 s.
  - Each camera operates at 4,800 fps while acquiring full resolution images
  - Exposure time for each image is 40×10^{-6} s
  - Maximum synchronization offset between cameras $\simeq 1 \times 10^{-6}$ s
- Rigid cross member for mounting both cameras.
  - Translation stage to rigidly move both cameras towards or away from the specimen without altering the relative position or orientation of the cameras.
- Matching f200 Nikon F1.4 ($\mathcal{N} = 1.4$) lenses with 2X attachment and extender tubes for viewing specimen with modest magnification from an extended distance.
  - Depth of field $\simeq$25 mm
- Stationary lighting with fiber optics and high intensity halogen lamps.

7.3.4 Calibration of the Stereovision System

The procedure outlined in Section 7.2.4 is used to calibrate the high speed stereovision system. Table 7.3 provides the calibration parameters for this application. The measured standard deviation of residuals for all images in a single camera is less than 0.043 pixels and 0.058 pixels as a stereo-rig. Due to the physical configuration of the relatively small specimen within the loading structure, the angle (β) between the optical axes of the two cameras is relatively small ($\simeq 20°$) and the image scale factor for this setup is $\simeq$15 pixels/mm.

Table 7.3 Calibration parameters

Parameter	Camera 1	Camera 2	Camera 1 → 2 transformation	
Center x (pixels)	392.836	372.231	Alpha (degrees)	0.1167
Center y (pixels)	300.285	293.171	Beta (degrees)	19.9905
$f S_x$ (pixels)	14,531.9	14,462.7	Gamma (degrees)	0.38223
$f S_y$ (pixels)	14,531.9	14,462.7	T_x (mm)	−327.71
Skew	0	0	T_y (mm)	0.68899
$\kappa_1^{normalized}$	3.32992	3.00058	T_z (mm)	61.4927
κ_1	15.77E-09	14.34E-09	Baseline (mm)	333.426
Residual	0.042543	0.022444	Residual after optimization	0.05647

7.3.5 Post-processing

The software VIC-3D again is used to analyze all images acquired by the calibrated stereovision system and obtain 3D displacement fields. All analyses were performed with 19×19 subsets with a step size of 2 between subset centers. All displacement components, (U,V,W), are converted to a global coordinate system located at the original crack tip in the reference configuration. The crack-tip centered coordinate system is shown in Fig. 7.15, where the X-direction is along the crack line, the Y-direction is perpendicular to the nominally straight fatigue crack and the Z-direction is perpendicular to the planar reference configuration. The imaged region ranged from -8 mm $\leq X \leq 32$ mm to -15 mm $\leq Y \leq 15$ mm in these experiments [298].

Strains at each time during the event were obtained in the reference configuration using the procedures defined in Section 7.2.7. The partial derivatives of the displacement field are computed from a quadratic polynomial approximation to the computed displacement field in a local neighborhood using a 9×9 array of measurements, and the strain tensor is defined by Eq. (A.5).

Crack tip positions also were determined using the procedures in Section 7.2.7. The resulting time history of crack tip locations is converted into a time history for the crack tip velocity (relative to the material reference frame) using a combination of forward and central difference formulae.

7.3.6 Experiments

After completing calibration of the stereovision system, the specimen is attached to the mixed mode I/III grips and placed in the impact loading frame. The drop weight is raised to the appropriate height of 0.050 m to achieve the terminal velocity $V_0 = 1$ m/s and all instrumentation is initialized.

Triggering of the synchronized cameras is performed using the output signal of an accelerometer mounted to the striker bar (loading anvil). The two high speed Phantom® cameras acquire full-resolution images at 4,800 fps with a maximum temporal offset of 1×10^{-6} s. The time-varying reaction force is estimated during the dynamic mixed mode I/III test using two load cells mounted in the load frame between the vertical top bar and the compression side members.

The signals from the accelerometer(s), both load cells and the triggering circuit are sampled, acquired and transmitted using software developed in Labview 7.0 Express. The Labview-based data acquisition system is triggered using a TTL signal, which is obtained from an infrared detector that transmits a 5V to 0V drop when the drop weight passes a pre-specified location during its descent towards impact.

7.3.7 Experimental Results

Figure 7.17 shows the plastically deformed shape of an impact fractured specimen, with $\varphi = 60°$. Figure 7.17 also shows a fractured specimen that underwent

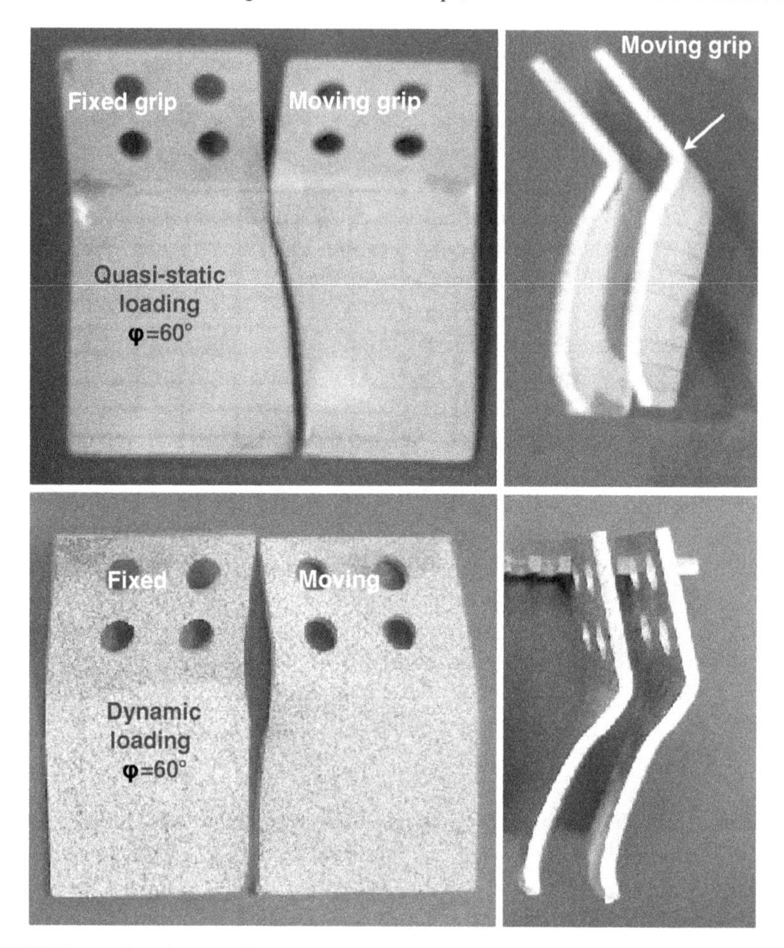

Fig. 7.17 Comparison between impact at $V_0 = 1$ m/s and quasi-static fractured AL6061-T6 specimens in mixed mode I/III experiments, $\varphi = 60°$. Crack path generally straight during impact fracture, and curvilinear during quasi-static fracture. Permanent out of plane deformation during impact or quasi-static fracture $\simeq$20 mm

quasi-static loading for comparison purposes, demonstrating that the overall shape is not affected appreciably by the corresponding strain rate increase [298].

Following the procedures described in Section 7.2.7 to locate the position of the growing crack tip in the material configuration, Fig. 7.18 shows both the crack path and also the velocity as a function of time and crack extension. As expected, the material-based crack path is nearly straight, with the crack velocity exceeding 1.5 m/s after 10 mm of stable tearing, before decreasing as the crack continued to extend.

Figure 7.19 presents the complete 3D displacement and surface Lagrangian strain fields (see Eq. (A.5)) at 11,259 μs after impact. The displacement field contours for the U and V fields are complex and without any apparent symmetry. The W

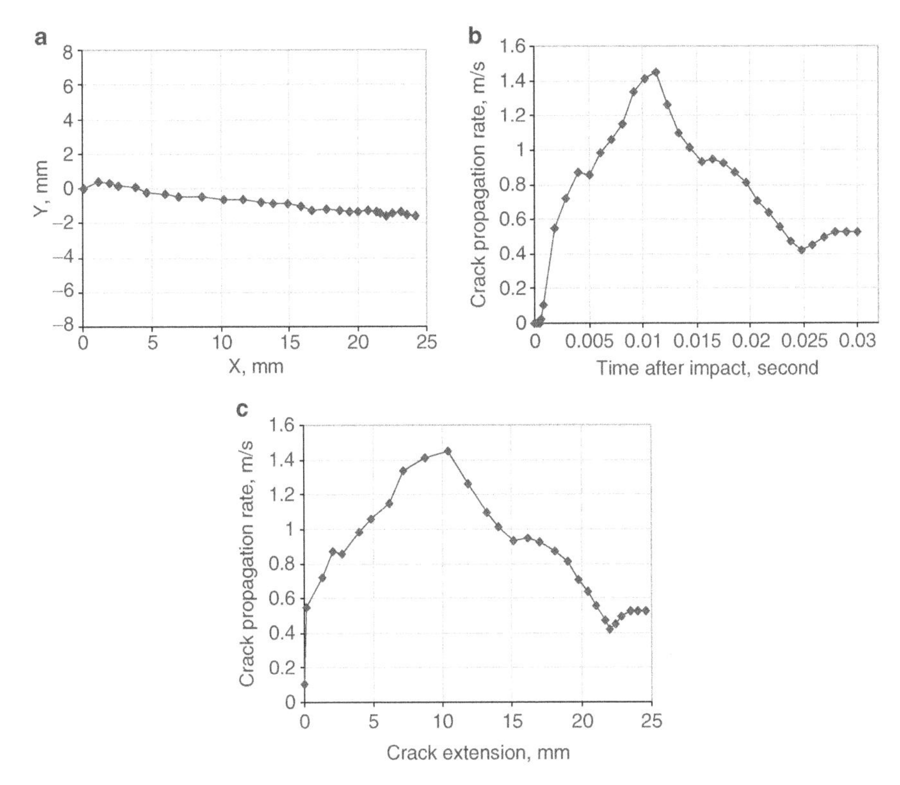

Fig. 7.18 Measured **a** crack path, **b** crack propagation rate as a function of time and **c** crack propagation rate as a function of crack extension. All data relative to the original configuration of AL6061-T6 specimen in dynamic mode I/III experiment, $V_0 = 1$ m/s, $\varphi = 60°$

displacement field is more understandable, with the right side of the specimen displacing downward ($W < 0$) relative to the fixed grip on the left side; the specimen has incurred up to 15 mm of relative out-of-plane motion at this time.

The in-plane surface strain fields induced during the large deformation fracture process show very specific trends.

- Crack tip region has a well defined in-plane shear strain field, $\varepsilon_{xy} > 0$, on the impact side of the crack, with the maximum shear strain shifted slightly towards the impact side.
- Strain ε_{xx} is approximately antisymmetric across the crack line, tensile on the impact side.
- Strain $\varepsilon_{yy} > 0$ throughout the crack tip region, with a maximum on the order of 0.16 shifted towards the fixed side of the specimen.
- Residual normal plastic strains ε_{xx} and ε_{yy} have opposite signs across the crack line, with $\varepsilon_{xx} > 0$ and $\varepsilon_{yy} < 0$ on the impact side of the crack and $\varepsilon_{xx} < 0$ and $\varepsilon_{yy} > 0$ on the fixed side of the crack.

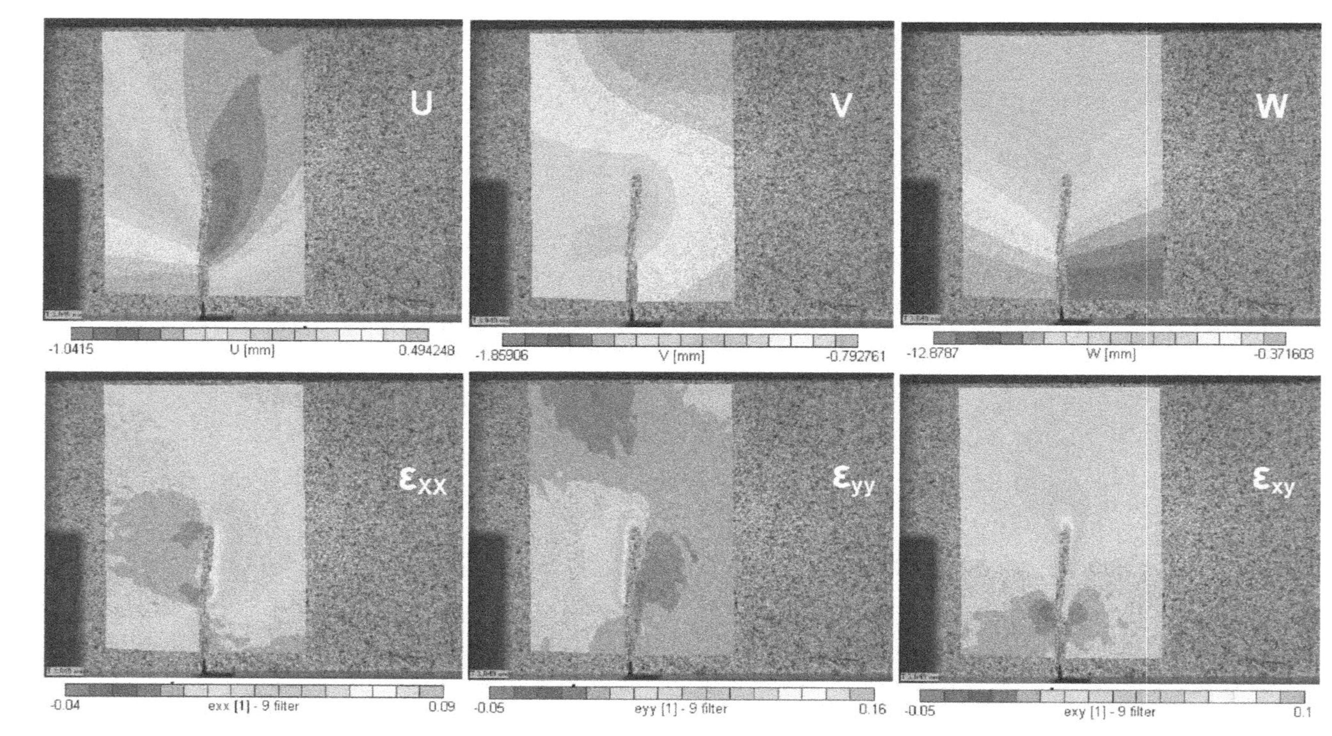

Fig. 7.19 Contours for all displacement components (U,V,W) and all in-plane strain components $(\varepsilon_{xx}, \varepsilon_{yy}, \varepsilon_{xy})$. Time after impact is 11,259 µs, crack extension is 10.46 mm, and tension–torsion loading angle is $\varphi = 60^{\circ}$

The complex displacement and strain fields induced by the relatively large three-dimensional specimen motions suggests the need for a hybrid experimental-computational model to estimate the through-thickness stress and strain fields in the crack tip region and improve our understanding of crack growth conditions under nominally tension-torsion conditions.

7.3.8 Discussion and Practical Considerations

An essential aspect of using a 3D digital image correlation system is to position cameras so that the baseline distance is adequate for accurate out-of-plane motion measurements and the specimen is clearly visible throughout the loading process. Also, since the specimen will be tilted away from the cameras by 30° or 60° to apply both tension and torsion, the issues highlighted in Fig. 7.14 for the bending application in Section 7.2 are also relevant for the tension–torsion experiments. In this application, the physical size of the grips and loading fixture made it impossible to view the full region of interest if the cameras are oriented on a horizontal plane and have a reasonable baseline distance. As shown in Fig. 7.15, the decision was made to place the cameras in a vertical plane. This approach allowed (a) full view of the specimen throughout the impact process, (b) increased baseline distance between cameras, thereby improving accuracy of the measured out-of-plane motion and (c) a modest angle between the normal to the specimen surface and the optical axis of each camera.

The choice of long focal length lenses and 2X extenders to increase viewing distance was required by safety considerations. All drop tower experiments must be performed within a protective steel enclosure to reduce the likelihood of injury in the event of fragmentation and high rate expulsion of material. In this application, the minimum physical set-back distance is 2 m, necessitating the use of f200 lenses.

To ensure adequate contrast in the images, lighting can be applied by a combination of stationary sources and/or high intensity strobes. Both the exposure time and the synchronization offset between images must be considered when evaluating the 3D data. The exposure time controls temporal averaging during imaging of a dynamic event. The synchronization offset may introduce relative motions between two stereo camera views. A discussion of the practical issues related to exposure time and image averaging for this application is presented in Section 10.1.4.

7.4 Development and Application of 3D Digital Image Correlation Principles in Stereomicroscopy for Microscale Shape and Deformation Measurements

Development of a modified, two-camera, stereomicroscope system for three-dimensional deformation measurements on miniature specimens is described. Baseline results are presented to validate its capability. Application of the system to make quantitative micro-scale measurements on biological materials is presented. Details regarding use of stereomicroscopy for biological material deformations are provided in a recent reference [281].

7.4.1 Problem Description: Shape and Deformation Measurements for Small, Soft Specimens Subjected to Mechanical Loading

There is an increasing interest in quantifying the properties of soft materials at reduced length scales (e.g., polymer layers, bio-materials). Since any form of attachment has the potential to affect the measurement, the issue has produced a need for novel non-contacting measurement systems. One area is in the study of MEMS devices and materials. Another example is in the characterization of mechanical properties of bio-components such as blood vessels where the local strain response can provide important information regarding the effect of variations in extra-cellular matrix composition for a variety of conditions.

The remainder of this section describes details regarding (a) the stereo microscope imaging system, (b) pattern application at reduced length scale, (c) preliminary validation experiments and (d) measurement of deformations during mechanical loading of a small mouse carotid.

7.4.2 Specimen Geometry and Preparation

Three types of objects were imaged in this study, and are shown in Fig. 7.20.

The first object is a specially designed and lithographically manufactured pattern on a glass substrate that contains both (a) several standard grid patterns to perform calibration of the stereo-microscope system and (b) several sizes of random patterns to perform a priori distortion removal.

The second is a small cannula with a cylindrical section having a diameter $\simeq$0.45 mm that is designed for connection to small vessels. To obtain a high contrast random pattern, the milky white cannula is lightly sprayed with flat white, slow-drying enamel paint. Prior to drying, several light dustings of Xerox toner powder are used to obtain an appropriate random pattern for the high magnification experiments. This specimen is used to validate the measurement system for use in the bio-material studies.

The third specimen is a mouse carotid, with outer diameter $\simeq$0.50 mm and internal diameter $\simeq$0.40 mm. Several methods have been used to apply a random pattern for high contrast imaging. The first method, dusting the specimen with Xerox toner powder, provides an excellent pattern for damp specimens. However, in a liquid environment, the hydrophobic powder tends to float off of the surface. In a liquid environment, a conjugation method has been developed to attach fluorescent beads to the specimen surface. To visualize the beads, a fluorescent illumination attachment

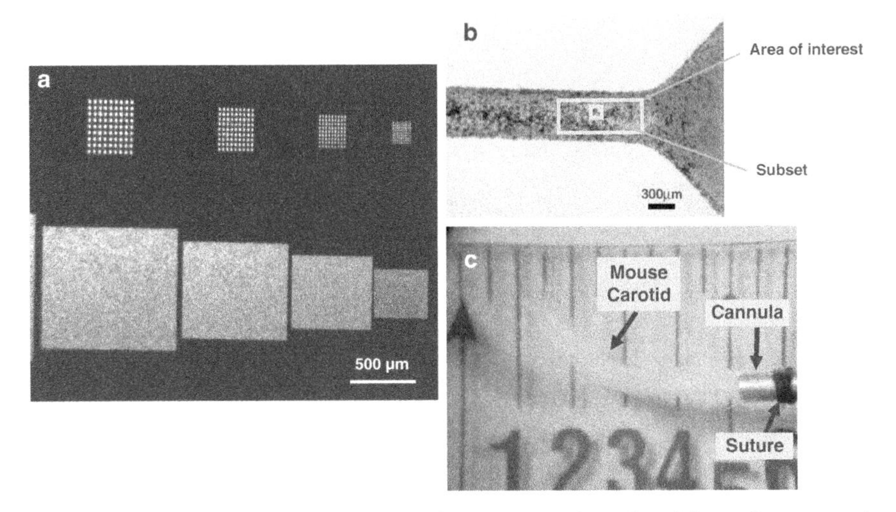

Fig. 7.20 **a** Specially-designed combination of patterns at various sizes. The random pattern is used for distortion correction and the grid pattern is used for calibration. Pattern can be used for calibration and distortion removal for fields of view ranging from 5 to 0.5 mm^2. **b** Twenty-five gage cannula used in bio-material studies. **c** Mouse carotid specimen

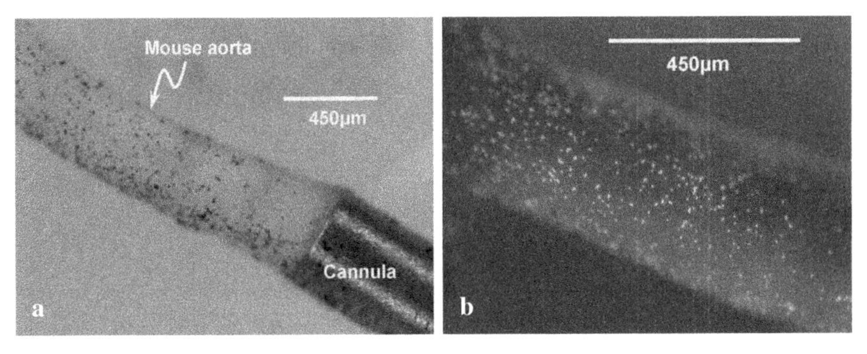

Fig. 7.21 Speckle pattern on mouse carotid using **a** Xerox toner powder with white light illumination and **b** fluorescent polystyrene microspheres (10 μm) attached using fibronectin, an extracellular matrix protein which binds to several components of the arterial wall. Illumination is via UV attachment, with bandpass optical filter on collection lens to recover micro-sphere emission

and bandpass filter are added to the microscope. Figure 7.21 shows images of a toner powder pattern and a fluorescent bead pattern on a mouse carotid. Both methods have been used successfully, though the fluorescent beads are preferred since the pattern is stable in a fluid environment so that long-term experiments can be performed. In this study, Ricoh toner powder with a size distribution 3 $\rightarrow$ 10 μm was used to develop a high contrast pattern. Specimens were spritzed every 1 min to maintain moisture on the outer surface throughout the estimated 30 min experimental time.

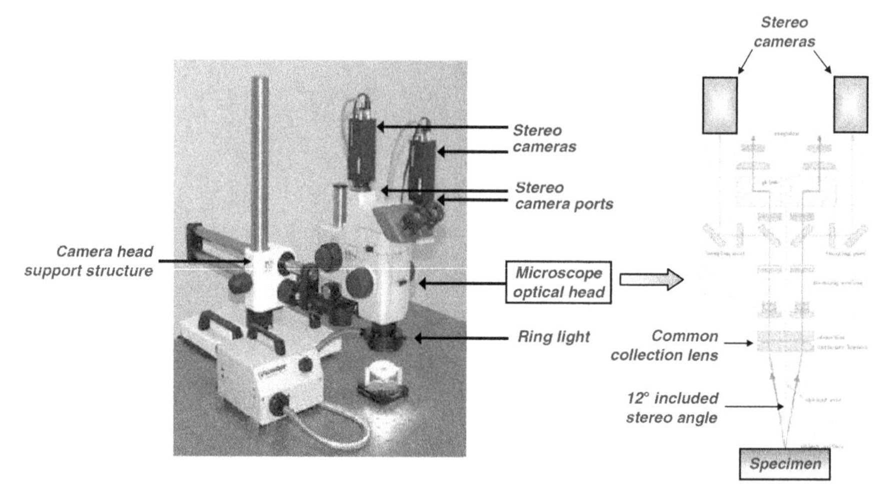

Fig. 7.22 Nikon SMZ-U microscope with dual camera attachment for stereo-imaging. Ring light provides uniform downward illumination for imaging. Complex internal optics in microscope optics provides compact stereo viewing arrangement

7.4.3 Optical Components in Stereo-Microscope and Experimental Considerations

The stereo-microscope measurement system shown in Fig. 7.22 uses the following system components:

- Nikon SMZ-U light microscope arranged for 2×2 mm field of view
- Two monochrome Q-Imaging QICAM 8-bit cameras with data acquisition and synchronization components (see Section 7.3.3)
- Nikon FOI-150 fiber optic ringlight for illumination
- Calibration target frame and 3D rotational mount (see Fig. 7.20)
- UV Illuminator with bandpass filter for imaging fluorescent microspheres
- Translation stage and random pattern on target for distortion correction

7.4.4 Distortion Correction and Stereo Calibration

Following the procedures outlined in Sections 3.1.2.3–3.1.2.4 and [254], the slide with the combined calibration and random speckle patterns shown in Fig. 7.20 is mounted onto a translation stage. Then, the slide is viewed by the stereo-microscope system. By moving the specimen small distances along a calibration path (see Fig. 7.23), the distortion correction fields for the horizontal and vertical directions shown in Fig. 7.23 are determined. The distortion fields are used to define corrected positions for each pixel in the sensor plane.

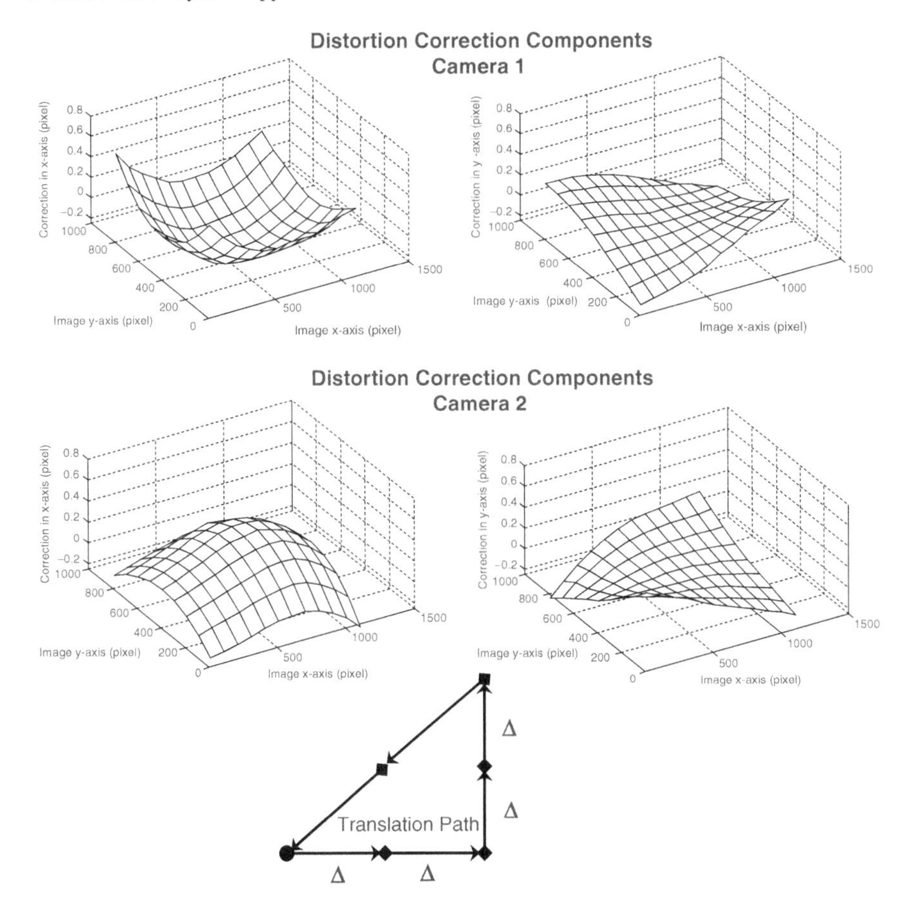

Fig. 7.23 Distortion correction fields for cameras in stereo-microscope system. Distortions computed using random pattern image pairs obtained along in-plane translation path. Total motion should be less than 10% of image size to minimize loss of area for measurements. Step size, Δ, is on order of 20 pixels in this application

By acquiring a series of stereo vision images of one of the grid pattern on the combined calibration-random pattern slide in Fig. 7.20, camera parameters for each path in the stereo vision system are determined and shown in Table 7.4. In this work, a total of 41 image pairs are acquired as the target undergoes in-plane translation, in-plane rotation and out-of-plane tilt. After correcting all images for distortion, the images are used to complete the camera calibration process. Here, the maximum out-of-plane tilt during calibration is limited to 30°[3] to maintain focus throughout the target area.

[3] Software is developed by Correlated Solutions, Incorporated to use a grid pattern instead of motions of a random pattern for the camera calibration process. Based on using distortion-corrected locations for the grid pattern, calibration of the system is conducted using the traditional bundle-adjustment method.

Table 7.4 Calibration Parameters

Parameter	Camera 1	Camera 2	Camera 1 → 2 transformation	
Center x (pixels)	−4,259.2	4,161.0	Alpha (degrees)	−0.5643
Center y (pixels)	483.3	553.5	Beta (degrees)	0.1231
fS_x (pixels)	19,022.3	18,929.3	Gamma (degrees)	3.4942
fS_y (pixels)	19,190.0	18,831.6	T_x(mm)	−19.1402
Skew	−0.9715	69.4	T_y (mm)	−1.5479
$\kappa_1^{normalized}$	0	0	T_z (mm)	−0.7756
κ_1	0	0	Baseline (mm)	19.2184
Residual	0.04456	0.03958	Residual after optimization	0.05679

7.4.5 Post Processing

The software VIC-3D is used to analyze all images acquired by the calibrated stereovision system and obtain 3D displacement fields. An approximate transverse magnification factor is 463 pixels/mm for both the validation and mouse carotid experiments.

For the validation experiments (Section 7.4.6), matching was performed with a subset size of 43×43 pixels^2 with a step size of 5 pixels.

For the mouse carotid experiments (Section 7.4.7), matching used large subsets, 101×101 pixels^2, with a small step size of 5 pixels.

Strains in both case were obtained in the reference configuration using the procedures defined in Section 7.2.6. The partial derivatives of the displacement field are computed from a quadratic polynomial approximation to the computed displacement field in a local neighborhood using a 9×9 array of measurements, and the strain tensor is defined by Eq. (A.5).

7.4.6 Validation Experiment

Previous studies have shown that the strain variability of a fully-corrected stereo-microscope system viewing a planar object is on the order of 200 μs when imaging planar surfaces. In order to demonstrate the accuracy of the microscopic 3D-DIC system when estimating strains on miniature cylindrical vessels, the calibrated and distortion-corrected system is used to record images of the cannula shown in Fig. 7.20 as it undergoes three-dimensional rigid body motion. The measurement area is $\simeq 290 \times 770$ μm^2 (130×350 pixels), with rigid body motions not exceeding 600 μm.

As shown in Table 7.5, the maximum variability in strain is <800 μs throughout the field of view, confirming that the method is sufficiently accurate for quantifying deformations in soft biological materials which typically undergo appreciable deformation.

Table 7.5 Standard deviation in surface strain fields during 3D translation experiments

Strain components	Standard deviation × 10^{-6}
ε_{xx}	350
ε_{yy}	790
ε_{xy}	360

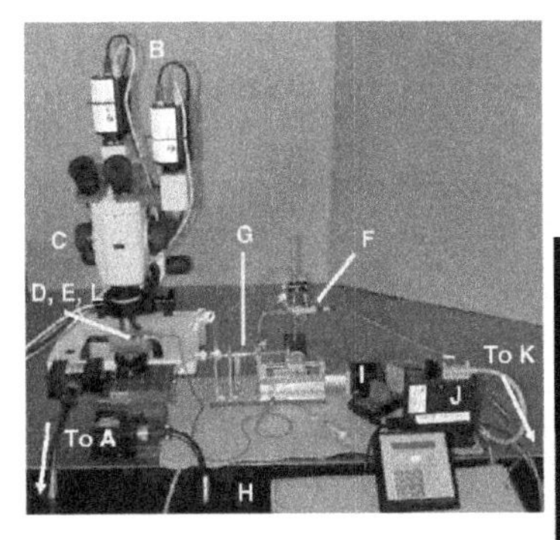

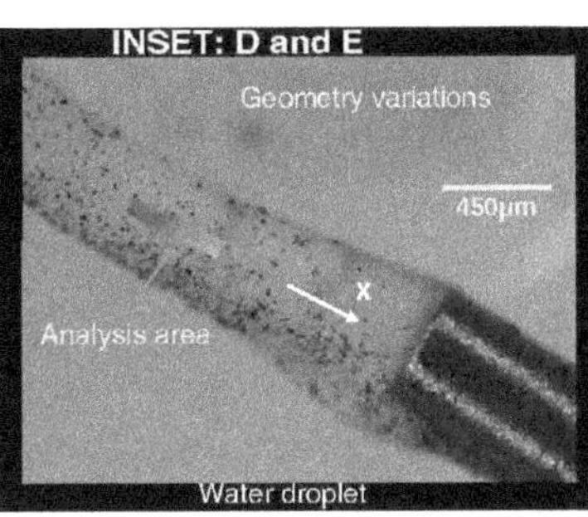

Fig. 7.24 Experimental setup for pressurizing a mouse carotid artery. *A*: Computer with two image capture cards; *B*: two cameras; *C*: microscope; *D*: mouse carotid artery and *E*: 25 gage luer; *F*: pressure transducer; *G*: syringe; *H*: holder; *I*: step motor; *J*: step motor controller; *K*: voltage amplifier; *L*: specimen holder. INSET: Mouse carotid undergoing pressurization in system with p = 151 mm Hg. Geometric variations clearly visible. The region of interest is outlined by the black-gray-white contour box

7.4.7 Mouse Carotid Experiments

After sacrificing a mouse via carbon dioxide asphyxiation, the carotid was cannulated prior to removal. Following cannulation, the carotid specimen is dissected free of the underlying structures and transferred to buffered physiological saline supplemented with protease inhibitors (Sigma) in preparation for pressure experimentation.

Figure 7.24 shows the relatively complex pressurization arrangement used in this study.

Each mice carotid specimen is connected to a pressure sensor's vertical end (P23, Gould Statham) at the proximal end via the 25 gage luer stub. The Gould transducer outputs a voltage signal from a traditional Wheatstone bridge circuit. The signal is sent to the amplifier and a cap-module with an inlet and outlet to allow for media to pass over the sensor. The amplifier (Gould universal console) outputs an analog signal from 0 to 5 V for all gain levels with 5 V corresponding to 250 mm Hg. It is noted that, during experimentation, the experimental apparatus was firmly attached to a floating optical bench (Newport) since the transducer is highly sensitive to vibrations at the high gain settings used in this work. The amplifier and computer were not placed on the table due to the presence of cooling fans that interfered with operation of the pressure sensor.

At this stage, the carotid artery specimen is positioned on the stereomicroscope stage over a white background. The pressure sensor's slanting end is connected to a computer-controlled step motor which drives a syringe. The vessel is flushed with PBS to remove any bubbles trapped in the lumen. For pressure testing, the distal end of the vessel is clamped shut with a microaneurysm clip (Harvard Apparatus) or with suture just proximal to the carotid bifurcation. At this point, the toner powder was misted onto the surface.

The pressurization system was programmed to cyclically pressurize and depressurize the artery in steps of 25 mm Hg from 0 to 240 mm Hg with a pressure rate of 1 mm Hg/s. At each pressurization step, synchronized images are acquired from both cameras for 3D-DIC analysis.

7.4.8 Results

Figure 7.24 also shows an image of the specimen with pressure, $p = 151$ mm Hg. Also shown is the region where image correlation was performed. Spanning approximately $50 \times 250\ \mu\text{m}^2$ in physical dimension, the region was selected to highlight the effect on axial strain of an observed geometric variation in the artery. Here, the ring-like region exhibited more than twice the axial strain as the surrounding carotid.

Figure 7.25 presents the average strains ε_{xx}, ε_{yy} and ε_{xy} in the region of interest as a function of applied pressure. Results indicate that the circumferential strain and the shear strain follow a nearly linear trend with minimal hysteresis, whereas the axial strain exhibits highly non-linear behavior with considerable hysteresis.

7.4.9 Discussion

When combined with 3D digital image correlation, stereo-microscopes offer unique advantages for non-contacting measurements on miniature specimens. As a full-field measurement method, averaged data can be obtained from the full field data, while also quantifying local variations in response that are important when studying non-homogeneous materials such as vascular tissues. Additional advantages include (a) good spatial resolution, on the order of 1/100th of the field of view for the displacement measurements. Thus, if a $1 \times 1\ \text{mm}^2$ area is imaged, then the spatial resolution is $\simeq 10\ \mu\text{m}$, (b) ability to make measurements on curved (e.g. cylindrical, spherical) or planar surfaces), providing the ability to make accurate measurements on blood vessels, (c) ability to make measurements on extremely small components (e.g. $100 \times 100\ \mu\text{m}$), (d) point-to-point strain accuracy down to 200×10^{-6} in some applications and full-field average strain accuracy on the order of 50×10^{-6},

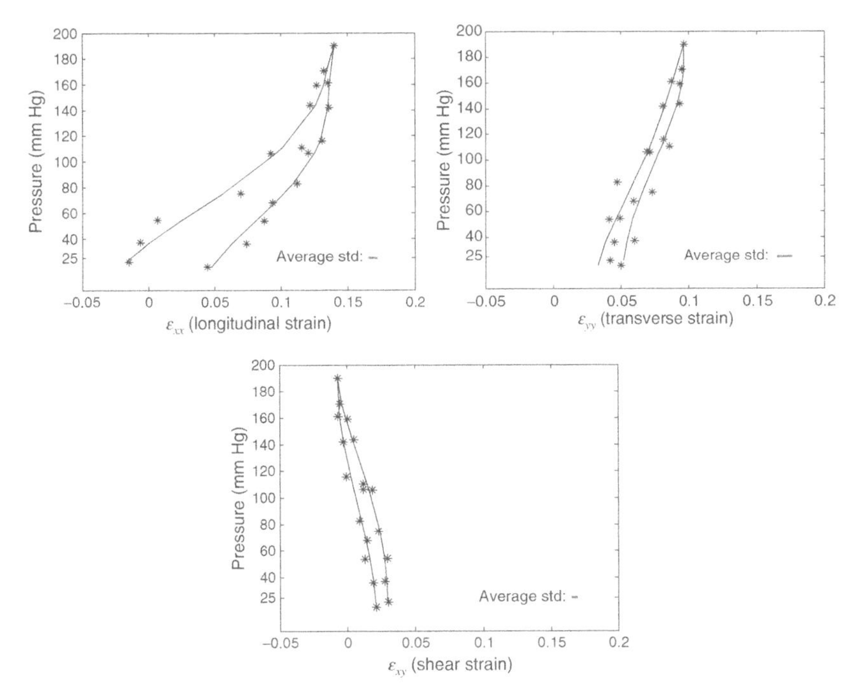

Fig. 7.25 Average longitudinal normal strain ε_{xx}, average transverse normal strain ε_{yy}, and average shear strain ε_{xy} as a function of pressure in a unfixed mouse common carotid artery during the fifth pressurization cycle. Data represented by symbols. Lines are least squares data fits

providing the ability to study both small and large deformation responses in a single specimen, (e) measurements can be made using either white light or light falling within a wide spectral range, ranging from infrared through ultraviolet, depending upon the type of camera sensor being used to record the images and (f) suitability for both static and high speed measurements at rates up to 20,000 frames/s so that quantitative measurements can be made during transient variations in environment and/or loading.

Inspection of Table 7.4 shows that the value for the skew parameter defined by Eq. (3.7) is quite large. The value shown is entirely consistent with the fact that, prior to performing calibration of the pinhole model, spatial distortions were removed from the images using the procedures outlined in Section 3.1.2.3. The distortion corrected images are not expected to have orthogonality between the modified row and column directions (see Figs. 3.2 and 3.17). In this application, using the definition for skew and the data in Table 7.4, the estimated angle between the distortion corrected x_s and y_s directions in camera 2 is $\cong 90.21^\circ$, indicating a relatively small deviation even though the skew parameter is large.

Regarding the relatively large variability noted in Table 7.5, the primary sources for increased variability lie with the pattern; neither particle density nor size are ideal, requiring a large correlation subset size to maintain accuracy. As noted by

Schreier [252, 253], such patterns can lead to local errors in the measurements that will propagate into the strain computations. Since the strain ε_{yy} has the largest variability, where the y-direction is nominally oriented in the local circumferential direction, the approach used to define strain may be a factor. Here, local strain is defined in the least squares best fit plane. Since the specimen will rotate during the rigid body motion process, it is likely that inaccuracies in the planar surface fits in the circumferential direction may increase strain errors.

Inspection of the Inset image in Fig. 7.24 shows the presence of a water droplet. If image correlation is performed to match subsets in this region, large position and strain errors will occur due to the effects of refraction at the curved air-droplet interface. In such cases, the only reasonable choices are to either (a) immerse the entire specimen and develop methods for properly correcting for refraction effects at a fixed interface or (b) remove surface droplets prior to imaging the specimen surface.

Chapter 8
Volumetric Digital Image Correlation (VDIC)

In all of the previous sections, images of an object surface that are acquired by a sensor array are combined with a calibrated imaging model to convert the image sensor information into 2D or 3D surface position data. Typically, the imaging model is a form of perspective transformation, with functions defined to correct the model predictions.

In recent years, a series of advances in imaging, not only of surfaces but also for material volumes, have resulted in new opportunities for quantitative internal measurement of 3D positions throughout an entire volume. Computed Tomography (CT) scanners [26,27,30,155,156,188,265], Confocal Imaging Microscope architectures (CIM) [80] and Magnetic Resonance Imaging (MRI) [219,324,325] systems are examples of modern hardware that are now able to obtain and store massive amounts of digital volumetric image data, oftentimes exceeding the ability of the hardware or software to process and analyze the data.

8.1 Imaging and Discrete Pattern Recording

The process used to acquire volumetric images is a function of the type of system being used. Computed tomography (CT), originally known as computed axial tomography, had its beginnings in the field of medicine. From its independent invention by Hounsfield and McCormack in the late 1960s, modern CT scanning systems have undergone several generations of improvement. Radiation sources are part of the evolution, beginning with thin X-ray beams that gave way to wide fan X-rays and helical X-ray system, eventually including high energy electron beam systems.

For a typical CT scan system, a planar X-ray image (known as a slice) is obtained by either rotating the specimen or the X-ray unit about a single axis, while passing an X-ray beam through the specimen about a common axis of rotation, with each slice being separated by a angle of rotation, $\Delta\theta$. Tomographic reconstruction using methods such as filtered back-projection [33,209] are used to extract local values for

M.A. Sutton et al., *Image Correlation for Shape, Motion and Deformation Measurements: Basic Concepts, Theory and Applications*, DOI: 10.1007/978-0-387-78747-3_8,

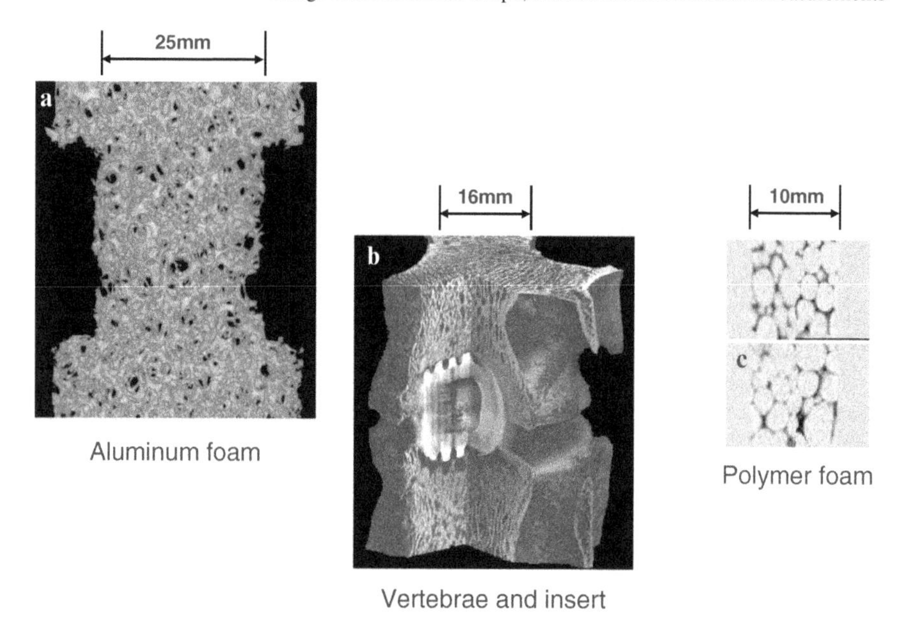

Fig. 8.1 Volumetric CT images for **a** metallic honeycombs, **b** bone structures and **c** polymeric foams. Images obtained using micro-CT X-ray source with CCD detector and 1,024 × 1,024 array for each slice. (Courtesy Professor B.K. Bay)

the material's "radiodensity", which is a measure of the ability (or lack thereof) to block the radiation that passes through the material. The feature data from individual slices is then combined with similar data from adjacent slices to obtain a continuously varying, volumetric representation for the specimen. Today, CT systems allow the resulting image data to be viewed as either a volumetric (3D) representation for the structure or as a series of individual 2D slices.

Figure 8.1 shows typical volumetric CT images for (a) metallic honeycombs, (b) bone structures and (c) polymeric foams. All images were performed using a micro-CT X-ray source and a CCD detector integrated with a materials testing system for imaging under load.[1] Voxel size in all cases was 0.05 mm; 2 × 2 binning of the polymer foam images reduced spatial resolution to 0.10 mm in these images.

Voxel-level CT data is known to have reduced signal to noise levels relative to optical methods. Additional error sources can include (a) beam hardening (increased beam attenuation near object center), (b) ring patterning and (d) partial volume blurring. Fortunately, software with image processing and system modifications is effective in accounting for typical CT scanning artifacts, though the effect of these image modifications on matching of sub-volumes remains an area of active investigation.

[1] Volumetric imaging and analysis performed in the Biomechanics Imaging and Measurements Laboratory at Oregon State University, under the direction of the laboratory developer, Professor Brian K. Bay.

Confocal Imaging Microscopy (CIM), also known as confocal laser scanning microscopy, uses a combination of (a) laser beam rastering using rotating mirrors and lenses to focus the beam in discrete volumes, (b) fluorescent species within a nominally transparent specimen that emit light when excited by incident laser illumination, (c) optical elements to collect the light emitted by the material due to laser illumination, (d) photo-multiplier devices to detect the resulting light emission and output an electrical signal and (e) A/D conversion for digital storage. Typically, the scanning process is performed sequentially to obtain $M \times N$ digital intensity values in each horizontal layer (slice). By shifting the process a vertical distance, Δz, additional slices, known as z-stacks, are obtained. The digital data obtained from all z-stacks is assembled to generate a volumetric reconstruction.

As an example, Fig. 8.2 shows three z-section images or slices (labelled 1, 10 and 21 in top right corner in Fig. 8.2) from a stained region of a mouse small intestine. All imaging was performed in a Zeiss LSM 510 Meta CLSM, which is a single photon spectral imaging systems. The red region in Fig. 8.2 resulted from DAPI stain of the nuclei. The green region was stained with f-actin. Each section corresponds to emission of light over 1 μm thickness. Also shown in Fig. 8.2 is the integrated image along the z-direction containing all 21 z-sections.

Voxel level CIM data has several error sources including (a) positioning errors for the laser beam, (b) laser intensity variability, (c) photo-multiplier noise and (d) variability in specimen emission. Relative to (d), when performing CIM using fluorescence for contrast, reductions in contrast are expected due to photo-bleaching of the fluorescent species under continued laser excitation.

Magnetic resonance imaging (MRI) has primarily been used to construct images that reflect variations in the nuclear magnetic resonance of hydrogen atoms in a specimen. MRI systems have been used to acquire high resolution volumetric representations for regions in the human body, where images are constructed based on magnetic interactions with water and fatty tissues.

As with all other volumetric imaging methods, voxel level MRI data has several error sources. These include all of the following (a) object motion during imaging process, (b) unanticipated variations in the magnetic field during imaging, (c) ring artifacts similar to those seen in CT images after processing, (d) hardware errors/inaccuracies during imaging, (e) variations in magnetic response of materials and (f) variations in shielding during imaging.

8.2 System Calibration for VDIC

In a manner that is directly analogous to the definitions used for 2D and 3D digital image correlation, for VDIC we can define the reference volumetric feature representation function as $I(P,t)$, where P is any three-dimensional position inside the component and t is a measure of time. Here, $I(P,t)$ is oftentimes known as the "intensity field". Since modern CT, MRI or CIM systems generally provide direct

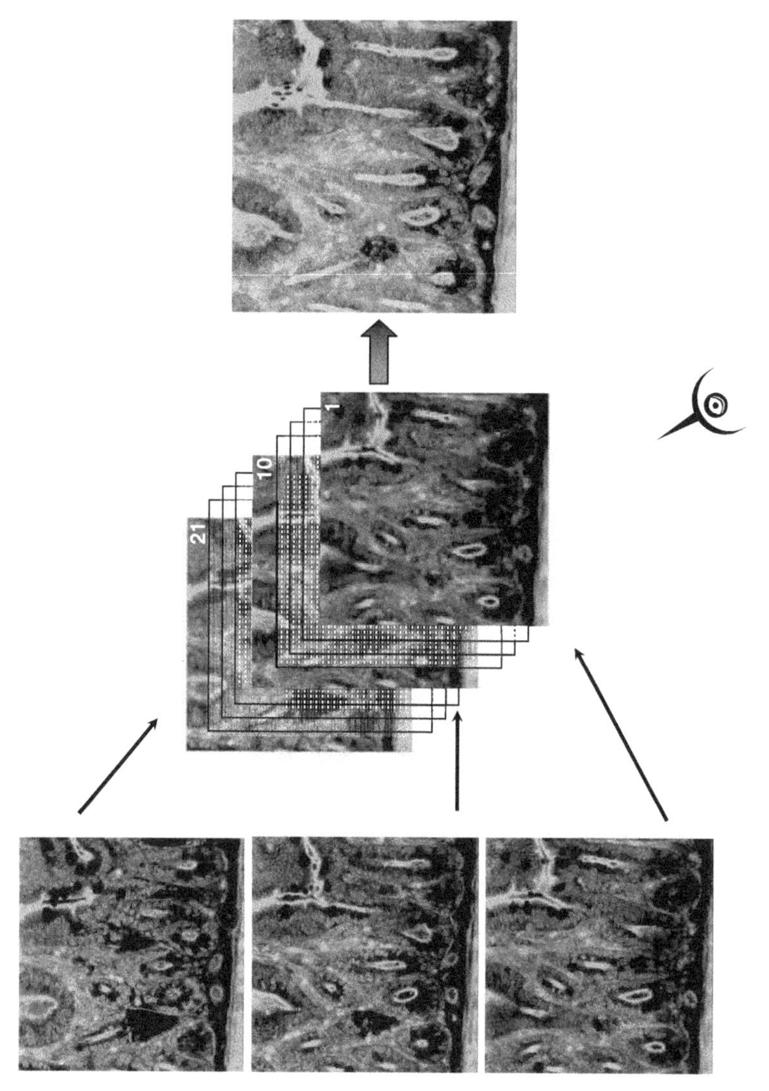

Fig. 8.2 Slice and integrated volumetric images of mouse aorta obtained using Confocal Imaging Microscopy with blue diode and helium neon laser illumination. Each slice is $0.289 \times 0.289 \times 0.001$ mm thick. (Courtesy Professor Robert Price)

output measurement of $I(P,t)$, *in principle* the calibration process for a VDIC system is considerably simpler than for 3D-DIC.[2] Specifically, these systems:

- Have simplified models relating sensor plane locations to object positions, minimizing complexity of the calibration process

[2] Methodologies for accurate calibration of volumetric imaging systems are actively being developed.

- Eliminate the requirement to relate multiple independent imaging systems for extraction of accurate 3D positional information
- Remove the requirement for synchronization of multiple imaging systems during calibration process

Given these simplifications, there are two effects to be quantified during a calibration process of a volumetric imaging system:

- Scale factors to convert sensor positions (hence reconstructed volume) into metric distances
- Distortion of measurement volume during the image acquisition and intensity data generation processes

To obtain scale factors, approaches that one may use include (a) acquisition of images of a calibration standard and (b) performance of known motions of an object to extract distortion fields.

8.3 Volumetric Shape and Deformation Measurements

Assuming that calibration has been completed and all 3D positions in the reconstructed object are corrected for distortions, image matching is performed using VDIC to locate specific sub-volumes in a series of images with optimal accuracy. Similar to the description in Section 5.4 and Appendix D, optimal matching is performed through minimization or maximization of a metric function that quantifies the level of similarity in sub-volumes between the reference and deformed volumetric images.

Figure 8.3 shows a rectangular prism region of an object that is imaged. Defining one of the volumetric reconstructions as the "reference" and the remaining as "deformed" images, the most common metric used to quantify the level of agreement is the least square form.

$$\sum_{k=1}^{N}\sum_{j=1}^{N}\sum_{i=1}^{N}\left[I'(x,y,z)-I(x_{ijk},y_{ijk},z_{ijk})\right]^2$$

where:

$$I(x_{ijk},y_{ijk},z_{ijk}) = \text{value of object feature function recorded at reconstructed integer object location } (x_{ijk},y_{ijk},z_{ijk}) \text{ in undeformed image} \tag{8.1}$$

$I'(x,y,z)$ = continuous representation for object feature function at arbitrary position (x,y,z), obtained by fitting digitized integer values $I'(x_{ijk},y_{ijk},z_{ijk})$ in the deformed image

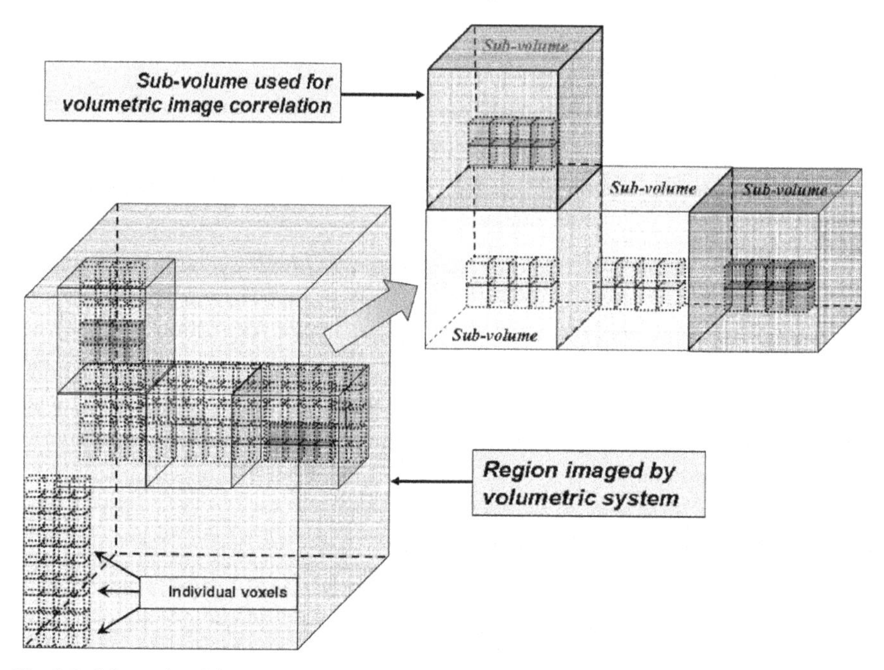

Fig. 8.3 Schematic with object volume, sub-volumes and individual voxel elements. Sub-volumes shown as disjoint but may overlap to increase data density

Equation (8.1) may be termed as the volumetric sum of squared differences (VSSD), a direct analogue to the SSD formula in Eq. (5.41) that is used for both 2D and 3D computer vision formulations.

One reason for selecting a least squares algorithm (and not a normalized cross correlation function), is the reduction in computation time that is accrued for the large volumetric images. For example, suppose that slice images are acquired every 0.6° during full rotation of the object, with each slice sampled at 520×520 locations. If the range of intensity values is 8 bits, then each image is 150 Mb, a size that is two orders of magnitude larger than used in 3D stereo vision.

8.4 Volumetric Shape Functions

As shown in Fig. 8.4, the location of the center-point of the deforming sub-volume is estimated during the matching process.

Since each sub-volume undergoes both translation and deformation, a cube in the undeformed image will deform into a more complex shape. If a cubic sub-volume is centered at object point P with image coordinates (X^P, Y^P, Z^P), then the intensity values for point P and an arbitrary point Q located at position $(X^P + \Delta X, Y^P + \Delta Y, Z^P + \Delta Z)$ respectively, can be written in terms of subset-based coordinates $\gamma^Q = X^Q - X^P, \eta^Q = Y^Q - Y^P$ and $\xi^Q = Z^Q - Z^P$ as follows:

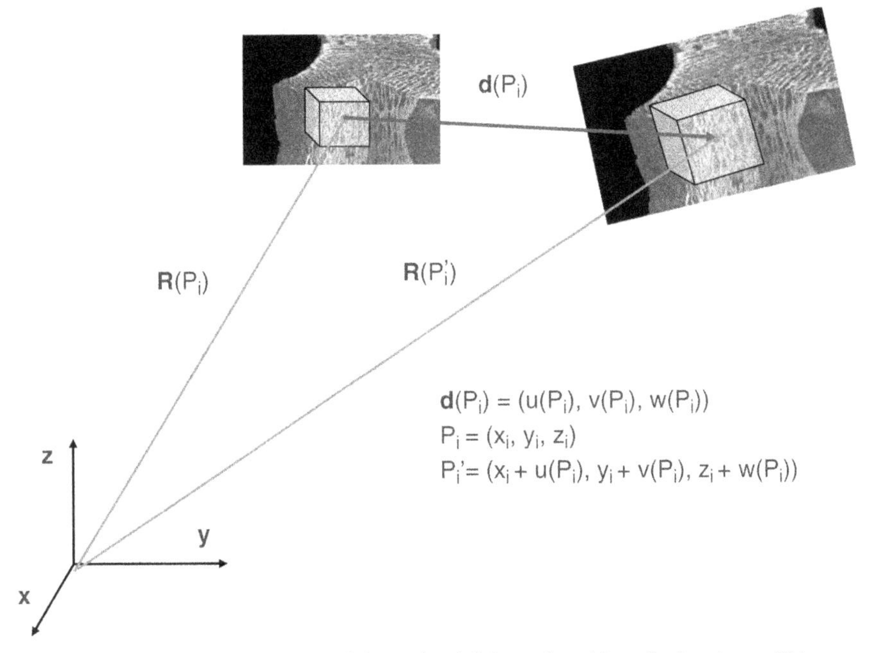

Fig. 8.4 Schematic demonstrating undeformed and deformed position of sub-volume. Object motions may result in translations, rotations and deformations

$$\begin{aligned}&(\gamma^Q,\eta^Q,\xi^Q) = (X^Q - X^P, Y^Q - Y^P, Z^Q - Z^P)\\ &\Rightarrow I(P) = I(\gamma^P,\eta^P,\xi^P) = I(0,0,0)\\ &\quad I(Q) = I(\gamma^P+\gamma^Q,\eta^P+\eta^Q,\xi^P+\xi^Q) = I(\gamma^Q,\eta^Q,\xi^Q)\end{aligned} \tag{8.2}$$

If the coordinates for Q are integer pixel values, then no interpolation of the image is required to perform the matching process.

Assuming that the intensity pattern recorded after deformation is related to the undeformed pattern through a continuous deformation field, subset-based coordinates for an arbitrary point can be written $\gamma = X - X^P$, $\eta = Y - Y^P$ and $\xi = Z - Z^P$. Thus, the displacement vector field for a general point in the sub-volume can be written in the form $\mathbf{d}(\gamma,\eta,\xi)$ and, the resulting intensity values at an arbitrary point q in the deformed image can be written:

$$\begin{aligned}I'(q) = I\big(&X^P+\gamma^Q+d_1(\gamma^Q,\eta^Q,\xi^Q), Y^P+\eta^Q+d_2(\gamma^Q,\eta^Q,\xi^Q),\\ &Z^P+\xi^Q+d_3(\gamma^Q,\eta^Q,\xi^Q)\big)\end{aligned}$$

$$\cong I\left(X^P + d_1(0,0,0) + \left(1 + \frac{\partial d_1}{\partial \gamma}(0,0,0)\right)\Delta\gamma + \frac{\partial d_1}{\partial \eta}(0,0,0)\,\Delta\eta + \frac{\partial d_1}{\partial \xi}(0,0,0)\,\Delta\xi,\right.$$

$$Y^P + d_2(0,0,0) + \frac{\partial d_2}{\partial \gamma}(0,0,0)\,\Delta\gamma + \left(1 + \frac{\partial d_2}{\partial \eta}(0,0,0)\right)\Delta\eta + \frac{\partial d_2}{\partial \xi}(0,0,0)\,\Delta\xi,$$

$$\left. Z^P + d_3(0,0.0) + \frac{\partial d_3}{\partial \gamma}(0,0,0)\,\Delta\gamma + \frac{\partial d_3}{\partial \eta}(0,0,0)\,\Delta\eta + \left(1 + \frac{\partial d_3}{\partial \xi}(0,0,0)\right)\Delta\xi\right)$$

$$= I(X^P, Y^P, Z^P, \Delta\gamma, \Delta\eta, \Delta\xi, B_1, B_2, B_3, B_4, B_5, B_6, B_7, B_8, B_9, B_{10}, B_{11}, B_{12})$$

where:

$$B_1 = d_1(0,0,0) \ ; \ B_2 = \tfrac{\partial d_1}{\partial \gamma}(0,0,0) \ ; \ B_3 = \tfrac{\partial d_1}{\partial \eta}(0,0,0) \ ; \ B_4 = \tfrac{\partial d_1}{\partial \xi}(0,0,0)$$

$$B_5 = d_2(0,0,0) \ ; \ B_6 = \tfrac{\partial d_2}{\partial \gamma}(0,0,0) \ ; \ B_7 = \tfrac{\partial d_2}{\partial \eta}(0,0,0) \ ; \ B_8 = \tfrac{\partial d_2}{\partial \xi}(0,0,0)$$

$$B_9 = d_3(0,0,0) \ ; \ B_{10} = \tfrac{\partial d_3}{\partial \gamma}(0,0,0) \ ; \ B_{11} = \tfrac{\partial d_3}{\partial \eta}(0,0,0) \ ; \ B_{12} = \tfrac{\partial d_3}{\partial \xi}(0,0,0) \quad (8.3)$$

As shown in Eq. (8.3), the relationship of points P and Q is approximated using a first order Taylor's series approximation, resulting in twelve subset-level parameters. To further simplify the expression, the form of the displacement function assumed within the subset is required. If a linear variation in displacement is assumed within the subset,[3] then:

$$\mathbf{d}(\gamma,\eta,\xi) = \{d_1(\gamma,\eta,\xi), d_2(\gamma,\eta,\xi), d_3(\gamma,\eta,\xi)\}$$
$$= \begin{Bmatrix} a_{10} + a_{11}\,\gamma + a_{12}\,\eta + a_{13}\,\xi \\ a_{20} + a_{21}\,\gamma + a_{22}\,\eta + a_{23}\,\xi \\ a_{30} + a_{31}\,\gamma + a_{32}\,\eta + a_{33}\,\xi \end{Bmatrix} \quad (8.4)$$

In this form, the terms in Eq. (8.4) are assumed to be constant within each subset, with the following physical interpretation:

- $\{a_{10}, a_{20}, a_{30}\} = \{B_1, B_5, B_9\}$ is the three-dimensional displacement of the subset center point due to the deformation process.
- $\frac{1}{2}(a_{21} - a_{12}) = \frac{1}{2}(B_6 - B_3)$ is an estimate for a local rotation component.
- $\frac{1}{2}(a_{12} + a_{21}) = \frac{1}{2}(B_6 + B_3)$ is an estimate for a local shear strain component.
- $\frac{1}{2}(a_{32} - a_{23}) = \frac{1}{2}(B_{11} - B_8)$ is an estimate for a local rotation component.

[3] Assuming each subset is sufficiently small, then the displacement gradients are approximately constant within a subset so that it undergoes rigid body rotation and uniform strain, resulting in a parallelogram shape.

- $\frac{1}{2}(a_{32}+a_{23})=\frac{1}{2}(B_{11}+B_8)$ is an estimate for a local shear strain component.
- $\frac{1}{2}(a_{13}-a_{31})=\frac{1}{2}(B_4-B_{10})$ is an estimate for a local rotation component.
- $\frac{1}{2}(a_{13}+a_{31})=\frac{1}{2}(B_4+B_{10})$ is an estimate for a local shear strain component.
- $a_{11}=B_2$ is an estimate for uniform subset strain ε_{xx}.
- $a_{22}=B_7$ is an estimate for uniform subset strain ε_{yy}.
- $a_{33}=B_{12}$ is an estimate for uniform subset strain ε_{zz}.

Similarly, if we define $\mathbf{d}(\gamma,\eta,\xi)$ as follows:

$$\mathbf{d}(\gamma,\eta,\xi)=\{d_1,d_2,d_3\}$$

where:

$$\begin{aligned}
d_1(\gamma,\eta,\xi)&=a_{10}+a_{11}\gamma+a_{12}\eta+a_{13}\xi+a_{14}\gamma\eta+a_{15}\eta\xi+a_{16}\xi\gamma+a_{17}\gamma^2\\
&\quad+a_{18}\eta^2+a_{19}\xi^2\\
d_2(\gamma,\eta,\xi)&=a_{20}+a_{21}\gamma+a_{22}\eta+a_{23}\xi+a_{24}\gamma\eta+a_{25}\eta\xi+a_{26}\xi\gamma+a_{27}\gamma^2\\
&\quad+a_{28}\eta^2+a_{29}\xi^2\\
d_3(\gamma,\eta,\xi)&=a_{30}+a_{31}\gamma+a_{32}\eta+a_{33}\xi+a_{34}\gamma\eta+a_{35}\eta\xi+a_{36}\xi\gamma+a_{37}\gamma^2\\
&\quad+a_{38}\eta^2+a_{39}\xi^2
\end{aligned} \tag{8.5}$$

then the quadratic form allows the sub-volume to undergo translation, rigid body rotation and non-uniform change in shape, allowing lines to deform into curved shapes.

If a form such as given in Eq. (8.5) is used to describe the local deformation experienced by a sub-volume, then the deformed intensity pattern can be written in functional form $I'(a_{ij},x,y,z)$ and the metric coefficient will be a function of the unknown deformation parameters, $\zeta(a_{ij})$. The optimization process is used to determine the parameters a_{ij} that provide the best match between the deformed and undeformed intensity patterns.

Two distinct disadvantages of a quadratic shape function in VDIC are noted. First, the requirement to solve for 30 parameters during the subset-level matching process defined in Eq. (8.1) increases computation time substantially. Second, the need to use larger sub-volumes so that the curvature parameters can be reasonably estimated for each subset will substantially decrease spatial resolution. For these reasons, it is common to assume either (a) local translations only or (b) locally linear transformation including strain and rotation, thereby minimizing the number of parameters.

8.5 Volumetric Image Reconstruction

Since general deformations of the sub-volumes will introduce non-integer displacements, improved accuracy when matching undeformed and deformed patterns requires that positions in the reference image be registered with deformed locations with sub-voxel accuracy.

To optimize the accuracy in this process, the discrete volumetric image data in each deformed image must be converted into a continuous functional representation, typically by use of a robust, three-dimensional interpolation scheme. A wide range of approaches have been used for volumetric interpolation. One approach that has been used successfully to interpolate CT images for VDIC analysis employs cubic Hermitian polynomials [26, 265]. The basis functions for each direction, β, in the 3D image have the form:

$$\begin{aligned} H_0 &= (2\beta+1)(\beta-1)^2 \\ H_1 &= \beta(\beta-1)^2 \\ H_2 &= \beta^2(3-2\beta) \\ H_3 &= \beta^2(\beta-1) \end{aligned} \tag{8.6}$$

where β is defined locally between voxel measurements on a normalized interval $[0,1]$. Thus, for a coordinate $(X_i+\gamma, Y_i+\eta, Z_i+\xi)$, where γ, η, ξ represent the sub-voxel coordinates in the range $[0,1]$, the Hermitian interpolated intensity value can be obtained from

$$I(X_i+\gamma, Y_i+\eta, Z_i+\xi) = \sum_{k=0}^{3}\sum_{j=0}^{3}\sum_{i=0}^{3} A_{ijk} H_i(\gamma) H_j(\eta) H_k(\xi) \tag{8.7}$$

The coefficients A_{ijk} are determined using the following information; (a) the measured three-dimensional intensity field at voxel locations, (b) estimates for the first derivatives at voxel locations and (c) selected second derivatives at each of the eight sub-voxel locations. A schematic of a central difference scheme used to estimate the first derivatives $\partial I/\partial\gamma$, $\partial I/\partial\eta$ and $\partial I/\partial\xi$ at voxel location (x_i, y_j, z_k) is shown in Fig. 8.5.

8.6 Case Study: Quantifying Micro-damage in Polymeric Foam Undergoing Uniaxial Compression

8.6.1 Background

It is well known that the effects of stress constraint can alter the mechanical behavior of materials. Since constraint is typically highest in the central sections of

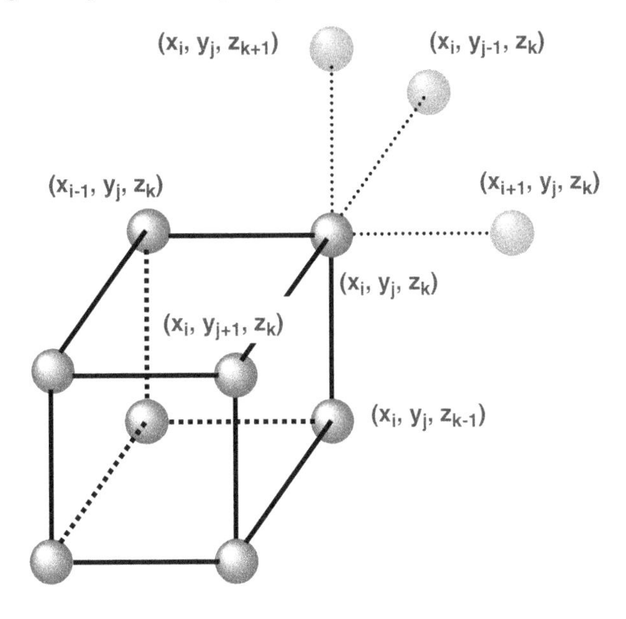

$$\partial I/\partial x\,(x_i, y_j, z_k) = [I\,(x_{i+1}, y_j, z_k) - I\,(x_{i-1}, y_j, z_k)]\,/\,2\Delta x$$
$$\partial I/\partial y\,(x_i, y_j, z_k) = [I\,(x_i, y_{j+1}, z_k) - I\,(x_i, y_{j-1}, z_k)]\,/\,2\Delta y$$
$$\partial I/\partial z\,(x_i, y_j, z_k) = [I\,(x_i, y_j, z_{k+1}) - I\,(x_i, y_j, z_{k-1})]\,/\,2\Delta z$$

Fig. 8.5 Graphical representation describing development of derivative expressions through finite difference

a specimen, the ability to track and quantify the extent of micro-damage within materials would allow for an increased understanding of material behavior under monotonic or cyclic loading.

Rigid polymeric foam materials are of interest since the complex closed-cell structure has mechanical properties that are being considered for bio-mechanical applications. Specifically, data has shown that the viscoelastic behavior and structural similarity of selected rigid foams to trabecular bone make it a viable option. Since polymer foams are "radioluscent", they can be used in micro-CT X-ray tomography systems to obtain high contrast, volumetric images that are appropriate for volumetric imaging using VDIC for post-processing. This study demonstrates the viability of VDIC in the analysis of the structural effects of mechanically induced internal damage in such foam materials.

8.6.2 Specimen and Experimental Considerations

All compression experiments were performed using a 500 N Enduratec Bose Elf 3200 mechanical loading system at a strain rate of 0.001 s^{-1} up to the experimental load of interest. Figure 8.6 shows the shape of the cylindrical closed-foam specimen.

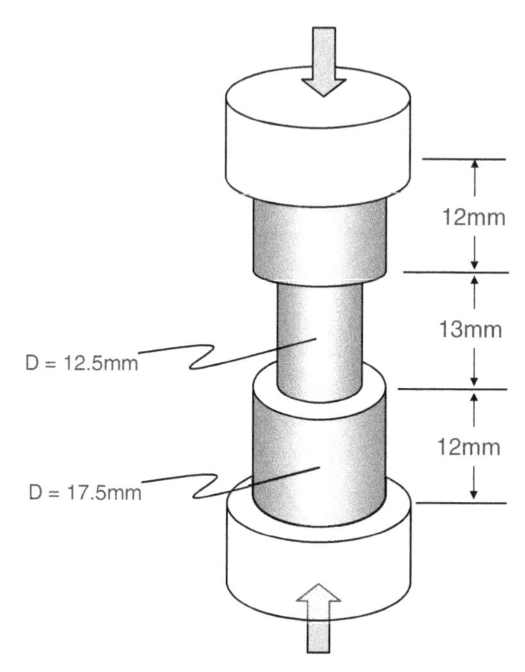

Fig. 8.6 Geometry of reduced diameter closed foam cylindrical specimen

Four cylindrical sections, approximately measuring 37 mm in length with a diameter of 17.5 mm, were removed from a single block of Sawbones 10 pcf rigid cellular polyurethane foam using a drill press and a coring tool. Each specimen was mounted in polycarbonate platens using pre-polymerized PMMA as the bonding agent. To ensure that the maximum load in the specimen was below the capacity of the Bose system, a lathe and a parting tool were used to reduce the diameter of the middle 13 mm in each sample to $\approx$12.5 mm. It is noted that considerable effort was taken during the entire specimen preparation process to ensure that the samples remained undamaged during this process.

Prior to performing the compression damage-evolution experiments, a stress–strain curve was generated by compressing a single sample at a strain rate of 0.001 s^{-1} to failure. Results show that the material response is nominally linear during the loading process with (a) E = 95 MPa, (b) yield stress of 1.78 MPa, (c) a maximum axial stress $\approx$1.85 MPa at a strain of 0.024 and (d) a failure strain $\approx$0.06. The maximum stress corresponds to a load of 230 N.

Prior to subjecting the remaining foam specimens to damage through cyclic compressive loading, each specimen was placed in the Bose system and compressively loaded to three different levels: (a) 10 N, (c) 50 N and (d) 100 N. While under load, each specimen is imaged in a micro-CT tomographic system. The CT-system parameters used during imaging are given in Table 8.1. It is noted that frame averaging is

Table 8.1 Micro-CT X-ray scan parameters

KVP	41
Micro-Amps	196
Exposure time	260 ms
Image binning	2 × 2
Frames averaged	4
Slice angle	0.48°/slice
Distance from X-ray source-to-object	150 mm

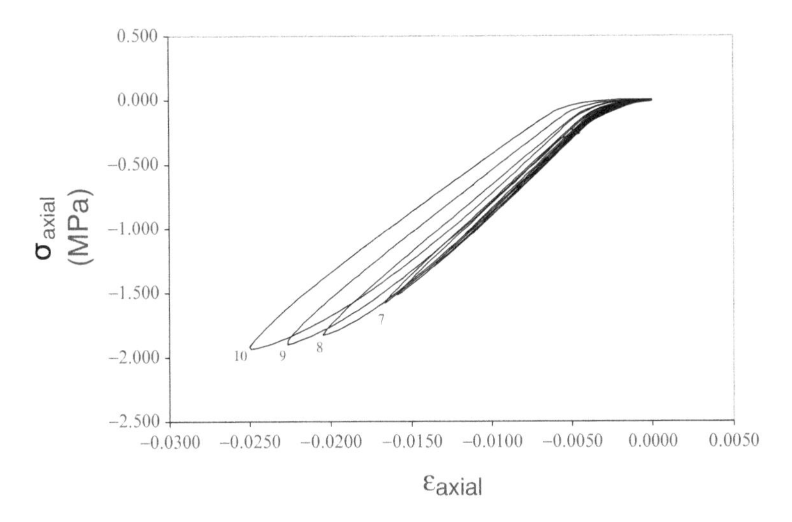

Fig. 8.7 Cyclic compressive stress–strain data for closed cell polymer foam specimen. (Courtesy Professor B.K. Bay)

common in CT-scanning to reduce voxel level noise and increase the accuracy of displacement and strain measurements when performing VDIC and matching sub-volumes.

To induce micro-damage within each cylinder, each sample was subjected to ten cycles of compressive loading modeled after accepted damaging protocol. The sequence consists of:

- Constant uniaxial compression strain rate at 0.001 s^{-1}
- Incremental increases in maximum compressive displacement from 0.17 to 0.92 mm over 10 cycles, which corresponds to global maximum strain values 0.005 – 0.025 over 10 cycles

Figure 8.7 shows examples of the applied engineering stress–strain cyclic loading sequence for each specimen.

8.6.3 CT Imaging and VDIC Parameters

After cyclic loading is completed, each sample is compressively loaded to 10, 50 and 100 N and CT scanned at fixed load using the same micro-CT system while retaining all imaging parameters in Table 8.1.

To perform VDIC, the first step is to define the outer surfaces of the specimen volume. Using the CT scan images to identify specimen boundaries, the finite-element analysis package PATRAN is used to construct a region of interest (ROI) for detailed motion measurements. VDIC is performed using the linear transformation in Eq. (8.3). Designated CDI and GDS, these custom-designed computer codes implement algorithms based on concepts outlined in Sections 8.3–8.5. Displacement results are obtained using sub-volumes with $30 \times 30 \times 30$ voxels.

Two sets of results are presented. For the pre-cyclic loading case (e.g., undamaged), results are obtained by comparing the 50 and 100 N images with the 10 N image from the pre-cyclic loading set. To quantify the effect of cyclic loading on the foam specimens, the 50 and 100 N images for specimens that were subjected to ten cycles of compressive loading are compared with the 10 N image from the post-cyclic loading set, providing a direct measure of the increase in damage that occurs when cyclically compressing the foam at higher loads.

For each case outlined above, VDIC was performed to obtain the full-field, volumetric 3D displacement field. Once the 3D displacement field is determined within the specimen volume, the equations in Appendix A and a volumetric extension of the local fitting shown in Fig. D.1 are used to define the volumetric strain field components, $\varepsilon_{ij}(x,y,z)$.

8.6.4 Experimental Results

Figure 8.8 presents the maximum principal strain fields in the center plane of the specimen for two specimens and two nominal load levels.

Though not clearly discernible in the specimen shape, the principal strain fields show that:

(a) Localization initiated within the central region in both damaged and undamaged specimens
(b) Damaged specimens exhibited much higher principal strains and hence an accumulation of larger permanent strains, than seen in the undamaged material
(c) Consistent with (b), damaged specimens also exhibited a progressive reduction in the initial modulus during the latter cycles of compression

It is important to emphasize that the measured strain localization in Fig. 8.8 is indicative of "micro-damage in the closed cell strut structures", rather than macro-scale yielding, large deformation of the foam and complete fracture of the foam structure. Micro-scale damage was achieved using the specified maximum

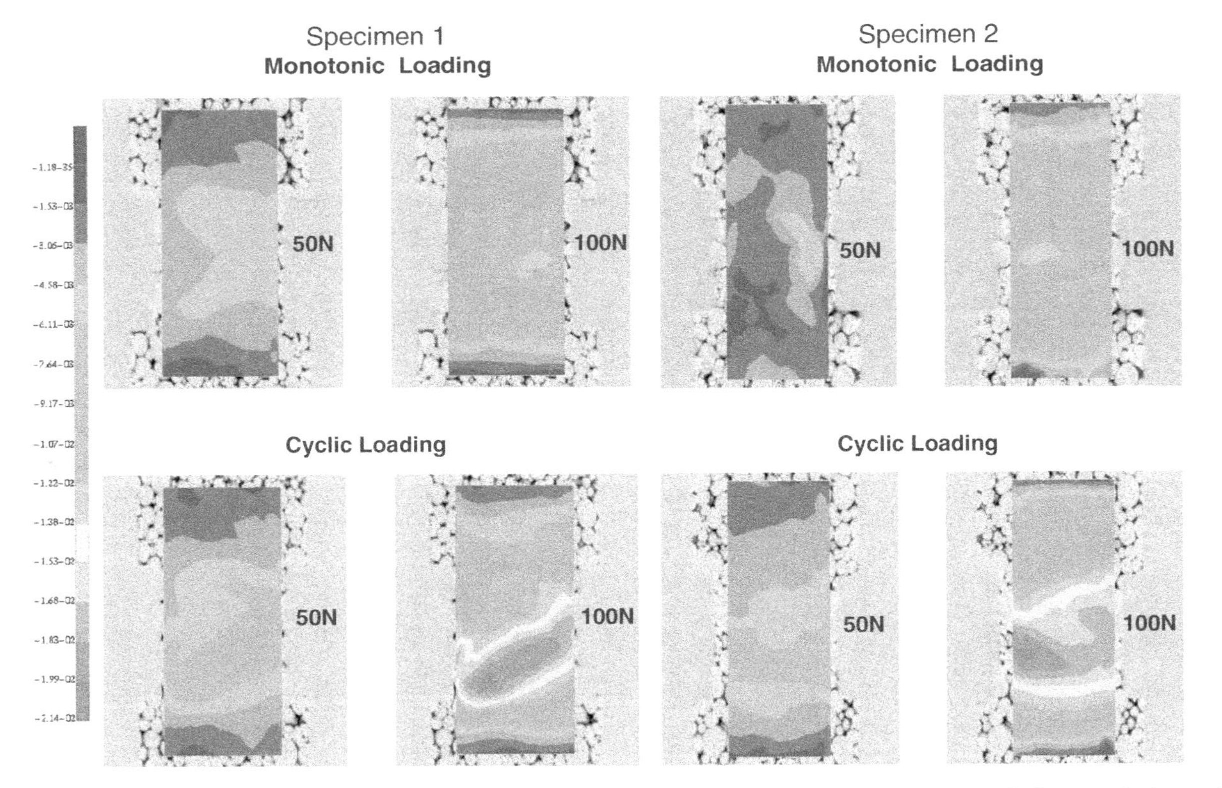

Fig. 8.8 Principal strain distribution in both undamaged and damaged polymer foam specimens. Micro-damage processes in foam results in consistent spatial localization. (Courtesy Professor B.K. Bay)

displacements during cyclic compression; larger maxima in the displacements could lead to overall yielding and overall foam deformation would dominate.

It is also noted that, during the cyclic compressive loading sequence, no cracking was heard for either sample, indicating that the samples' strain magnitudes were kept below their respective yield strains and that macro-scale damage to the cell struts was avoided. Additional visual investigation confirmed this inference.

Finally, as shown in Fig. 8.8, there are strong similarities in the relative distribution of strain between load steps, particularly in the post-damage cases. Comparison of the post-damage 50 and 100 N load steps for each sample illustrate the relatively consistent strain fields, indicating that the same regions account for the largest strain magnitudes. This would be expected under the condition of induced micro-damage, given that each micro-damaged strut would exhibit a larger strain response to the applied stress.

8.6.5 CT-Imaging Discussion

In this study, no effort was made to (a) develop or calibrate a CT-imaging model, (b) correct for spatial distortions during imaging, (c) determine the magnitude of time-varying distortions (e.g., drift) on the voxel-level intensity data or (d) assess the magnitude and form of individual voxel intensity noise. Rather, the intensity data obtained when integrating all 750 slices to construct the measurement volume is assumed to be located at the position defined by the hardware/software used to process the raw CT-scans.

This use of such $4\times$ averaged CT-data is entirely appropriate for situations where the data can be use to quantify the presence (or absence) of differences in material response. Here, the approach was used successfully to identify the clear differences in strain magnitude and distribution that occurs when a material is damaged, with the success of the method attributed to the large strain in the damaged specimen (1.32%) that can be reasonably estimated in this manner.

Detailed calibration and correction studies would provide estimates for the magnitude of errors expected in the micro-CT system. Once such studies are completed, error estimates could be obtained for the micro-CT displacement data and used to assess the range of applied strains for which the method is appropriate to give accurate results.

Chapter 9
Error Estimation in Stereo-vision

Error estimation in stereo-vision systems has been an area of research for many years [14, 59, 75, 81, 104, 124, 183, 228, 335]. The enclosed developments, based on the fundamental equations given in Chapters 3 and 4, are meant to provide an introduction to their application for estimation of both 3D position bias and variability. Consider a two-camera stereo vision system. As shown in Fig. 4.7, one point in space is in direct correspondence with two sensor locations, one in each camera system. Since the transformations between the world system and the camera 1 are known after calibration has been completed, it is assumed that the world coordinate system is transferred to the camera 1 pinhole location. Thus, if the common point is **P**, after the transformation is completed (a) the 3D coordinates are defined in the camera 1 system (e.g., see pinhole system in Fig. 3.1) and (b) the transformation given in Eq. (3.9) for the camera 1 view is greatly simplified, since $[\mathbf{R}]_1 = [\mathbf{I}]$ and $\{\mathbf{t}\}_1 = \{\mathbf{0}\}$.

9.1 Sub-optimal Position Estimator

Using terminology for the camera parameters shown in Eq. (3.33), we define the camera parameters for each view by $\boldsymbol{\beta_i}$, with $i = 1,2$ as follows:

$$\begin{aligned} \boldsymbol{\beta_1} &= \{\mathbf{f_{x1}}, \mathbf{f_{y1}}, \mathbf{c_{x1}}, \mathbf{c_{y1}}, \mathbf{f_{s1}}\} \\ \boldsymbol{\beta_2} &= \{\mathbf{f_{x2}}, \mathbf{f_{y2}}, \mathbf{c_{x2}}, \mathbf{c_{y2}}, \mathbf{f_{s2}}; \\ &\quad \mathbf{R_{11}}, \mathbf{R_{12}}, \mathbf{R_{13}}, \mathbf{R_{21}}, \mathbf{R_{22}}, \mathbf{R_{23}}, \mathbf{R_{31}}, \mathbf{R_{32}}, \mathbf{R_{33}};\ \mathbf{t_x}, \mathbf{t_y}, \mathbf{t_z}\} \end{aligned} \tag{9.1}$$

where $[\mathbf{R}]_1$ and $\{\mathbf{t}\}_1$ are omitted from $\boldsymbol{\beta_1}$ (since the camera 1 system is assumed to be the reference), and $[\mathbf{R}] = [\mathbf{R}]_2$, $\{\mathbf{t}\} = \{\mathbf{t}\}_2$ in $\boldsymbol{\beta_2}$.

Letting (x_{s1}, y_{s1}) and (x_{s2}, y_{s2}) be the sensor coordinates for the two camera views for the common point **P**, and assuming that both cameras have skew, using Eq. (3.9) we can write the following matrix relationship for both views:

M.A. Sutton et al., *Image Correlation for Shape, Motion and Deformation Measurements: Basic Concepts, Theory and Applications*, DOI: 10.1007/978-0-387-78747-3_9,

$$[\mathbf{M}] \bullet \{\mathbf{y}\} = \{\mathbf{b}\}$$

where

$$\mathbf{M} = \begin{bmatrix} f_{x1} & f_{s1} & c_{x1} - x_{s1} \\ 0 & f_{y1} & c_{y1} - y_{s1} \\ R_{11} f_{x2} + R_{21} f_{s2} + R_{31}(c_{x2} - x_{s2}) & R_{12} f_{x2} + R_{22} f_{s2} + R_{32}(c_{x2} - x_{s2}) & R_{13} f_{x2} + R_{23} f_{s2} + R_{33}(c_{x2} - x_{s2}) \\ R_{21} f_{y2} + R_{31}(c_{y2} - y_{s2}) & R_{22} f_{y2} + R_{32}(c_{y2} - y_{s2}) & R_{23} f_{y2} + R_{33}(c_{y2} - y_{s2}) \end{bmatrix}$$

$$\mathbf{y} = \begin{bmatrix} X_p \\ Y_p \\ Z_p \end{bmatrix}$$

$$\mathbf{b} = \begin{bmatrix} 0 \\ 0 \\ -(t_x f_{x2} + t_y f_{s2} + t_z(c_{x2} - x_{s2})) \\ -(t_y f_{y2} + t_z(c_{y2} - y_{s2})) \end{bmatrix} \tag{9.2}$$

Assuming that β_1 and β_2 are known through the stereo-calibration process (e.g., see Section 3.2 or 4.2), and that cross-camera image correlation has been performed to optimally determine the corresponding sensor positions, then Eq. (9.2) provides an over-determined set of equations to solve for the common 3D position. Least squares can be performed directly on Eq. (9.2) to solve for the 3D positions using the measured sensor positions

$$\{\mathbf{y_0}\} = \left([\mathbf{M}]^{\mathbf{T}}[\mathbf{M}]\right)^{-1}[\mathbf{M}^{\mathbf{T}}]\{\mathbf{b}\} \tag{9.3}$$

where $[\mathbf{M}]$ and $\{\mathbf{b}\}$ are computed using the as-measured sensor positions and the expectations for each camera parameter obtained via the calibration process. As discussed in the following paragraph, the value obtained generally is a sub-optimal estimate for the 3D position.

9.2 Optimal Position Estimator

As shown in Sections 5.6 and 5.7 digital image correlation matching will result in both bias and random variability in the sensor plane measurements. Thus, a more robust approach for determining the optimal estimates would employ a least squares approach to obtain $\{\mathbf{y}\}_{\text{opt}}$ that minimizes the least square variation between the measured sensor plane positions and model predictions using an inverted form of Eq. (9.2). The corresponding functional form may be written in the following form:

$$\begin{aligned}\chi^2 = & \left[x_{s1}^{\text{measured}} - x_{s1}^{\text{model}}(X_p, Y_p, Z_p; \beta_1, \beta_2)\right]^2 \\ & + \left[y_{s1}^{\text{measured}} - y_{s1}^{\text{model}}(X_p, Y_p, Z_p; \beta_1, \beta_2)\right]^2 \\ & + \left[x_{s2}^{\text{measured}} - x_{s2}^{\text{model}}(X_p, Y_p, Z_p; \beta_1, \beta_2)\right]^2 \\ & + \left[y_{s2}^{\text{measured}} - y_{s2}^{\text{model}}(X_p, Y_p, Z_p; \beta_1, \beta_2)\right]^2\end{aligned}$$

where (9.4)

$$\begin{bmatrix} x_{s1}^{\text{model}}(X_p, Y_p, Z_p; \beta_1, \beta_2) \\ y_{s1}^{\text{model}}(X_p, Y_p, Z_p; \beta_1, \beta_2) \\ x_{s2}^{\text{model}}(X_p, Y_p, Z_p; \beta_1, \beta_2) \\ y_{s2}^{\text{model}}(X_p, Y_p, Z_p; \beta_1, \beta_2) \end{bmatrix} = \begin{bmatrix} f_{x1} & f_{s1} & 0 & 0 & c_{x1} \\ 0 & f_{y1} & 0 & 0 & c_{y1} \\ 0 & 0 & f_{x2} & f_{s2} & c_{x2} \\ 0 & 0 & 0 & f_{y2} & c_{y2} \end{bmatrix} \begin{bmatrix} \frac{X_p}{Z_p} \\ \frac{Y_p}{Z_p} \\ \frac{R_{11} X_p + R_{12} Y_p + R_{13} Z_p + t_x}{R_{31} X_p + R_{32} Y_p + R_{33} Z_p + t_z} \\ \frac{R_{21} X_p + R_{22} Y_p + R_{23} Z_p + t_y}{R_{31} X_p + R_{32} Y_p + R_{33} Z_p + t_z} \\ 1 \end{bmatrix}$$

The concepts described in Appendix D may be used to develop a Levenberg–Marquardt algorithm to iteratively solve the nonlinear least square function in Eq. (9.4). To initialize the search process, a preliminary estimate for the 3D position is needed. An obvious choice is the "sub-optimal" solution obtained from Eq. (9.3). The right hand side vector in Eq. (9.4) is initialized using this solution, designated $\{\mathbf{y_0}\}$, and the iterative search process is performed. When performing the iteration process, the mean values are used for the camera parameters for all calculations. The resulting least square solution provides a reasonable approximation for the expectation in the sensor plane positions, $E(x_{s1})$, $E(y_{s1})$, $E(x_{s2})$ and $E(y_{s2})$. Hence, the optimal $\{\mathbf{y}\}$ is an estimate for the expectation in the 3D position, $\{E(X_p), E(Y_p), E(Z_p)\}$.

With regard to the relationship between Eqs. (9.3) and (9.4), the following points are noted. First, if the measured sensor positions input into Eq. (9.3) are the true expectations, then the least square result corresponds to the optimal solution obtained in Eq. (9.4). Second, if the measured sensor positions do not correspond to the expectations, then Eq. (9.4) provides a method that (a) iteratively estimates an optimal solution for the 3D position while also (b) estimating the expectations for the measured sensor positions.

9.3 Variance in 3D Position

Equation (9.2) can be written as a general nonlinear equation $\mathbf{y} = \mathbf{g}(\mathbf{X}, \boldsymbol{\Gamma})$, where $\{\mathbf{X}\}$ contains sensor plane position $(x_{s1}, y_{s1}, x_{s2}, y_{s2})$, and $Var(\mathbf{X})$ is the covariance matrix (see Appendix E) of $\{\mathbf{X}\}$ which is obtained from image matching.

The variance of the 3D position, $Var(\mathbf{y})$, is determined using Eq. (E.14) to give:

$$Var(\mathbf{y}) \approx [\mathbf{J}] \bullet Var(\mathbf{X}) \bullet [\mathbf{J}]^{\mathrm{T}} \tag{9.5}$$

where $[\mathbf{J}]$ is the Jacobian of the transformation g(.) with respect to $\{\mathbf{X}\}$ (see Eq. (E.14)). It is noted that $[\mathbf{J}]$ is evaluated using the $E(\mathbf{X})$ and $E(\boldsymbol{\Gamma})$.

9.4 Discussion

As noted previously, throughout the estimation process for $E(\mathbf{y})$ and $Var(\mathbf{y})$, the expectation in the camera parameters was used. This is consistent with the first order estimation process outlined in Appendix E, and hence does not use the variability in camera parameters.

This approach is advantageous since the camera parameters are obtained from calibration through a non-linear solution process involving the sensor positions (see Section 3.2). Since the sensor plane positions used in calibration are obtained from a series of reference grid images, the estimated camera parameter variances (obtained from calibration) are a function of both the expectation and variance of sensor plane positions and the reference grid and its positions.

Therefore, if the computed variances in the camera parameters are used to predict the variance in 3D positions, the result is related to the reference grid positions during calibration as well as the robustness of the reference grid extraction process. By utilizing only the expectations in the camera parameters when determining 3D position variance, the effect of the grid motion and extraction process is minimized.

Chapter 10
Practical Considerations for Accurate Measurements with DIC

As shown in the several 2D examples in Chapter 6, the 3D examples in Chapter 7 and the volumetric imaging study in Chapter 8, digital image correlation has been used to make accurate measurements in a wide range of physically relevant experiments, ranging from the micro-scale to structural scale.

As with any measurement method, effective use in laboratory or field conditions requires that the experimentalist exercise appropriate judgment when:

- Selecting lenses and other optical components for specific experiments
- Identifying and employing appropriate digital cameras
- Configuring and locating each component in the imaging system
- Arranging the lighting components
- Determining the appropriate patterning approach and speckle size
- Choosing appropriate exposure time

In Section 10.1, practical suggestions regarding general imaging considerations are provided. These include (a) depth of field and field of view, (b) image artifacts that may affect measurement accuracy, (c) subset patterning and (d) exposure time.

In Section 10.2, issues related to 2D-DIC measurements that are discussed include (e) out-of-plane motion (f) high magnification and depth of field and (g) measurement accuracy.

In Section 10.3, issues related to 3D-DIC measurements that are discussed include (h) camera selection, (i) system configuration and specimen positioning, (j) calibration and (k) measurement accuracy.

10.1 General Imaging Considerations

10.1.1 Depth of Field and Field of View

As discussed in Section 2.1.1.1, equations exist to quantify the depth of field (Eq. (2.13)) and field of view (Eq. (2.16)) for an imaging system in terms of key

M.A. Sutton et al., *Image Correlation for Shape, Motion and Deformation Measurements: Basic Concepts, Theory and Applications,* DOI: 10.1007/978-0-387-78747-3_10,

optical system parameters such as the lens focal length, numerical aperture, focusing factor, image format and image distance. The following examples describe how to determine the DOF and FOV for specific applications.

10.1.1.1 Example

Consider a 35 mm camera with a standard lens having a focal length of $\overline{f} = 50$ mm and an entrance pupil with diameter of 6.25 mm. The dimensions of the 35 mm image format are $v = 24$ mm (vertically) by $h = 36$ mm (horizontal), giving a diagonal $d \approx 43.3$ mm.

The viewing angles and other imaging parameters are as follows:

- Horizontally, $\alpha_h = 2 \arctan \frac{h}{2\overline{f}} = 39.6^\circ$
- Vertically, $\alpha_v = 2 \arctan \frac{v}{2\overline{f}} = 27.0^\circ$
- Diagonally, $\alpha_d = 2 \arctan \frac{d}{2\overline{f}} = 46.7^\circ$
- $\mathcal{N} = 8$

The field of view (in meters) at a given camera-to-subject distance D can be derived from the angle of view or directly from the focal length and the image sensor format $(h \times v)$:

- Horizontal FOV: $\text{FOV}_h = Dh/\overline{f} = 2D\tan(\alpha_h/2)$
- Vertical FOV: $\text{FOV}_v = Dv/\overline{f} = 2D\tan(\alpha_v/2)$

Assuming that the circle of confusion is limited to 1/10th of the size of an image plane sensing element, D_s, in pixel units we have:

- Nearest location of DOF:

$$D_N = \frac{\frac{10\overline{f}^2 D}{\mathcal{N} D_s}}{\frac{10\overline{f}^2 D}{\mathcal{N} D_s} + D}$$

- Farthest location of DOF:

$$D_F = \frac{\frac{10\overline{f}^2 D}{\mathcal{N} D_s}}{\frac{10\overline{f}^2 D}{\mathcal{N} D_s} - D}$$

For a camera-to-subject distance, $s = D = 1$ m and a pixel size in the sensor plane, $D_s = 6.25 \times 10^{-6}$ m, the field of view and depths of field are as follows:

- Horizontally: 72 cm
- Vertically: 48 cm
- Diagonally: 86 cm
- Closest distance where object is still in focus: 0.998 m
- Furthest distance where object is still in focus: 1.002 m

Thus, an object with physical size of 720 by 480 mm can be imaged with magnification of $\simeq$ 1/20th. Furthermore, the object can move towards and away from the camera by 2 mm and still have images that appear well focused, with image motions less than 1/10th of the 6.25 μm sensing elements.

10.1.2 Image Artifacts

As shown in Fig. 6.23, recorded images may contain unwanted artifacts. Here, the artifacts are due to water spots, dust or imperfections in a component within the imaging elements. Most often, well-focused spots such as these are due to dust or other contaminants on the glass elements in front of the sensor plane. Since the image of such artifacts will remain at a fixed location in the sensor plane, motions of the object estimated by using image subsets containing "fixed points" will be shifted (biased) due to their presence.

To identify such artifacts in an image, the following approach is recommended.

I. Defocus object, resulting in a uniform gray background intensity.
   (a) Contaminants on the glass elements near sensor plane will remain visible.

II. If lens is attached using a C-mount, hold the lens in place and unscrew camera body up to 90°.
   (a) Fixed points located on image-path elements within the camera will remain in the same image position as the camera is rotated.
   (b) Image positions of fixed points located on the lens will rotate as the camera is rotated.

III. If the lens is attached to the camera body with a bayonet mount, carefully loosen bayonet mount and rotate camera body slightly.
   (a) Fixed points located on image-path elements within the camera again will remain in the same image position as the camera is rotated.
   (a) Image positions of fixed points located on the lens will rotate as the camera is rotated.

If the source of the fixed points is located within the camera, then it is most likely due to contaminants on the glass cover (or IR filter) of the sensor plane and careful cleaning of the glass is required. If the source of fixed points is located in the lens train, the lenses should be removed and the front of the camera body capped to protect the sensor plane glass. Again, careful cleaning of the lens elements is recommended to remove contaminants.

An additional type of "fixed point artifact" occurs when a pixel or a cluster of pixels do not respond to incident light, typically recording a 0 value for the intensity. Known as "dead pixels", these artifacts are readily identified by imaging a gray card, recording N images and averaging the images to obtain a "mean" image. By searching through the image, either visually or via an array search algorithm, the locations of all dead pixels can be determined.

If the camera manufacturer has not excluded these pixels via on-board camera processing prior to displaying their values, it is recommended that all such pixels be excluded from the image correlation process; inclusion of such points when performing image matching will introduce local measurement errors. Exclusion of dead pixels can be achieved in a variety of ways. A common approach is to maintain a database containing the dead pixel locations, using this information to exclude intensity comparisons at these locations in any subset that contains the dead pixel.

Finally, image artifacts due to local reflections on the object surface should be mentioned. In general, reflective locations on an object have image points that either have minimal intensity or are fully saturated. The presence of saturation is readily observed as outliers at maximum intensity in an image histogram, such as the one in Fig. 10.1 for the foam specimen used in the inverse analyses in Section 6.5.

Here, the saturated pixels due to reflections will be to the far right side of the histogram. Since specimen deformations and motions can cause sharp changes in image intensity at these locations, measurement errors will occur due to bias in the image correlation process.

As such, it is always recommended that experiments be configured to minimize the potential for reflections. Since reflections are generally introduced by use of glossy paint or the presence of a reflective surface on the object that has not been coated, the use of flat paints and/or anti-reflection coatings is recommended. For situations where such methods cannot be used, specimen lighting should be arranged

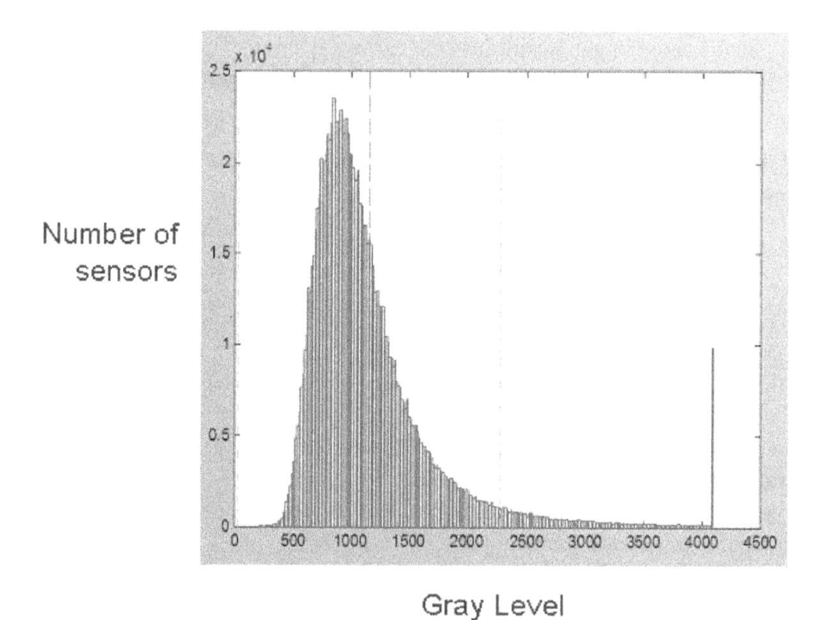

Fig. 10.1 Histogram for polymer foam image

so that local bright spots in the image due to reflections are minimized. If reflections are identified in images, eliminating these regions from the correlation process is the prudent choice whenever possible.

10.1.3 Subset Patterning

10.1.3.1 Methods to Estimate Appropriate Pattern Density for Accurate Motion Measurements

As noted in Section 5.1, accurate reconstruction of the underlying continuous intensity pattern requires accurate identification of features in a series of deformed images, with speckle patterns having a distinct advantage relative to periodic grid structures by having variations in contrast along all directions. The derivation of confidence margins for displacement measurements in Section 5.6 indicates that in order to minimize the noise in displacement measurements, it is important to maximize the term $\sum(\nabla F)^2$. Indeed, the covariance matrix in Eq. (5.98) can be used to rapidly assess the suitability of a speckle pattern for image correlation and to determine the required subset size to achieve a desired confidence margin for the displacement estimates for a given level of gray value noise. However, this simple approach does not always work in practice. The maximization of gradients dictates the use of the smallest features to obtain as many edges as possible in a small window. There is, however, a practical limit, and if the feature size is too small, aliasing will occur and severely bias displacement estimates. To obtain high levels of subpixel accuracy when performing pattern matching, it is therefore required to over-sample the features in each speckle pattern. Obviously, there is a need to obtain an estimate for the speckle size so that the level of over-sampling can be determined and optimized.

One approach for estimating the average feature size that must be over-sampled is to employ a form of *normalized autocorrelation*, using a definition such as the following:

$$A = \frac{\int_{-\infty}^{\infty} I(x)\,I(x-u)dx}{\int_{-\infty}^{\infty} [I(x)]^2\,dx} \cong \frac{\sum\limits_{i=1}^{M} I(x_i)\,I(x_i-u)}{\sum\limits_{i=1}^{M} [I(x_i)]^2} \tag{10.1}$$

Figure 10.2 shows a graphical representation for the normalized autocorrelation function for a one-dimensional intensity pattern, where the summation is performed over a pre-defined discrete sampling set for the intensity pattern.

As shown in Fig. 10.2, an estimate for the average feature size, FS, (in pixels), is oftentimes associated with the width of the autocorrelation function at a pre-specified

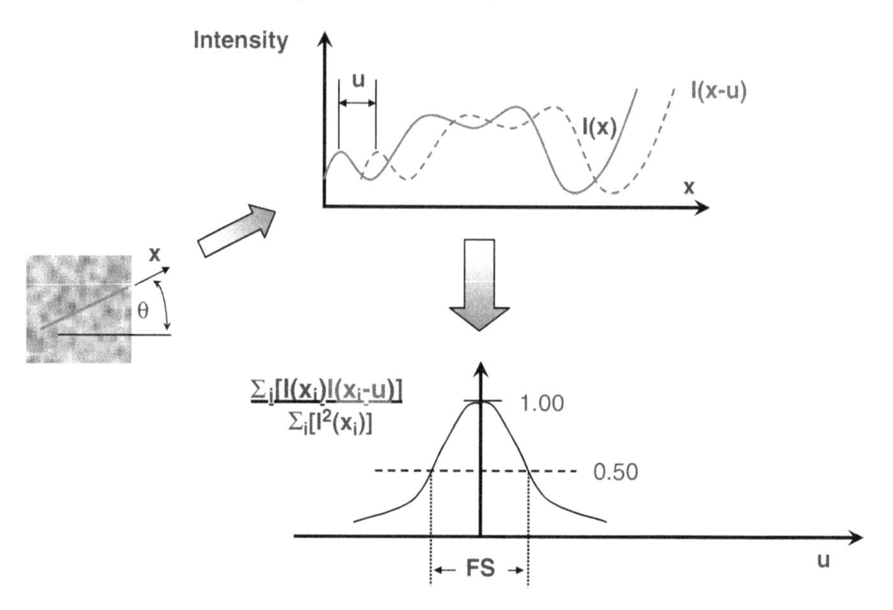

Fig. 10.2 Graphical representation of the normalized autocorrelation function for a one-dimensional intensity pattern. Width of autocorrelation peak is a measure of the average speckle size along θ direction

value, typically when $A = 0.500$. Here, if the autocorrelation process indicates that $1 < FS < 3$, then there is minimal over-sampling of the pattern along the direction being shown. In such cases, the displacement values obtained when comparing subsets will likely suffer from localized bias due to aliasing and hence there is increased probability of higher errors in the measurements. To increase accuracy in the measurements, either a higher magnification with the existing pattern or a new pattern with larger features at the same magnification would generally be recommended.

If $3 \leq FS \leq 6$, then there is a reasonable amount of over-sampling in the pattern and it is appropriate for accurate reconstruction with suitable interpolation filters. If $FS > 6$, then the pattern is well over-sampled.[1] In this case, a reduction in feature size may be needed if increased spatial resolution is required for the problem of interest, e.g., the determination of the strain field in regions of stress concentration.

Figure 10.3a–c show three examples representing near-optimal sampling, slight over-sampling and under-sampling, respectively, of a fixed portion of the object speckle pattern defined by the specified subset region.

Analysis of the sampling process shown in Fig. 10.3 indicates (a) the selected subset contained portions of 20 speckles and (b) the speckles are sampled by

[1] Image feature size is important since sensor arrays have a fixed size (e.g. $1{,}024 \times 1{,}024$) and oversampling is essential for accurate reconstruction. Thus, control of feature size during pattern application is necessary so that excessive over-sampling is not required, which will result in a requirement for larger subset sizes during matching and a reduction in the spatial resolution that is achievable.

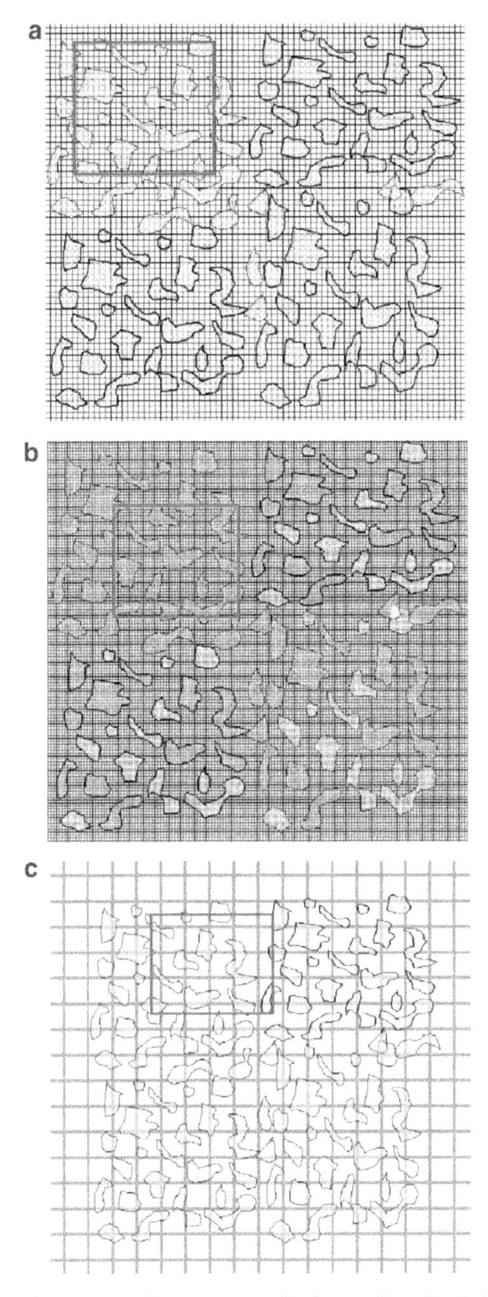

Fig. 10.3 Three examples representing **a** near-optimal sampling, **b** slight over-sampling and **c** under-sampling, of a fixed portion of the object speckle pattern defined by the specified subset region

Table 10.1 Basic sampling data for the patterns shown in Fig. 10.3a–c

Figure	Subset sampling (pixels)^2	Object speckles sampled	Samples per speckle	Characteristic sampling length (pixels)	Characteristic of sampling	Remarks
10.3a	23×23	20	10.5	3.25	Near-optimal	Subset and speckle sizes are appropriate
10.3b	45×45	20	42	6.5	Slightly over-sampled	De-magnify image, consider reducing subset size
10.3c	5×5	20	0.5	0.7	Under-sampled	Magnify image and increase subset size

approximately (b-1) 210 pixels for the optimally sampled image, (b-2) 840 pixels for the over-sampled image and (b-3) 10 pixels for the under-sampled image. Assuming a spatially isotropic pattern distribution, the corresponding characteristic sampling lengths are 3.2, 6.5 and 0.7 pixels for the optimally sampled, slightly over-sampled and severely under-sampled patterns, respectively. Table 10.1 presents the basic sampling data for the patterns shown in Figure 10.3a–c.

Of particular note is the under-sampled pattern shown in Fig. 10.3c. In this case, the number of pixel samples for the speckles exceeds the number of pixels in the subset. Such situations are only possible when the speckles are too small relative to the sampling medium so that a single pixel is sampling more than one speckle at the same time. For CCD or CMOS sensors that integrate light over the entire area of each pixel, such severely under-sampled images will result in low contrast. More problematic, however, is the case where aliasing is introduced by a sensor with an active area smaller than the area of the pixel, i.e., a sensor with a low fill-factor. In either situation, it will be difficult, if not impossible, to accurately identify the position of a deformed subset during image matching using digital image correlation. If the speckle size cannot be altered, increased magnification of the image is recommended, and the use of interpolation filters with low phase errors is necessary to minimize measurement bias.

Engineering Approach Relating Speckle Size to Error Estimates 2D Computer Vision

Based on the previous discussion, it is possible to provide an engineering approach for estimating appropriate parameters for image matching in 2D computer vision. The following are assumed to be the primary parameters for this approach.

Field of View (FOV): L by L (m^2)
Recording Resolution: N by N (pixels^2)
Magnification Factor: M_T (pixels/m)

Subset Size: m by m (pixels^2)
Image displacement accuracy: β_I (pixels)
Image speckle dimension: η_I (pixels)
Object speckle dimension: η_O (m)
Object dimension: l_O (m)

With these definitions, it is possible to develop various useful factors. For example, the object displacement accuracy, β_O, can be defined as follows:

$$\beta_O = \frac{L(\mathrm{m})}{N(\mathrm{pixels})} \cdot \beta_I(\mathrm{pixels}) = \frac{\beta_\mathrm{I}}{\mathrm{M_T}}(\mathrm{pixels})$$

With these parameters set, two simple precepts (rules of thumb) provide the basis for the approach. The first precept specifies a minimum level of image plane oversampling for accurate matching.

Rule 1 *For accurate matching, image plane speckles should be sampled by at least a 3 by 3 pixel array to ensure minimal oversampling and reasonable intensity pattern reconstruction via interpolation.*

The second precept specifies a minimum number of speckles in a subset to maintain reasonable accuracy in the matching process.

Rule 2 *Each image plane subset of size N by N should contain at least 3 by 3 speckles to ensure reasonable accuracy and isotropy in the subset matching process.*

First Example

A displacement accuracy of 10^{-6} m is required for a specific experiment, with a recording resolution of $1{,}000 \times 1{,}000$ pixels^2 and an image displacement accuracy, $\beta_I = 0.01$ pixels. (a) What is the required maximum field of view? (b) What is the minimum speckle size on the object? (c) What is the minimum possible subset size?

A solution to the problem proceeds as follows. For part (a), we use the object displacement accuracy equation to define the parameter

$$\beta_O = \tfrac{L}{N} \cdot \beta_I$$

$$\Rightarrow 10^{-6} = \frac{L}{1000} \cdot 0.01$$

$$\Rightarrow L = 0.10\ \mathrm{m}$$

Thus, the field of view should be smaller than 0.10×0.10 m to maintain the required object-based displacement accuracy.

For part (b), the minimum object speckle size is obtained by using Rule 1 for the minimum allowable image plane speckle size and the magnification factor.

$$\eta_O = \eta_I \left(\tfrac{L}{N}\right) \geq 3\ \mathrm{pixels} \left(\tfrac{0.10\ \mathrm{m}}{1000}\right)$$

$$\Rightarrow \eta_I \geq 3 \times 10^{-5}\ \mathrm{m}$$

Thus, the field of view should have speckles that are no smaller than 30 μm to maintain accurate subset matching.

For part (c), the minimum subset size is obtained using Rule 2 for the minimum subset size in terms of the minimum image speckle dimension.

$$N \geq 3 \cdot \eta_{Imin} = 3 \cdot 3$$

$$\Rightarrow N \geq 9 \text{ pixels}$$

$$\Rightarrow (N \times N)_{min} = 9 \times 9 \text{ pixels}^2 = 81 \text{ pixels}^2$$

A 9×9 subset is generally too small for practical use, as variability in the distribution of particles across the entire $1{,}000 \times 1{,}000$ image may result in too few particles in some subsets. Thus, a more conservative approach would be to select a subset size of 15×15 pixels^2.

Second Example

Measurements must be obtained in a region that is 10 by 10 mm on an object. The recording resolution should be $1{,}000 \times 1{,}000$ pixels^2 and an image displacement accuracy, $\beta_I = 0.01$ pixels is to be maintained. (a) What is the object displacement accuracy? (b) What is the minimum speckle size on the object? (c) What is the minimum subset size for reasonable accuracy and associated object region?

For part (a), we again employ the object displacement accuracy equation to give the following:

$$\beta_O = \tfrac{L}{N} \cdot \beta_I$$

$$\Rightarrow |(\beta_O)_{min}| = \left(\frac{0.01 \text{ m}}{1000 \text{ pixels}}\right) \cdot 0.01 \text{ pixels}$$

$$\Rightarrow |(\beta_O)_{min}| = 1 \times 10^{-7} \text{ m}$$

In this case, the best possible accuracy in displacement accuracy that is achievable is ± 100 nm.

For part (b), the minimum object speckle size is obtained by using Rule 1 for the minimum allowable image plane speckle size and the magnification factor.

$$\eta_O = \eta_I \left(\tfrac{L}{N}\right) \geq 3 \text{ pixels} \left(\tfrac{0.10 \text{ m}}{1000}\right)$$

$$\Rightarrow \eta_O \geq 3 \times 10^{-5} \text{ m}$$

Here, the required object speckle is about the size of larger toner powders, which may be appropriate for this case.

For part (c), the minimum subset size is obtained using Rule 2 for the minimum subset size in terms of the minimum image speckle dimension.

$$N \geq 3 \cdot \eta_{Imin} = 3 \cdot 3 = 9 \text{ pixels}$$

$$\Rightarrow (N \times N)_{min} = 9 \times 9 \text{ pixels}^2 = 81 \text{ pixels}^2$$

$$\Rightarrow (l_0)_{min} = \frac{9 \text{ pixels}}{M_T} = 9 \times 10^{-5} \text{ m}$$

If the minimum size of 9 pixels is selected for a correlation subset, then the corresponding object dimension, 90 μm, is a measure of the minimum spatial resolution achievable with image correlation.

10.1.3.2 Pattern Application

Once the appropriate pattern density has been determined, application of the pattern is the next concern. In many experiments, the process is performed using spray paint in the following manner:

- Lightly coat the surface with a light (dark) paint
- Overspray the coated specimen with a dark (light) mist
- Continue misting and re-misting until the density matches the required size

Though conceptually simple, the process described above can fail for a variety of reasons. For example, this process was used to pattern metallic specimen for the impact experiments in Section 7.3, with delamination of the pattern occurring in high strain regions.

In this application, the authors initially thinly coated specimens with a standard flat white enamel. However, it was quickly discovered that the enamel delaminated in the crack tip region during impact loading. Several surface modifications, including machining, sand-blasting and light chemical etching, did not improve adherence. Eventually, the use of standard enamel paints was abandoned and the Rustoleum™High Heat paint introduced in Section 7.2.2 was employed successfully in dynamic experiments with both aluminum and steel specimens that were lightly sanded to roughen the surface.

Based on experience, when using paint to coat the specimen the best results have been obtained as follows:

- Light roughening of surface whenever possible
- Cleaning of surface to remove oils, grit and other contaminants
- Applying paint just prior to the experiment to maintain maximum adherence to the specimen in high strain regions

When specimens have been painted one or two days prior to an experiment, the authors have noted that adherence is reduced, especially in high strain regions around crack tips or stress concentrators.

Since patterns with a range of sizes are applied to components ranging from meters to microns, a brief summary of methods that have been used successfully are noted in Tables 10.2 and 10.3, along with the approximate size range where applicable.

Table 10.2 Random pattern methods, appropriate size range and typical pattern images

Specimen \ Topic	Patterning method	Speckle size (mm)	Typical pattern image	Remarks
Aluminum 2024-T3 as rolled	Filtered 11 μm toner powder misted onto wet enamel paint (see Section 6.4.4). Specimen roughened with 200 grit metal sandpaper and cleaned with alcohol.	Speckle size ranges from 0.08 to 0.15 mm. Magnification is 150 pixels/mm.		Ideal subset is 400×400 μm which is $\approx 60 \times 60$ pixels at magnification of 150 pixels/mm.
AA8009 aluminum	Lithographically applied tantalum (see Section 6.4.4). Specimen polished prior to coating.	Individual squares are 8 μm. Illumination via in-line lighting for full-field microscope.		Ideal subset size is 100×100 μm which is $\approx 25 \times 25$ pixels at magnification of 250 pixels/mm.

Polymer composite sheet	Surface lightly roughened.	Light coating with white enamel spray paint.	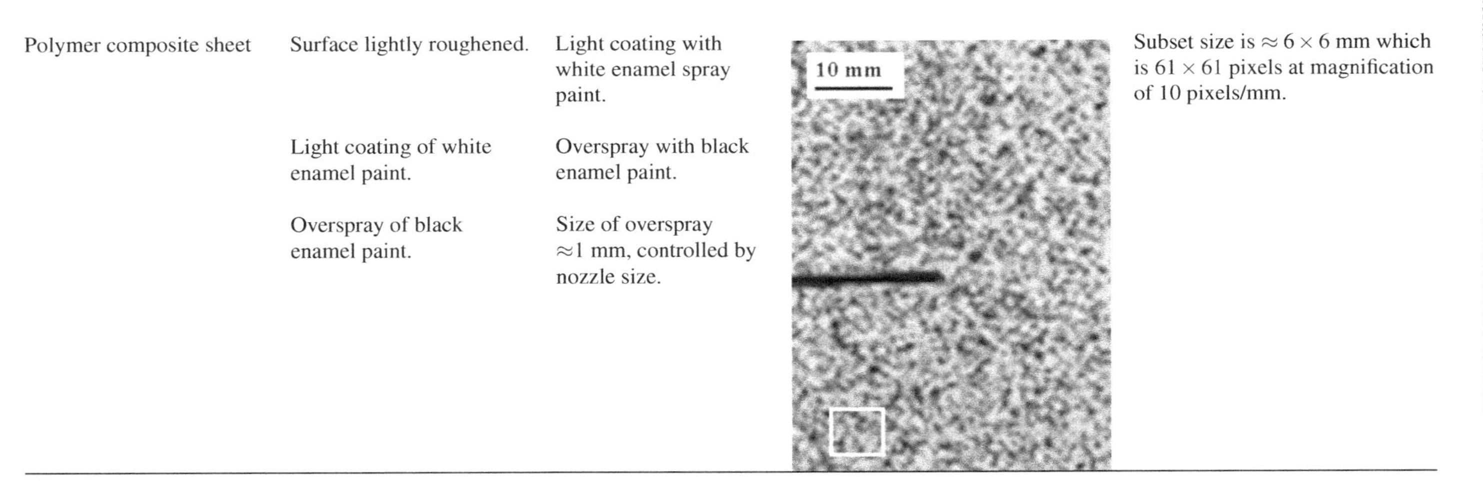	Subset size is $\approx 6 \times 6$ mm which is 61×61 pixels at magnification of 10 pixels/mm.
	Light coating of white enamel paint.	Overspray with black enamel paint.		
	Overspray of black enamel paint.	Size of overspray $\approx$1 mm, controlled by nozzle size.		

Table 10.3 Random pattern methods, appropriate size range and typical pattern images

Specimen \ Topic	Patterning method	Speckle size (mm)	Typical pattern image	Remarks
Polymer composite sheet	Surface lightly roughened. Pattern printed on vinyl sheet with adhesive backing. Adherence through high strength epoxy released when rolled onto surface.	Average speckle size is 5 mm	10mm	Ideal subset size is $\approx 15 \times 15$ mm which is 31×31 pixels at image magnification of $\approx$2 pixels/mm
Palmetto trunk wood	Sawed surface lightly sanded to be planar.	Natural pattern with size of 0.80 mm. Visual pattern related to internal transport mechanism structures.	20 mm 5 mm	Ideal subset size is $\approx 10 \times 10$ mm which is 31×31 pixels at image magnification of $\approx$3 pixels/mm

The natural pattern in the palm tree trunk shown in Table 10.3 requires additional discussion. First, if the pattern and 31×31 subset are used for image matching, then deformations of individual features in the pattern will be averaged to extract the subset motions. Thus, effects such as relative deformations between the matrix and the fiber-like features will not be measured by the image correlation process. To extract such information, a higher magnification image will be required along with an appropriate pattern at the higher magnification.

Secondly, if the entire 0.5×0.5 m trunk area is to be viewed during the experiment, then the size of the pixel array required to maintain sufficient resolution of the naturally-occurring pattern for accurate image analysis is $2{,}048 \times 2{,}048$. Such large-format cameras are available today, though somewhat more expensive for high quality sensor arrays that have a minimum of dead pixels.

10.1.4 Exposure Time

When a digital camera acquires an image, one may interpret each intensity value as the average over a specified time, the *image exposure time*. For quasi-static situations where the object is essentially stationary during image acquisition, the primary effect of increasing exposure time is to reduce the required object illumination for adequate image contrast, allowing an investigator to perform experiments under a wider range of laboratory and field conditions.

However, for situations where object motions occur during the image acquisition time (e.g. fatigue loading [Section 6.4], impact loading [Section 7.3], time-varying pressurization [Section 7.4], inadvertent slip), the role of exposure time is far more significant. To demonstrate this issue, the studies described in Section 7.3 are used.

Considering exposure time, if one has a vertical velocity, v (mm/s), an approximate image magnification of M_T (pixels/mm of object motion) and an exposure time of t_e (s), then the average image displacement during the exposure time is $\approx M_T \cdot v \cdot t_e$. The value obtained is generally compared to the estimated "pixel variability"[2] for the system to determine whether it is significant. For example, in this application v = 1 m/s, $M_T \simeq 15$ pixels/mm and $t_e = 40 \times 10^{-6}$ s. Thus, the image will be slightly "blurred" along the vertical direction due to averaging $\simeq$0.6 pixels of motion. This value is larger than the calibration variability (see Table 7.3). Hence, regions having high spatial displacement gradients or possibly significant velocity gradients may incur slight increases in error due to averaging of the motions.

Similarly, the relative pixel motion between two views due to synchronization offset, t_d, is $M_T \cdot v \cdot t_d$. Here, the magnitude of 0.015 pixels is smaller than system variability, and hence is generally considered negligible.

[2] The image plane residuals for each camera obtained during the calibration process provides one quantitative estimate for pixel-level variability.

10.2 Practical Considerations for 2D-DIC Measurements

10.2.1 Out-of-Plane Motion

Without question, the issue of most concern to investigators using a single camera to make quantitative measurements is the potential for out-of-plane motion and how to account for its presence.

This issue is the focus of all the developments in Section 6.3, with selected issues discussed briefly in Sections 6.3.6 and 6.3.7. From a practical standpoint, the following points are noted.

First, *arrange the camera and lens so that the object distance from the front of the lens is maximized.* This is most easily achieved using long focal length lenses and, if necessary, an appropriate magnification lens or extenders to view the region of interest.

Second, it is important to maintain perpendicularity between viewing direction and object surface. Use of a carpenter's square, tape measure, laser pointer or similar devices may be helpful in achieving reasonable alignment. One approach is to use a small glass mirror with double-sided tape and temporarily attach it to the specimen surface. Then, a laser pointer can be aligned with the lens axis and the camera system adjusted until the incident and reflected beams are approximately coincident.

Third, ensure specimen loading is aligned so that the tendency for out-of-plane deformations (i.e., bending effects) and motions is minimized. If accuracy is a priority, this can be verified prior to the test by using a dummy specimen and a dial indicator to measure the out-of-plane motion, w, with appropriate modifications to ensure that the pseudo-strains are small compared to the expected strains using the following equation:

$$\frac{100\left[\frac{w}{z}\right]}{\varepsilon_{nn}} < \delta \tag{10.2}$$

where:

$$\delta = \text{acceptable percent error due to out-of-plane motion}$$

$$\varepsilon_{nn} = \text{level of normal strain to be measured}$$

$$z = \text{object distance}$$

For many tests, the problem of out-of-plane motions can be greatly reduced by applying a small pre-load to the sample. This is particularly helpful for compliant samples that have a tendency to buckle at small compressive loads that may be induced by wedge grips even if load control is used during the clamping process.

10.2.2 Depth of Field at High Magnification

As noted in Section 6.4.10, high magnification systems have small depths of field. Using simple approximations, Eq. (2.13) can be shown to imply that DOF $\propto (M_T)^{-2}$, decreasing rapidly at higher magnification. In this case, small out-of-plane motions can cause defocus of the pattern and lead to increased measurement errors.

To compensate for this effect, it is possible to develop an automated procedure to re-position the entire high-magnification imaging system to maintain the same level of focus throughout the experiment. This can be accomplished using an iterative, image-based assessment of the level of blurring with motions towards and away from the specimen until optimal sharpness is achieved in the image. This procedure has been used effectively by the authors for the far-field microscope studies (see Section 6.4). In this case, the small depth of field is actually advantageous, since it can be used to accurately maintain the same distance between camera and object by refocusing.

10.2.3 Measurement Accuracy and Intensity Noise

Application of Eqs. (5.89), (5.95) and (5.97) suggests that baseline measurement errors can be minimized using any/all of the following whenever practicable; (a) reducing intensity noise to reduce variability and bias in measurements; (b) use of optimized intensity interpolation methods to minimize bias in the measurements; (c) increasing image contrast to maximize intensity pattern gradients and decrease both bias and variability in measurements.

10.2.3.1 Intensity Noise Reduction

Since much of the background noise in a camera signal is due to thermal fluctuations, one way to reduce noise is to use "cooled" digital cameras that minimize thermal noise. For quasi-static cases where multiple images can be acquired at each stage, image averaging is a straight-forward approach for eliminating random intensity noise, especially when noise levels are high.

10.2.3.2 Intensity Interpolation

In a broad sense, higher order interpolating functions are recommended to decrease measurement bias. Our studies have shown that linear interpolation is less accurate than cubic polynomials, which are not as robust as cubic splines. The strong impact of the interpolation method on measurement bias cannot be over-emphasized.

In practice, however, a trade-off must be found between computation time and accuracy, particularly in real-time applications. As shown in Sections 5.6.1.2 and 5.6.2, polynomial interpolation schemes should generally be avoided due to their extremely poor performance. For the same computational cost, greatly improved results can be obtained using filters designed to minimize phase and amplitude errors. Unless real-time analysis is required for only a few data points, interpolation methods based on recursive pre-filters and a convolution mask size of six pixels (see Section 5.6.1.2) are recommended.

10.2.3.3 Intensity Pattern Contrast

A general approach for increasing contrast is to expand the range, while controlling the noise. This is achievable by increasing the quantization from 8 to 10, 12 or higher bits through camera selection. Though clearly advantageous for reducing both bias and noise, a distinct disadvantage of this choice is the added expense of such cameras and the increased storage (2 bytes per pixel) requirements.

If camera selection is restricted to eight bits, the remaining choice is to ensure that the applied pattern is brightly illuminated to maintain the maximum range possible. A range from 20 to 230 out of 255 would be nearly ideal, since saturation (values of 255) should be avoided so that the pattern can be reconstructed with sub-pixel accuracy.

10.2.4 Thermal Effects at Moderate Temperatures

An area of increasing interest is the measurement of deformations on a specimen at elevated temperatures. In these cases, the authors have shown that the local temperature increase in the vicinity of the specimen can introduce index of refraction changes in the surrounding air, causing a lens-like effect that warps the optical path and distorts the image.

An approach that has been used effectively by one of the authors [176] is to mix the surrounding gases using a fan or similar device, thereby randomizing the refraction changes and introducing a nominally random fluctuation in the image-based measurements. Another approach that has been used effectively is to heat a specimen in a vacuum, thereby mitigating local environmental concerns, and viewing the heated specimen through an appropriate transparent window.

10.2.5 Other Factors

Even after constructing a measurement system that minimizes out-of-plane motion and optimizes the factors outlined in the previous sections, additional factors may play a major role and increase measurement error. Typical factors that must be considered include mechanical vibrations and camera stability. If such effects are of potential concern, baseline experiments can be performed to isolate specific aspects and determine the relative importance of these factors in the measurements.

10.3 Practical Considerations for 3D-DIC Measurements

Stereo vision systems, with the additional geometric complexity associated with multi-camera configurations, are inherently more complex. As such, our efforts in this section will be to provide general suggestions regarding (a) camera and lens selection, (b) system configurations, (c) calibration procedures and (d) measurement procedures.

10.3.1 Camera and Lens Selection

From a practical standpoint, whenever the camera-lens combinations are identical (or nearly the same) for the stereo system, the entire measurement process is simplified:

- Camera control software is the same, making integration simpler
- Sensor sensitivity is similar so that uniform lighting will give similar image contrast
- Sensor size and sensor plane sampling density is similar, so that the same magnification in each camera gives similar spatial resolution
- Identical lenses allow user to employ similar aperture and focus settings to attain similar image pattern contrast and sharpness

Therefore, unless there is a specific reason to use mixed camera-lens systems, it is recommended that similar optics and cameras be used in a multi-camera stereovision system.

10.3.1.1 Camera Selection

Typical features of modern digital cameras used in practical stereovision application include features such as (a) adjustable gain function to balance intensity across cameras, (b) square pixels, (c) CCD or CMOS sensors, (d) array sizes ranging from 800 by 600 to 4,000 by 4,000, (e) image acquisition rate from 5 to 30 frames per second and (f) a minimum of dead pixels in the array.

Typical features of modern high speed digital cameras used in practical stereovision application include features such as (a) adjustable gain function, (b) square pixels, (c) CCD or CMOS sensors, (d) maximum array sizes ranging from 800 by

600 to 1,024 by 1,024 and (e) image acquisition rates at full frame size from 5,000 frames per second with storage of over 5,000 images to 200,000,000 frames per second with a maximum of 16 images.

10.3.1.2 Lens Selection

Since most modern scientific grade cameras can be configured using an adapter to accept either bayonet or screw-mount attachments, there are a range of high quality lenses with good light transmission, focal lengths from 19 to 200 mm and internal apertures to control depth of field that can be used. The authors have used multi-element, fixed focal length, Nikon, Schneider, Canon, Minolta and Yashika lenses successfully.

Even though long focal length lenses have been used successfully in stereo-vision applications (see Section 7.3) their lower light transmission and potential for non-uniqueness in the image center calculation indicate they are best used in 2D applications where the effect of out-of-plane motion must be minimized. A more robust option is the use of modest focal length lenses whose imaging characteristics are well approximated by the pinhole imaging model in Section 3.1.1, and where the DOF can be increased through aperture reduction and corresponding increases in $\mathcal{N}$ (see Eq. (2.13)).

10.3.2 Stereo-vision System Configuration

In many cases, the arrangement of cameras and lighting for image acquisition is dictated by the physical situation; limited space and restricted access oftentimes lead to few options for the investigator. Thus, the following recommendations should be viewed as guidelines for improving the overall measurement process and not as mandatory requirements.

First, it is noted that the angle between t_{0-1} and t_{0-2} in Fig. 4.8 (oftentimes designated the *pan angle* or *stereo angle*) greatly impacts the depth-resolution, with larger stereo angles providing increased resolution in the out-of-plane direction. This consideration might indicate that large stereo angles are generally desirable to maximize measurement system performance. For the vast majority of deformation measurement applications, this is however not the case and relatively small stereo angles are desirable in order to maximize in-plane sensitivity and to reduce the matching bias that may occur with larger stereo angles. For example, consider two points on a planar object that are a distance of 10 mm apart. For simplicity, we assign the first point as the origin of our coordinate system, and assume that the second point is located at $\mathbf{P}_2 = (10\,\text{mm}, 0, 0)$. If we further assume that only the second point is affected by measurement noise, we can express the strain error induced by a noise vector in the displacement measurement $\mathbf{U}$ as

$$\Delta\varepsilon = \frac{|\mathbf{P}_2 + \mathbf{U}|}{10\text{mm}} - 1 \,. \tag{10.3}$$

If we analyze an in-plane noise vector $\mathbf{U}_x = (0.1 \text{ mm}, 0, 0)$ and an out-of-plane noise vector $\mathbf{U}_z = (0, 0, 0.1 \text{ mm})$, we find the corresponding strain errors as

$$\begin{aligned} \Delta\varepsilon_x &= 1\% \\ \Delta\varepsilon_z &= 0.005\% \,, \end{aligned} \tag{10.4}$$

i.e., the same amount of noise in the in-plane direction causes 200 times more error than if the noise occurs in the out-of-plane direction. In most practical strain measurement applications, the out-of-plane sensitivity should therefore not be maximized at the expense of in-plane sensitivity by using large stereo angles, unless the terms $1/2(\partial W/\partial X)^2$ or $1/2(\partial W/\partial Y)^2$ in the Lagrangian strain equations become significant. A recommended stereo angle range is between 10° to 30°, providing good sensitivity to out-of-plane motion while maintaining modest perspective differences between corresponding image subsets (see Fig. 7.14). However, if large stand-off distances are required and out-of-plane sensitivity is of importance, stereo angles of 60° or higher can be employed at the expense of in-plane accuracy.

Second, similar to the situation shown in Fig. 7.14, it is recommended that the object surface normal nominally bisect the pan angle, minimizing relative perspective differences while ensuring good accuracy in the out-of-plane measurements. Another advantage of this approach is that the camera aperture, DOF and FOV of both cameras can be set equal, simplifying the experimental process.

Third, if the direction and magnitude of motion is known, then the system should be configured so that the specimen remains within the DOF and FOV of all cameras during the experiment. An example is shown in Fig. 7.6, where the specimen is placed near one side of the DOF limit so that it moves into the well-focused region during mechanical loading.

Fourth, care must be taken to ensure that various camera parameters defined in Eq. (4.2.3.3) are unaltered after calibration. These include the following:

- Focal length
- Relative position of cameras in a stereo system

Altering these parameters will degrade measurement quality, requiring recalibration to maintain measurement accuracy. Conversely, the overall position of the entire stereo-vision system can be changed, as this does not alter the measurement. If lens distortions are small, then the effective lens diameter can be altered without affecting stereo system parameters.

Fifth, whenever possible the remaining angles orienting one camera relative to another should be minimized; the tilt and swing angles, β and γ, should be small. In such cases, identification of corresponding subsets in each stereo camera is relatively easy since the image subsets are not severely distorted and the initial motion estimation process simplified.

10.3.3 Individual Camera Calibration

The authors have used a wide range of camera calibration methods successfully. Two types of approaches are discussed here. The first type assumes (a) known target motions or (b) known target spacing. These have included:

- 3D specimen translations by specified amounts of a speckle target, with image correlation used to obtain full field motions
- 3D camera translations by specified amounts of a speckle target, with image correlation used to obtain full field motions
- 3D translations of precision line grids, with line extraction required to determine the location of all grid intersections for distortion estimation
- 3D translation of precision dot grids, with dot centroid extraction required to determine the location of all grid intersections for distortion estimation (see Fig. 3.17)

When the grid spacing or motion(s) are assumed to be known, then Eq. (3.41) is modified to exclude re-estimation of the specified grid, dot or speckle motion data. When using this approach, it is recommended that translations (a) towards and away from the camera and (b) in-plane be performed over the range expected in the experiment to ensure that system parameters are optimally estimated for this motion range.

The second type of calibration employs either grid or dot pattern targets and acquires images of the target as it undergoes unknown, arbitrary out-of-plane and in-plane rotations and translations. In this case, Eq. (3.41) is employed and a full bundle-adjustment is performed, with the initial grid spacing used as the initial spacing estimate. It is helpful to think of the calibration process as a shape measurement of the calibration target. To reconstruct the shape of the calibration target accurately, it is desirable to maximize the triangulation angles by tilting the target to obtain good out-of-plane resolution, while also taking some perpendicular views to achieve good in-plane resolution. Large tilt angles are of particular importance when the focal length or stand-off distance are high, since the magnification change in the image plane caused by an out-of-plane motion of the calibration target is small and oftentimes insufficient for accurate calibration. It is noted that in our experience, the bundle adjustment calibration process is by far the most robust and accurate calibration method.

10.3.4 Single Lens Stereo-vision System

There are cases where a compact vision system is needed, necessitating the use of a single collection lens for both optical paths; the stereo-microscope example in Section 7.4 is such a system. In this application, the camera parameters extracted from the calibration process require additional discussion.

Cursory inspection of the calibration data in Table 7.4 indicates two points that require elaboration:

- The center locations for the two paths are in opposite directions and far removed from the sensor centerline.
- The skew factor is non-zero.

Regarding the skew factor, using the definition in Eq. (3.6) and the calibrated values for $f\,S_x$, the skew angles for camera 1 and camera 2 are $\theta_1 = -0.00293°$ and $\theta_2 = 0.21°$, respectively. Consistent with the discussion in Section 7.4, the a priori correction process resulted in slightly skewed (non-orthogonal) coordinate axes after distortion removal. As to the large, sensor plane offset, Fig. 10.4 provides

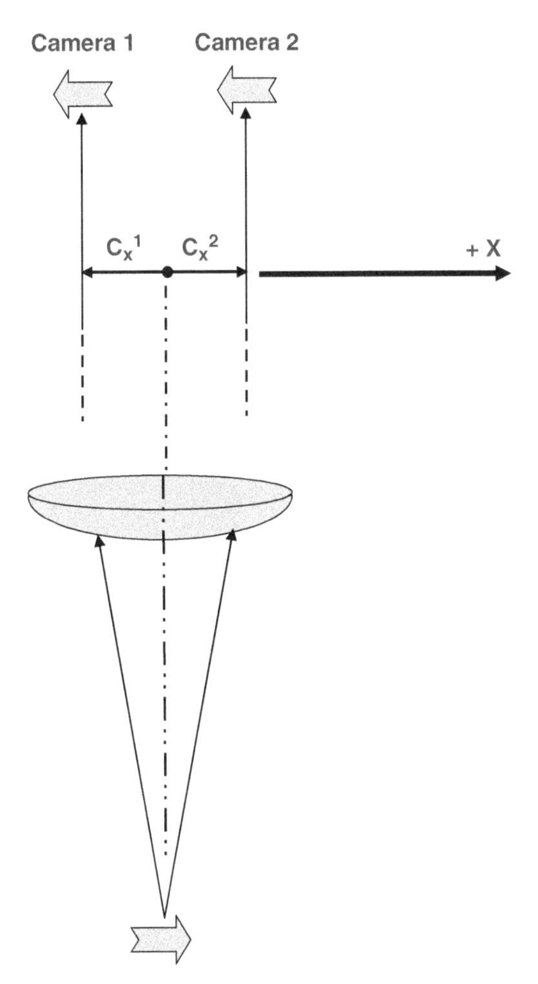

Fig. 10.4 Stereo-imaging through a common collimating lens. Individual paths use separate portions of the collection lens

a qualitative explanation for the observation. With all distortion corrections performed in the individual sensor planes, the optimization process is performed with all distortion parameters, $\kappa_i = 0$, so that the best fit of the sensor positions using a pinhole camera model is obtained when the center is offset to account for the use of a common collection lens.

10.4 Stereo-vision Measurement Process

Once the system has been configured, focused and calibrated, images are acquired for deformation measurements. Issues to be considered include (a) subset size selection, (b) initialization of matching process and (c) error assessment.

10.4.1 Subset Selection and Matching

Selection of an appropriate size for image correlation is based on the principles outlined in Section 10.1.3.1, and requires that the existing pattern be part of the size selection procedure.

Since cross-camera matching is required for stereo motion measurements, perspective differences will alter the shape of the matching subset. Thus, higher order subset shape functions are recommended for matching (e.g. homographic, quadratic) whenever they are available.

APPENDICES

Appendix A
Continuum Mechanics Formulation for Deformations

Figure A.1 shows the basic concept underlying the development of deformation equations. A set of differential line elements emanating from point P are selected. In this presentation, the elements are defined in an orthogonal, rectilinear coordinate system.

Letting $\mathbf{x}(X_1,X_2,X_3)$ be the deformed position of a general point, P, then the transformation of the differential line elements into a deformed state can be written as follows:

$$
\begin{aligned}
&\boldsymbol{x} = \boldsymbol{X} + \boldsymbol{d} \quad ; \quad x_i = X_i + d_i \\
&d\mathbf{x} = \mathbf{F}(\mathbf{X})\, d\mathbf{X} \\
&\mathbf{F}(\mathbf{X}) = \mathbf{x}\nabla \cdot d\mathbf{X} \\
&\mathbf{x}\nabla_{IJ} = \begin{bmatrix} \frac{\partial x_1}{\partial X_1}(X_1,X_2,X_3) & \frac{\partial x_1}{\partial X_2}(X_1,X_2,X_3) & \frac{\partial x_1}{\partial X_3}(X_1,X_2,X_3) \\ \frac{\partial x_2}{\partial X_1}(X_1,X_2,X_3) & \frac{\partial x_2}{\partial X_2}(X_1,X_2,X_3) & \frac{\partial x_2}{\partial X_3}(X_1,X_2,X_3) \\ \frac{\partial x_3}{\partial X_1}(X_1,X_2,X_3) & \frac{\partial x_3}{\partial X_2}(X_1,X_2,X_3) & \frac{\partial x_3}{\partial X_3}(X_1,X_2,X_3) \end{bmatrix}
\end{aligned}
\tag{A.1}
$$

Known as the deformation gradient tensor, $\mathbf{F}$ is the primary tensor relating undeformed and deformed states [181, 233]. Defining $\mathbf{x}(X_1,X_2,X_3) = (X_1,X_2,X_3) + \mathbf{d}(X_1,X_2,X_3)$, where $\mathbf{d}$ is the displacement vector, then the deformation gradient tensor can be written in the form:

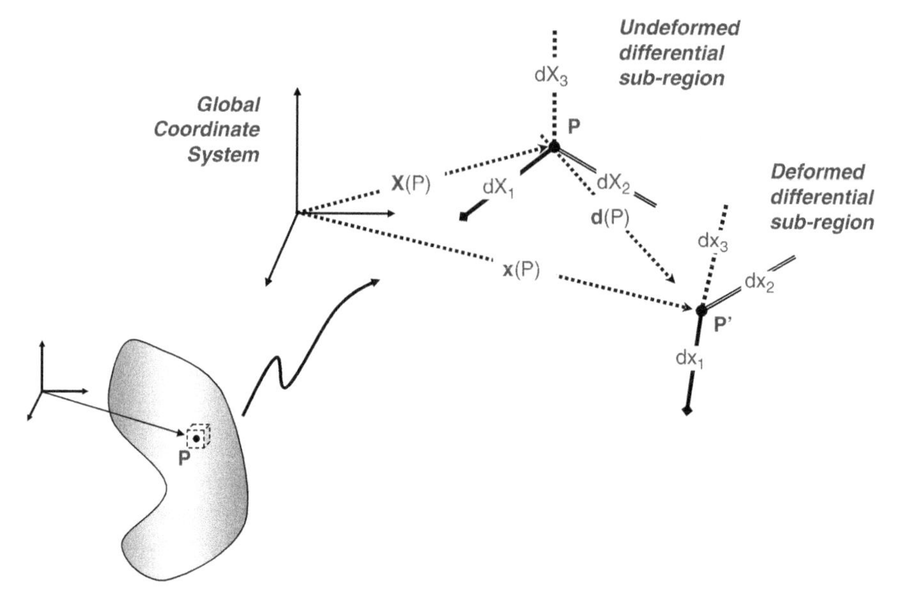

Fig. A.1 Schematic of deformation process with differential lines forming basis for continuum mechanics development of local strain metrics

$$\mathbf{F}(\mathbf{X}) = \mathbf{I} + \mathbf{d}\nabla$$
$$\mathbf{I}_{IJ} = \begin{bmatrix} 1 & 0 & 0 \\ 0 & 1 & 0 \\ 0 & 0 & 1 \end{bmatrix} \tag{A.2}$$
$$\mathbf{d}\nabla_{IJ} = \begin{bmatrix} \frac{\partial d_1}{\partial X_1}(X_1,X_2,X_3) & \frac{\partial d_1}{\partial X_2}(X_1,X_2,X_3) & \frac{\partial d_1}{\partial X_3}(X_1,X_2,X_3) \\ \frac{\partial d_2}{\partial X_1}(X_1,X_2,X_3) & \frac{\partial d_2}{\partial X_2}(X_1,X_2,X_3) & \frac{\partial d_2}{\partial X_3}(X_1,X_2,X_3) \\ \frac{\partial d_3}{\partial X_1}(X_1,X_2,X_3) & \frac{\partial d_3}{\partial X_2}(X_1,X_2,X_3) & \frac{\partial d_3}{\partial X_3}(X_1,X_2,X_3) \end{bmatrix}$$

In a similar manner, the appropriate transformation from deformed to undeformed state is written:

$$d\mathbf{X} = \mathbf{F}^{-1}(\mathbf{x}) \cdot d\mathbf{x}$$
$$\mathbf{F}^{-1}(\mathbf{x}) = \mathbf{X}\nabla$$
$$\mathbf{X}\nabla_{IJ} = \begin{bmatrix} \frac{\partial X_1}{\partial x_1}(x_1,x_2,x_3) & \frac{\partial X_1}{\partial x_2}(x_1,x_2,x_3) & \frac{\partial X_1}{\partial x_3}(x_1,x_2,x_3) \\ \frac{\partial X_2}{\partial x_1}(x_1,x_2,x_3) & \frac{\partial X_2}{\partial x_2}(x_1,x_2,x_3) & \frac{\partial X_2}{\partial x_3}(x_1,x_2,x_3) \\ \frac{\partial X_3}{\partial x_1}(x_1,x_2,x_3) & \frac{\partial X_3}{\partial x_2}(x_1,x_2,x_3) & \frac{\partial X_3}{\partial x_3}(x_1,x_2,x_3) \end{bmatrix} \tag{A.3}$$

A.1 Strain Tensors

Using these definitions, the difference in the squares of the deformed and undeformed differential line elements can be written in the form:

$$ds^2 - dS^2 = 2d\mathbf{X}\cdot\mathbf{E}\cdot d\mathbf{X} \qquad ds^2 - dS^2 = 2d\mathbf{x}\cdot\mathbf{E}^*\cdot d\mathbf{x}$$

$$\mathbf{E} = \tfrac{1}{2}(\mathbf{C}-\mathbf{I}) \qquad \mathbf{E}^* = \tfrac{1}{2}(\mathbf{I}-\mathbf{B}^{-1})$$

$$\mathbf{C} = \mathbf{F}^{\mathbf{T}}\mathbf{F} \qquad \mathbf{B}^{-1} = (\mathbf{F}^{-1})^{\mathbf{T}}\mathbf{F}^{-1}$$

$$(\mathbf{F}^{\mathbf{T}}\mathbf{F})_{IJ} = \sum_{k=1}^{3}\left(\frac{\partial x_k}{\partial X_I}\frac{\partial x_k}{\partial X_J}\right) \qquad \left[(\mathbf{F}^{-1})^{\mathbf{T}}\mathbf{F}^{-1}\right]_{IJ} = \sum_{k=1}^{3}\left(\frac{\partial X_k}{\partial x_I}\frac{\partial X_k}{\partial x_J}\right)$$

$$\mathbf{E}_{IJ} = \frac{1}{2}\left(\frac{\partial d_I}{\partial X_J}+\frac{\partial d_J}{\partial X_I}+\sum_{k=1}^{3}\left(\frac{\partial d_k}{\partial X_J}\frac{\partial d_k}{\partial X_J}\right)\right) \qquad \mathbf{E}^*{}_{IJ} = \frac{1}{2}\left(\frac{\partial d_I}{\partial x_J}+\frac{\partial d_J}{\partial x_I}-\sum_{k=1}^{3}\left(\frac{\partial d_k}{\partial x_J}\frac{\partial d_k}{\partial x_J}\right)\right) \tag{A.4}$$

Here, $\mathbf{E}$ is the *Lagrangian strain tensor* and $\mathbf{E}^*$ is the *Eulerian strain tensor*. In terms of the gradients of the displacement vector components, the components of each tensor are given as follows:

$$\mathbf{E_{11}} = \frac{\partial d_1}{\partial X_1}+\frac{1}{2}\left[\left(\frac{\partial d_1}{\partial X_1}\right)^2+\left(\frac{\partial d_2}{\partial X_1}\right)^2+\left(\frac{\partial d_3}{\partial X_1}\right)^2\right]$$

$$\mathbf{E_{22}} = \frac{\partial d_2}{\partial X_2}+\frac{1}{2}\left[\left(\frac{\partial d_1}{\partial X_2}\right)^2+\left(\frac{\partial d_2}{\partial X_2}\right)^2+\left(\frac{\partial d_3}{\partial X_2}\right)^2\right]$$

$$\mathbf{E_{33}} = \frac{\partial d_3}{\partial X_3}+\frac{1}{2}\left[\left(\frac{\partial d_1}{\partial X_3}\right)^2+\left(\frac{\partial d_2}{\partial X_3}\right)^2+\left(\frac{\partial d_3}{\partial X_3}\right)^2\right] \tag{A.5}$$

$$\mathbf{E_{12}} = \frac{1}{2}\left(\frac{\partial d_1}{\partial X_2}+\frac{\partial d_2}{\partial X_1}\right)+\frac{1}{2}\left[\left(\frac{\partial d_1}{\partial X_1}\frac{\partial d_1}{\partial X_2}\right)+\left(\frac{\partial d_2}{\partial X_1}\frac{\partial d_2}{\partial X_2}\right)+\left(\frac{\partial d_3}{\partial X_1}\frac{\partial d_3}{\partial X_2}\right)\right]$$

$$\mathbf{E_{23}} = \frac{1}{2}\left(\frac{\partial d_2}{\partial X_3}+\frac{\partial d_3}{\partial X_2}\right)+\frac{1}{2}\left[\left(\frac{\partial d_1}{\partial X_2}\frac{\partial d_1}{\partial X_3}\right)+\left(\frac{\partial d_2}{\partial X_2}\frac{\partial d_2}{\partial X_3}\right)+\left(\frac{\partial d_3}{\partial X_2}\frac{\partial d_3}{\partial X_3}\right)\right]$$

$$\mathbf{E_{31}} = \frac{1}{2}\left(\frac{\partial d_3}{\partial X_1}+\frac{\partial d_1}{\partial X_3}\right)+\frac{1}{2}\left[\left(\frac{\partial d_1}{\partial X_3}\frac{\partial d_1}{\partial X_1}\right)+\left(\frac{\partial d_2}{\partial X_3}\frac{\partial d_2}{\partial X_1}\right)+\left(\frac{\partial d_3}{\partial X_3}\frac{\partial d_3}{\partial X_1}\right)\right]$$

$$
\begin{aligned}
\mathbf{E}^*_{11} &= \tfrac{\partial d_1}{\partial x_1} - \tfrac{1}{2}\left[\left(\tfrac{\partial d_1}{\partial x_1}\right)^2 + \left(\tfrac{\partial d_2}{\partial x_1}\right)^2 + \left(\tfrac{\partial d_3}{\partial x_1}\right)^2\right] \\
\mathbf{E}^*_{22} &= \tfrac{\partial d_2}{\partial x_2} - \tfrac{1}{2}\left[\left(\tfrac{\partial d_1}{\partial x_2}\right)^2 + \left(\tfrac{\partial d_2}{\partial x_2}\right)^2 + \left(\tfrac{\partial d_3}{\partial x_2}\right)^2\right] \\
\mathbf{E}^*_{33} &= \tfrac{\partial d_3}{\partial x_3} - \tfrac{1}{2}\left[\left(\tfrac{\partial d_1}{\partial x_3}\right)^2 + \left(\tfrac{\partial d_2}{\partial x_3}\right)^2 + \left(\tfrac{\partial d_3}{\partial x_3}\right)^2\right] \\
\mathbf{E}^*_{12} &= \tfrac{1}{2}\left(\tfrac{\partial d_1}{\partial x_2} + \tfrac{\partial d_2}{\partial x_1}\right) - \tfrac{1}{2}\left[\left(\tfrac{\partial d_1}{\partial x_1}\tfrac{\partial d_1}{\partial x_2}\right) + \left(\tfrac{\partial d_2}{\partial x_1}\tfrac{\partial d_2}{\partial x_2}\right) + \left(\tfrac{\partial d_3}{\partial x_1}\tfrac{\partial d_3}{\partial x_2}\right)\right] \\
\mathbf{E}^*_{23} &= \tfrac{1}{2}\left(\tfrac{\partial d_2}{\partial x_3} + \tfrac{\partial d_3}{\partial x_2}\right) - \tfrac{1}{2}\left[\left(\tfrac{\partial d_1}{\partial x_2}\tfrac{\partial d_1}{\partial x_3}\right) + \left(\tfrac{\partial d_2}{\partial x_2}\tfrac{\partial d_2}{\partial x_3}\right) + \left(\tfrac{\partial d_3}{\partial x_2}\tfrac{\partial d_3}{\partial x_3}\right)\right] \\
\mathbf{E}^*_{31} &= \tfrac{1}{2}\left(\tfrac{\partial d_3}{\partial x_1} + \tfrac{\partial d_1}{\partial x_3}\right) - \tfrac{1}{2}\left[\left(\tfrac{\partial d_1}{\partial x_3}\tfrac{\partial d_1}{\partial x_1}\right) + \left(\tfrac{\partial d_2}{\partial x_3}\tfrac{\partial d_2}{\partial x_1}\right) + \left(\tfrac{\partial d_3}{\partial x_3}\tfrac{\partial d_3}{\partial x_1}\right)\right]
\end{aligned}
\tag{A.6}
$$

As shown in Eqs. (A.5) and (A.6), strain estimation at any point in a body requires determination of the deformation gradients. For methods that measure the displacement field, methods must be developed to optimally estimate the displacement gradient field in the region of interest.

A.2 Strain Rate Tensors

Since modern image correlation measurement methods are capable of acquiring full-field surface displacement data at discrete times, the resulting time-varying displacement fields can be used to estimate the velocity field. When reference positions (e.g., physical particle locations) are used to describe motion, then the form for both acceleration and velocity are written:

$$
\begin{aligned}
\mathbf{v}(X,Y,Z;t) &= \lim_{dt\to 0}\left[\tfrac{(\mathbf{d}(X,Y,Z;t+dt)-\mathbf{d}(X,Y,Z;t))}{dt}\right] \\
\mathbf{v}(X,Y,Z;t+\tfrac{\Delta t}{2}) &\cong \left[\tfrac{(\mathbf{d}(X,Y,Z;t+\Delta t)-\mathbf{d}(X,Y,Z;t))}{\Delta t}\right] \\
\mathbf{a}(X,Y,Z;t) &= \lim_{dt\to 0}\left[\tfrac{(\mathbf{v}(X,Y,Z;t+dt)-\mathbf{v}(X,Y,Z;t))}{dt}\right] \\
\mathbf{a}(X,Y,Z;t+\tfrac{\Delta t}{2}) &\cong \left[\tfrac{(\mathbf{v}(X,Y,Z;t+\Delta t)-\mathbf{v}(X,Y,Z;t))}{\Delta t}\right]
\end{aligned}
\tag{A.7}
$$

$(X,Y,Z) =$ position on object in reference configuration

When using this form, it is necessary to measure the location of a specific subregion on the object as it moves in space and time. Typically known as a Lagrangian description of motion, this approach is quite natural when used with modern image

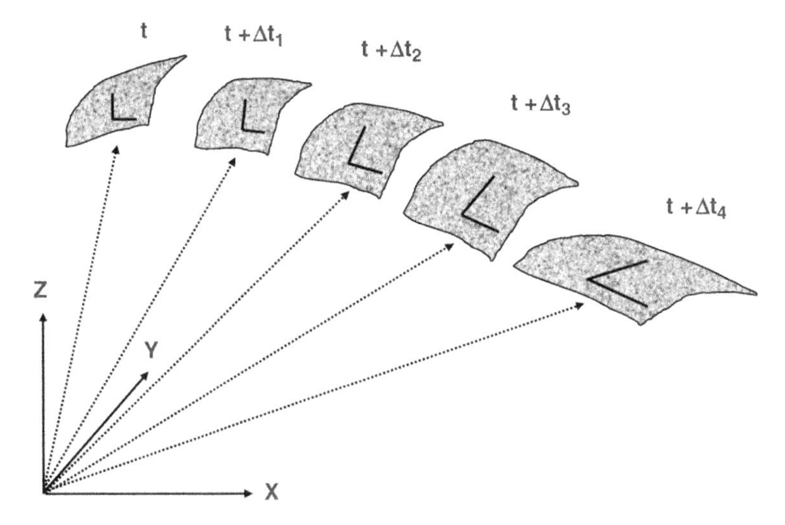

Fig. A.2 Lagrangian formulation: Motions of a fixed object region are measured for all time, t

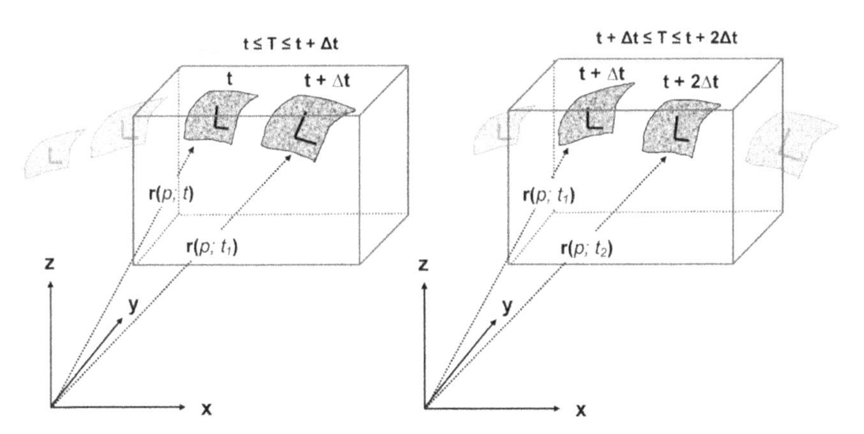

Fig. A.3 Eulerian formulation: Object motions past fixed spatial position, p, in a specified measurement volume are measured for all time, t

correlation measurements that perform matching to identify the location of a subregion throughout the loading process.

Figures A.2 and A.3 present a graphical comparison of Lagrangian and Eulerian formulations in continuum mechanics, respectively. Formulations in fluid mechanics and related areas are typically Eulerian, where physical measurements are performed in a specific measurement volume throughout time and not relative to fixed object positions. As shown in Fig. A.3, spatial position measurements in the Eulerian formulation provide relative motions of whatever physical system is currently present in the measurement volume. Hence, it is natural for the velocity field to be the primary variable in Eulerian formulations.

A.3 Lagrangian

$$
\begin{aligned}
&\frac{d(ds^2)}{dt} = 2d\mathbf{X}\cdot\dot{\mathbf{E}}\cdot d\mathbf{X} \\
&\dot{\mathbf{E}} = \tfrac{1}{2}\dot{\mathbf{C}} \\
&\dot{\mathbf{E}}_{IJ} = \tfrac{1}{2}\left(\frac{\partial v_I}{\partial X_J}+\frac{\partial v_J}{\partial X_I}\right)+\tfrac{1}{2}\sum_{k=1}^{3}\left(\frac{\partial v_k}{\partial X_I}\frac{\partial d_k}{\partial X_J}+\frac{\partial v_k}{\partial X_J}\frac{\partial d_k}{\partial X_I}\right) \\
&\mathbf{v} = \left(\frac{\partial \mathbf{d}}{\partial t}\right)\Big|_{(X,Y,Z)} = \left(\frac{\partial d_1}{\partial t}(X,Y,Z;t),\frac{\partial d_2}{\partial t}(X,Y,Z;t),\frac{\partial d_3}{\partial t}(X,Y,Z;t)\right)
\end{aligned}
\tag{A.8}
$$

To use Eq. (A.8) directly, the velocity and displacement fields must be determined and numerically differentiated in space at each point of interest. An alternative approach would be to compute the strain tensor field at each time of interest, $\mathbf{E}(X,Y,Z;t)$, using Eq. (A.5) and then numerically differentiating the strain field in time at each point of interest.

A.4 Eulerian

$$
\begin{aligned}
&\frac{d(ds^2)}{dt} = 2d\mathbf{x}\cdot\mathbf{D}\cdot d\mathbf{x} \\
&\mathbf{L} = \mathbf{v}\nabla;\mathbf{L}_{IJ} = \frac{\partial v_I}{\partial x_J}\cdots\text{Velocity gradient tensor} \\
&\mathbf{D} = \tfrac{1}{2}\left[\mathbf{L}+\mathbf{L}^{\mathbf{T}}\right];\mathbf{D}_{IJ} = \tfrac{1}{2}\left(\frac{\partial v_I}{\partial x_J}+\frac{\partial v_J}{\partial x_I}\right)\cdots\text{Rate of deformation tensor} \\
&\mathbf{v} = \left(\frac{\partial \mathbf{x}}{\partial t}\right)\Big|_{(X,Y,Z)} = (v_1(x_1,x_2,x_3;t),v_2(x_1,x_2,x_3;t),v_3(x_1,x_2,x_3;t)) \\
&\dot{\mathbf{E}} = \mathbf{F}^{\mathbf{T}}\cdot\mathbf{D}\cdot\mathbf{F};\dot{\mathbf{E}}_{IJ} = \sum_{m=1}^{3}\sum_{n=1}^{3}\left(\frac{\partial x_m}{\partial X_I}D_{mn}\frac{\partial x_n}{\partial X_J}\right)\cdots\text{Lagrangian strain rate tensor} \\
&\dot{\mathbf{E}}^* = \mathbf{D}-\left(\mathbf{E}^*\mathbf{L}+\mathbf{L}^{\mathbf{T}}\mathbf{E}^*\right) \\
&\quad = \mathbf{D}_{IJ}-\sum_{K=1}^{3}(\mathbf{E}^*_{\mathbf{IK}}\mathbf{L}_{\mathbf{KJ}})-\sum_{K=1}^{3}(\mathbf{E}^*_{\mathbf{KI}}\mathbf{L}_{\mathbf{KJ}})\cdots\text{Eulerian strain rate tensor}
\end{aligned}
\tag{A.9}
$$

If positions and displacements are acquired in a spatial coordinate system (i.e., incremental motions are measured at fixed positions), then the measurements can be converted to a spatial velocity field. Using Eq. (A.9), the strain rate tensor can be determined throughout the measurement volume as a function of time.

Appendix B
Elements of Linear Algebra

Let **A** be an $m \times n$ matrix that can be written in all three forms shown below.

$$[A] = \begin{bmatrix} A_{11} & A_{12} & A_{13} & \cdots & A_{1n} \\ A_{21} & A_{22} & A_{23} & \cdots & A_{2n} \\ A_{31} & A_{32} & A_{33} & \cdots & A_{3n} \\ \vdots \\ \vdots \\ A_{m1} & A_{m2} & A_{m3} & \cdots & A_{mn} \end{bmatrix} = \begin{bmatrix} \leftarrow r_1 \rightarrow \\ \leftarrow r_2 \rightarrow \\ \leftarrow r_3 \rightarrow \\ \vdots \\ \vdots \\ \leftarrow r_m \rightarrow \end{bmatrix} = \begin{bmatrix} \uparrow & \uparrow & \uparrow & \uparrow & \uparrow \\ c_1 & c_2 & c_3 & \cdots & c_n \\ \downarrow & \downarrow & \downarrow & \downarrow & \downarrow \end{bmatrix}$$

$$[A^T] = \begin{bmatrix} A_{11} & A_{21} & A_{31} & \cdots & A_{m1} \\ A_{12} & A_{22} & A_{32} & \cdots & A_{m2} \\ A_{13} & A_{23} & A_{33} & \cdots & A_{m3} \\ \vdots \\ \vdots \\ A_{1n} & A_{2n} & A_{3n} & \cdots & A_{mn} \end{bmatrix} = \begin{bmatrix} \leftarrow c_1 \rightarrow \\ \leftarrow c_2 \rightarrow \\ \leftarrow c_3 \rightarrow \\ \vdots \\ \vdots \\ \leftarrow c_n \rightarrow \end{bmatrix} = \begin{bmatrix} \uparrow & \uparrow & \uparrow & \uparrow & \uparrow \\ r_1 & r_2 & r_3 & \cdots & r_m \\ \downarrow & \downarrow & \downarrow & \downarrow & \downarrow \end{bmatrix} \tag{B.1}$$

As shown in (B.1), **A** can be viewed in terms of its m row vectors with n components and its n column vectors with m components. Assuming that there are r independent row vectors, the "row-space" spanned by the row vectors is r-dimensional and is designated $\mathbf{R_A}$. Thus, the $n - r$ remaining vectors span the "null space of **A**", $\mathbf{N_{RA}}$, and the vectors in $\mathbf{N_{RA}}$ are orthogonal to all vectors in $\mathbf{R_A}$ [268].

Since the first fundamental theorem of linear algebra states that the number of independent rows and columns are equal, there must be "r" columns that are linearly independent. Thus, the "column-space" spanned by the column vectors is r-dimensional and is designated $\mathbf{C_A}$. The remaining $m - r$ vectors span the "column null space", $\mathbf{N_{CA}}$.

By direct comparison, the column space and row space for the transpose $\mathbf{A^T}$ are related to the spaces of $\mathbf{A}$ by the equations $\mathbf{C_A^T} = \mathbf{R_A}$ and $\mathbf{R_A^T} = \mathbf{C_A}$. Clearly, the structure of a general matrix is contained in $\mathbf{C_A}$ and $\mathbf{R_A}$ defined by the columns (rows) of $\mathbf{A}$.

Consider the individual terms in the $n \times n$ and $m \times m$ square matrices defined by $\mathbf{A^T A}$ and $\mathbf{AA^T}$, respectively. The matrices are shown in (B.2). As shown in these matrices, the terms are related to the inner products $c_i \cdot c_j$ and $r_i \cdot r_j$ of the separate columns or rows of $\mathbf{A}$. If the columns (rows) are orthogonal, then $c_i \cdot c_j (r_i \cdot r_j) = 0$ for $i \neq j$ and all of the off-diagonal terms are zero. If the columns (rows) are orthonormal, then $c_i \cdot c_i (r_k \cdot r_k) = 1$ for all $i = 1, 2, \ldots, n$ and $k = 1, 2, \ldots, m$ and all of the off-diagonal terms also are zero. Because of their construction, the two square matrices provide a clear picture of the relationship between the columns (rows) that compose $\mathbf{A}$.

$$
\begin{aligned}
\left[A^T\right][A] &= \begin{bmatrix} \leftarrow c_1 \rightarrow \\ \leftarrow c_2 \rightarrow \\ \leftarrow c_3 \rightarrow \\ \vdots \\ \vdots \\ \leftarrow c_n \rightarrow \end{bmatrix}_{n\times m} \begin{bmatrix} \uparrow & \uparrow & \uparrow & \uparrow & \uparrow \\ c_1 & c_2 & c_3 & \cdots & c_n \\ \downarrow & \downarrow & \downarrow & \downarrow & \downarrow \end{bmatrix}_{m\times n} \\
&= \begin{bmatrix} c_1 \cdot c_1 & c_1 \cdot c_2 & c_1 \cdot c_3 & \cdots & c_1 \cdot c_n \\ & c_2 \cdot c_2 & c_2 \cdot c_3 & \cdots & c_2 \cdot c_n \\ & & c_3 \cdot c_3 & \cdots & c_3 \cdot c_n \\ & & & & \vdots \\ & \text{SYMMETRIC} & & & \vdots \\ & & & & c_n \cdot c_n \end{bmatrix}_{n\times n}
\end{aligned}
\tag{B.2}
$$

$$[A]\left[A^T\right] = \begin{bmatrix} \leftarrow r_1 \rightarrow \\ \leftarrow r_2 \rightarrow \\ \leftarrow r_3 \rightarrow \\ \vdots \\ \vdots \\ \leftarrow r_m \rightarrow \end{bmatrix}_{m\times n} \begin{bmatrix} \uparrow & \uparrow & \uparrow & & \uparrow \\ r_1 & r_2 & r_3 & \cdots & r_m \\ \downarrow & \downarrow & \downarrow & & \downarrow \end{bmatrix}_{n\times m}$$

$$= \begin{bmatrix} r_1\cdot r_1 & r_1\cdot r_2 & r_1\cdot r_3 & \cdots & r_1\cdot r_m \\ & r_2\cdot r_2 & r_2\cdot r_3 & \cdots & r_2\cdot r_m \\ & & r_3\cdot r_3 & \cdots & r_3\cdot r_m \\ & & & & \vdots \\ & \text{SYMMETRIC} & & & \vdots \\ & & & & r_m\cdot r_m \end{bmatrix}_{m\times m}$$

B.1 Orthonormal Bases for Column Space, $\mathbf{C_A}$

Consider the columns in **A** and the following construction of an orthonormal basis for m-dimensions, oftentimes known as the Gram-Schmidt process. Here, it is assumed that $m > n$. If "r" of the columns are linearly independent, then these columns can be used to form a basis for $\mathbf{C_A}$.

Let $\mathbf{u_1} = \mathbf{c_1}/|\mathbf{c_1}|$ be the first vector in the orthonormal basis. Then define $\mathbf{c}'_2 = \mathbf{c_1} - [\mathbf{u_1}\cdot\mathbf{c_2}]\cdot\mathbf{u_1}$. By construction, $\mathbf{u_2} = \mathbf{c}'_2/|\mathbf{c}'_2|$ is a unit vector that is orthogonal to $\mathbf{c_1}$. Since $\mathbf{u_1}$ and $\mathbf{u_2}$ are orthogonal, we can define $\mathbf{c}'_3$ by simultaneously subtracting both components from c_3 so that $\mathbf{c}'_3 = \mathbf{c_3} - [\mathbf{u_1}\cdot\mathbf{c_3}]\mathbf{u_1} - [\mathbf{u_2}\cdot\mathbf{c_3}]\mathbf{u_2}$. Since c'_3 is orthogonal to both $\mathbf{u_1}$ and $\mathbf{u_2}$ by construction, the unit vector is given by $\mathbf{u_3} = \mathbf{c}'_3/|\mathbf{c}'_3|$. Continuing the process, an orthonormal basis with "r" orthonormal vectors is constructed for the m-dimensional $\mathbf{C_A}$.

B.2 Orthonormal Bases for Row Space, $\mathbf{R_A}$

Following a similar procedure, consider the column entries in $\mathbf{A^T}$ and the following construction of an orthonormal basis for n-dimensions. Here, it is assumed that $m > n$. If the "r" independent columns of $\mathbf{A^T}$ are selected, then they can be used to form a basis for $\mathbf{R_A}$.

Let $\boldsymbol{v}_1 = \boldsymbol{r}_1/|\boldsymbol{r}_1|$ be the first vector in the orthonormal basis. Then, define $\boldsymbol{r}'_2 = \boldsymbol{r}_1 - [\boldsymbol{v}_1 \cdot \boldsymbol{r}_2]\boldsymbol{v}_1$. By construction, $\boldsymbol{v}_2 = \boldsymbol{r}'_2/|\boldsymbol{r}'_2|$ is a unit vector that is orthogonal to $\boldsymbol{v}_1$. Since $\boldsymbol{v}_1$ and $\boldsymbol{v}_2$ are orthogonal, we can define $\boldsymbol{r}'_3$ by simultaneously subtracting both components from $\boldsymbol{r}_3$ so that $\boldsymbol{r}'_3 = \boldsymbol{r}_3 - [\boldsymbol{v}_1 \cdot \boldsymbol{r}_3]\boldsymbol{v}_1 - [\boldsymbol{v}_2 \cdot \boldsymbol{r}_3]\boldsymbol{v}_2$. Since $\boldsymbol{r}'_3$ is orthogonal to both $\boldsymbol{v}_1$ and $\boldsymbol{v}_2$ by construction, the unit vector is given by $\boldsymbol{v}_3 = \boldsymbol{r}'_3/|\boldsymbol{r}'_3|$. Continuing the process, an orthonormal basis with "r" orthonormal vectors is constructed for the n-dimensional $\boldsymbol{R_A}$.

B.3 Matrices Constructed from Rotation and Warping Components

The matrix multiplication $\mathbf{A} \cdot \mathbf{b}$, where $\mathbf{b}$ is an n-dimensional vector, is a transformation that takes vectors in $\mathbb{R}^n$ into vectors in $\mathbb{R}^m$. Following the procedure outlined previously for matrix multiplication, the process can be viewed in two ways:

(a) A linear combination of columns in $\mathbf{A}$ with weighting factors corresponding to components in $\mathbf{b}$ to construct the transformed vector, $\mathbf{a} = \sum \mathbf{c_j} bj$ or
(b) An inner product of $\mathbf{b}$ with the rows of $\mathbf{A}$ to form the components of the transformed vector, $\mathbf{a}$, where $a_i = \sum r_{ij} b_j$

$$\boldsymbol{a} = [A]_{m\times n}\{b\}_{n\times 1} = b_1\{c_1\}_{m\times 1} + b_2\{c_2\}_{m\times 1} + b_3\{c_3\}_{m\times 1} + \cdots + b_n\{c_n\}_{m\times 1}$$
$$= \begin{Bmatrix} \boldsymbol{r}_1 \cdot \boldsymbol{b} \\ \boldsymbol{r}_2 \cdot \boldsymbol{b} \\ \boldsymbol{r}_3 \cdot \boldsymbol{b} \\ \vdots \\ \boldsymbol{r}_m \cdot \boldsymbol{b} \end{Bmatrix}_{m\times 1} \tag{B.3}$$

An important, yet oftentimes overlooked, fact concerning matrix multiplication is that a general transformation will result in a combination of (a) warping and stretching of the original vector and (b) rotation of the vector relative to its original orientation. For example, in two dimensions, the following multiplication results in a rotation and stretch as shown graphically below:

$$\begin{bmatrix} 1.1\cos\theta & \sin\theta \\ -\sin\theta & 0.7\cos\theta \end{bmatrix} \begin{Bmatrix} 1 \\ 1 \end{Bmatrix} = \begin{Bmatrix} 1.1\cos\theta + \sin\theta \\ -\sin\theta + 0.7\cos\theta \end{Bmatrix}$$
$$= \begin{bmatrix} \cos\theta & \sin\theta \\ -\sin\theta & \cos\theta \end{bmatrix} \begin{bmatrix} 1.1 & 0 \\ 0 & 0.7 \end{bmatrix} \begin{Bmatrix} 1 \\ 1 \end{Bmatrix} \tag{B.4}$$
$$= [U][\lambda] \begin{Bmatrix} 1 \\ 1 \end{Bmatrix}$$

Fig. B.1 Rotation and stretch

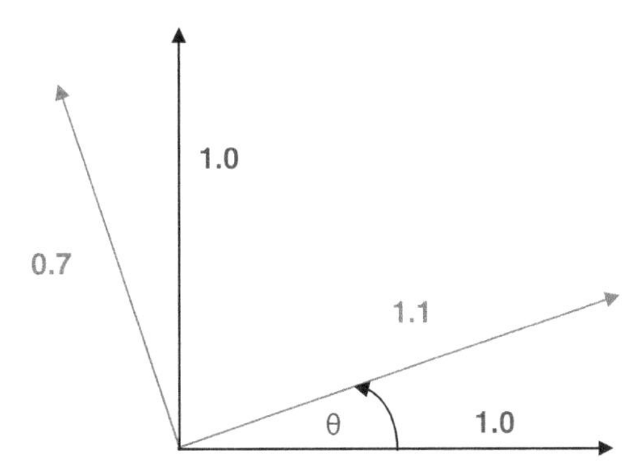

As shown in Eq. (B.4) and in Fig. B.1, the matrix multiplication process results in a stretch of the original components and a rotation of the two line segments by an angle θ. The process can be written in terms of the matrix multiplication of a rotation matrix, **U**, and a dilatation matrix, $\boldsymbol{\lambda}$.

The process outlined above can be used to demonstrate that all matrices can be decomposed into a combination similar to that shown in (B.4). Consider an $m \times n$ array, $[\mathbf{A}]$. Letting $\boldsymbol{a_1}, \boldsymbol{a_2}, \ldots \boldsymbol{a_n}$ be a set of orthonormal vectors that form an n-dimensional basis for the row space of $[\mathbf{A}]$, then we have the following:

$$
\begin{aligned}
&\boldsymbol{A}\boldsymbol{a_i} = \boldsymbol{f_i} \\
&\|\boldsymbol{A}\boldsymbol{a_i}\| = \lambda_i \\
&\boldsymbol{e_i} = \begin{cases} \frac{\boldsymbol{f_i}}{\lambda_i} & \lambda_i \neq 0 \\ O & \lambda_i = 0 \end{cases}
\end{aligned}
\tag{B.5}
$$

where each $\mathbf{e_i}$ is an m-dimensional unit vector within the column space of $[\mathbf{A}]$. Thus, $[\mathbf{A}]$ transforms vectors from $\mathbf{R_A}$ into $\mathbf{C_A}$. If $m > n$ (most common when solving an over-determined set of equations), then construction of the matrix is generally such that the f_i are non-zero. If $m < n$ (an under-determined set of equations), then it is more likely that one of the $\mathbf{f_i}$ is zero.

In any case, it is possible to construct $[\mathbf{A}]$ using $\mathbf{f_i}$ in the manner described in the following equation set. In this construction, $[\mathbf{A}]$ is obtained using a left matrix multiplication, a central multiplication by the diagonal matrix obtained in (B.5) and a right multiplication by a square matrix containing rows that contain an n-dimensional orthonormal basis.

$$[B] = \begin{bmatrix} \uparrow & \uparrow & \uparrow & \uparrow & \uparrow \\ e_1 & e_2 & e_3 & \cdots & e_n \\ \downarrow & \downarrow & \downarrow & \downarrow & \downarrow \end{bmatrix}_{m\times n} \cdot [\lambda_i \delta_{ij}]_{n\times n} \cdot \begin{bmatrix} a_1 \rightarrow \\ a_2 \rightarrow \\ a_3 \rightarrow \\ \vdots \\ a_n \rightarrow \end{bmatrix}_{n\times n}$$

$$\Rightarrow [B]\, a_i = \begin{bmatrix} \uparrow & \uparrow & \uparrow & \uparrow & \uparrow \\ e_1 & e_2 & e_3 & \cdots & e_n \\ \downarrow & \downarrow & \downarrow & \downarrow & \downarrow \end{bmatrix}_{m\times n} \cdot \begin{Bmatrix} 0 \\ 0 \\ 0 \\ \lambda_i \\ 0 \\ 0 \end{Bmatrix} = \lambda_i e_i = f_i = [A]\, a_i$$

$$\Rightarrow ([B] - [A])\, a_i = 0 \qquad \forall a_i \text{ that are vectors in the orthonormal basis for } n\text{-dimensions}$$

$$\Rightarrow [B] = [A] \tag{B.6}$$

where $\mathbf{a_i}$ is an n-dimensional orthonormal basis.

B.4 Singular Value Decomposition of a Matrix

Singular value decomposition (SVD) is a powerful technique in matrix analysis. Three reasons that SVD is a preferred method is that it (a) exists for each rectangular or square matrix of any size or sparseness and (b) it reveals the basic structure of a matrix, including its rank and (c) provides an optimal "approximation" to a matrix of rank "r". Due to its properties, the SVD is commonly used in computational methods because it reduces computational errors while providing a platform for optimal estimations. Today, the process of SVD is employed as a common tool when obtaining a least squares problem solutions to a set of linear equations.

Decomposition of an $m \times n$ matrix $[\mathbf{A}]$, as shown in Eq. (B.6), can be written in the following form:

$$[A]_{m\times n} = [U]_{m\times m}\, [\lambda]_{m\times n}\, [V]^T_{n\times n}$$

where :

$$[V]^T_{n\times n} = n\text{-dimensional orthonormal basis}$$

$$= \begin{bmatrix} a_1 & \cdots & \cdots & \cdots \\ a_2 & \cdots & \cdots & \cdots \\ \vdots & \vdots & \vdots & \vdots \\ a_n & \cdots & \cdots & \cdots \end{bmatrix}_{n\times n} \tag{B.7}$$

$[\boldsymbol{\lambda}]_{m \times n}$ = expanded matrix, including the n eigenvalues of $[V]$ and arbitrary $m-n$ additional values

$$= \begin{bmatrix} \lambda_1 & 0 & 0 & 0 \\ 0 & \lambda_2 & 0 & 0 \\ & & \ddots & \\ 0 & 0 & \lambda_n & 0 \\ \vdots & \vdots & \vdots \ \ddots & \vdots \\ 0 & 0 & 0 & \lambda_m \end{bmatrix}_{m \times n}$$

$[U]_{m \times m}$ = columns are vectors defined by Eq. (B.5)

$$= \begin{bmatrix} e_1 & e_2 & \cdots & e_{m-n+1} & \cdots & e_m \\ \vdots & \vdots & & \vdots & & \vdots \end{bmatrix}$$

For Eq. (B.7) to represent the singular value decomposition for $[\mathbf{A}]$, the orthonormal bases vectors, $\mathbf{a_i}$ in $[\mathbf{V}]^{\mathbf{T}}$, must be selected so that $\mathbf{e_i} \cdot \mathbf{e_j} = \lambda_\mathbf{i} \delta_\mathbf{ij}$, where $\mathbf{e_i} = [\mathbf{A}] \cdot \mathbf{a_i} / (\|[\mathbf{A}] \cdot \mathbf{a_i}\|)$.

Fundamental results from the spectral decomposition theorem can be used to show how appropriate selection for $\mathbf{a_i}$ is used to define $\mathbf{a_i}$ as the eigenvectors for $[\mathbf{A}]^{\mathbf{T}}[\mathbf{A}]$. The legitimacy of this approach can be shown by the following development that shows the general nature of the results.

$$\begin{aligned}
&\mathbf{e_i} = [\mathbf{A}]\mathbf{a_i} \qquad ; \qquad \mathbf{e_j} = [\mathbf{A}]\mathbf{a_j} \\
&\mathbf{a_j} \text{ is an eigenvector for } \left([\mathbf{A}]^{\mathbf{T}}[\mathbf{A}]\right) \\
&\mathbf{e_i} \cdot \mathbf{e_j} = ([\mathbf{A}]\mathbf{a_i}) \cdot ([\mathbf{A}]\mathbf{a_j}) \\
&\qquad\quad = \mathbf{a_i^T} \cdot \left([\mathbf{A}]^{\mathbf{T}}[\mathbf{A}]\right) \mathbf{a_j} \\
&\qquad\quad = \mathbf{a_i^T} \left\{ \left([\mathbf{A}]^{\mathbf{T}}[\mathbf{A}]\right) \mathbf{a_j} \right\} \\
&\qquad\quad = \mathbf{a_i^T} \lambda_j \mathbf{a_j} \\
&\qquad\quad = \lambda_j \left(\mathbf{a_i^T} \cdot \mathbf{a_j}\right) \\
&\qquad\quad = \lambda_j \delta_{\mathbf{ij}} \\
&\Rightarrow \mathbf{e_i} \text{ is an } n\text{-dimensional orthogonal basis} \\
&\mathbf{e_i} \text{ is an } n\text{-dimensional orthogonal basis} \\
&\qquad \lambda_j \delta_{\mathbf{ij}} = \mathbf{e_i} \cdot \mathbf{e_j} \\
&\lambda_j \mathbf{a_i^T} \cdot \mathbf{a_j} = ([\mathbf{A}]\mathbf{a_i}) \cdot ([\mathbf{A}]\mathbf{a_j}) \\
&\qquad\qquad = \left(\mathbf{a_i^T}[\mathbf{A}]^{\mathbf{T}}\right) \cdot ([\mathbf{A}]\mathbf{a_j}) \\
&\qquad\qquad = \mathbf{a_i^T} \left([\mathbf{A}]^{\mathbf{T}}[\mathbf{A}]\right) \mathbf{a_j} \\
&\qquad\qquad = \mathbf{a_i^T} \left\{ \left([\mathbf{A}]^{\mathbf{T}}[\mathbf{A}]\right) \mathbf{a_j} \right\} \\
&\mathbf{0_{ij}} = \mathbf{a_i^T} \left\{ \lambda_j \mathbf{a_j} - \left([\mathbf{A}]^{\mathbf{T}}[\mathbf{A}]\right) \mathbf{a_j} \right\} \\
&\lambda_j \mathbf{a_j} = \left([\mathbf{A}]^{\mathbf{T}}[\mathbf{A}]\right) \mathbf{a_j} \\
&\Rightarrow \mathbf{a_j} \text{ is an eigenvector for } \left([\mathbf{A}]^{\mathbf{T}}[\mathbf{A}]\right)
\end{aligned} \tag{B.8}$$

Appendix C
Method for Local Surface Strain Estimation

Consider a surface point in the reference configuration of the specimen, as shown in Fig. C.1.

Once the initial topology of the specimen has been measured (e.g., reference profile or surface shape) relative to a world coordinate system, then it is possible to define the orientation of the specimen surface at each point. Specifically, the surface normal can be defined at each point, as well as the two tangent vectors, thereby defining a local planar patch at each point.

If it is assumed that the surface can be presented by a function of the form $Z_{\text{surface}} = f(X,Y)$. Defining $G(X,Y,Z) = Z - f(X,Y)$, the normal is given as:

$$\begin{aligned}
&\mathbf{n} = \pm\frac{\nabla G}{|\nabla G|} \\
&\nabla G = \pm\left(-\frac{\partial f}{\partial X}, -\frac{\partial f}{\partial Y}, 1\right) \\
&|\nabla G| = \left[\left(\frac{\partial f}{\partial X}\right)^2 + \left(\frac{\partial f}{\partial Y}\right)^2 + 1\right]^{1/2}
\end{aligned} \tag{C.1}$$

Since the selection of tangent vectors is non-unique, a typical approach is to parameterize positions on the surface. One approach is to use the global coordinate system to define vector positions on the surface. If X is selected to parameterize the surface; then a tangent vector can be described as follows:

$$\begin{aligned}
&\mathbf{r} = (X, Y, f(X,Y)) \\
&\mathbf{t} \cong \frac{\frac{\partial \mathbf{r}}{\partial X}(X, Y=a, f(X,a))}{\left\|\frac{\partial \mathbf{r}}{\partial X}(X, Y=a, f(X,a))\right\|} \\
&\frac{\partial \mathbf{r}}{\partial X}(X, Y=a, f(X,a)) = \left(1, 0, \frac{\partial f}{\partial X}(X,a)\right) \\
&\left\|\frac{\partial \mathbf{r}}{\partial X}(X, Y=a, f(X,a))\right\| = \sqrt{1 + \left(\frac{\partial f}{\partial X}(X,a)\right)^2}
\end{aligned} \tag{C.2}$$

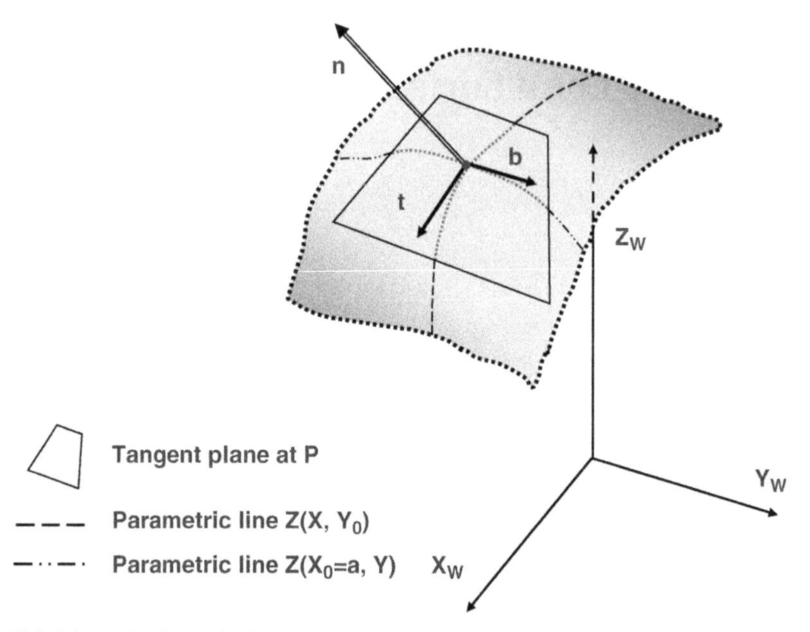

Fig. C.1 Schematic of a methodology for defining local coordinate system at point on non-planar reference surface. The vectors are the normal, tangent and bi-normal unit vectors defined by the surface topology

The procedure described in Eq. (C.2) could be repeated by parameterizing the surface using Y, and providing a second tangent vector for the surface. To ensure that perpendicular tangent directions are selected, one may also define the second tangent vector, designated the bi-normal, as follows:

$$\mathbf{b} = \mathbf{n} \times \mathbf{t}$$

$$\mathbf{b} = \begin{pmatrix} \mathbf{e}_1 & \mathbf{e}_2 & \mathbf{e}_3 \\ n_1 & n_2 & n_3 \\ t_1 & t_2 & t_3 \end{pmatrix}$$

$\mathbf{e}_i$ = unit vectors in the 1, 2 and 3 (i.e., X, Y and Z) directions, respectively

n_i = components of normal vector in the X, Y and Z directions

t_i = components of tangent vector in the X, Y and Z directions

(C.3)

One approach for using these concepts to estimate the local Lagrangian strain components (see Appendix A) is as follows. First, a small contiguous collection of 3D surface points (e.g., a 7×7 array with center-point P) is selected from the initial surface shape (profile). Next, a least square plane is fitted to this 3D position data set. The best fit plane equation can be combined with Eqs. (C.1, C.2, C.3) to obtain the normal, tangent and bi-normal unit vectors, where the tangent direction is typically selected along a preferred specimen direction (e.g., longitudinal axis of a tensile bar). This process is repeated for as many small data sets as needed to define a faceted representation for the initial specimen surface shape.

Using **t**, **n** and **b** as unit vectors for a local specimen coordinate system, $(X,Y,Z=Z_0)$, all displacement components for each of the specimen points are converted into the local system as the loading and deformation process proceeds. At each specified load level, a 7×7 array of displacement components (d_t,d_b,d_n) is obtained and written (d_X,d_Y,d_Z). The array of data points for each component of displacement is then fitted with a least squares function of the form $g(X,Y)$, where $g()$ is typically a complete quadratic function of X and Y.

Differentiating the functional fit for each displacement component in both the X and Y directions, and computing the values of all derivatives at P, the in-plane Lagrangian strains $E_{11}(P)(E_{xx})$, $E_{22}(P)(E_{yy})$ and $E_{12}(P)(E_{xy})$ can be computed using Eq. (A.5) in Appendix A.

The process outlined in the previous paragraphs is repeated for as many small data sets as necessary, resulting in a dense set of strain values throughout the specimen surface shape.

Appendix D
Methods of Non-linear Optimization and Parameter Estimation

D.1 Levenberg–Marquardt and Non-linear Optimization

A common approach for solving non-linear minimization problems is the straightforward gradient descent method. For problems with several variables, the approach can result in long convergence times due to the conflicting requirements for extremely small step sizes in regions of high gradient and large step sizes in relatively flat regions. Such issues are exacerbated when there are high gradients in some variables and low gradients in other variables during a given iteration, requiring a more sophisticated approach in computer vision applications.

D.1.1 Mathematical Background

Let E be a scalar function of the vector $\boldsymbol{\beta}$. Then the second order form for Taylor's series expansion can be written as follows:

$$E(\boldsymbol{\beta}) \cong E(\boldsymbol{\beta_0}) + \boldsymbol{\nabla}_{\boldsymbol{\beta}}\mathbf{E}(\boldsymbol{\beta_0}) \bullet (\boldsymbol{\beta} - \boldsymbol{\beta_0}) + \tfrac{1}{2}(\boldsymbol{\beta} - \boldsymbol{\beta_0}) \bullet \boldsymbol{\nabla\nabla}_{\boldsymbol{\beta}}\mathbf{E}(\boldsymbol{\beta_0}) \bullet (\boldsymbol{\beta} - \boldsymbol{\beta_0})$$

where:

$\boldsymbol{\beta} =$ vector of N variables $\beta_1, \beta_2, \ldots, \beta_N$

$E(\boldsymbol{\beta}) =$ scalar function of N variables $\beta_1, \beta_2, \ldots, \beta_N$ to be minimized

$\boldsymbol{\nabla}\mathbf{E}(\boldsymbol{\beta_0}) =$ gradient vector with components $\frac{\partial E}{\partial \beta_i}(\boldsymbol{\beta_0})$

$\boldsymbol{\nabla\nabla}\mathbf{E}(\boldsymbol{\beta_0}) =$ Hessian tensor with components $\frac{\partial^2 E}{\partial \beta_i \partial \beta_j}(\boldsymbol{\beta_0})$

(D.1)

Equation (D.1) can be viewed as a quadratic approximation for the scalar function E in the vicinity of β_0, where the gradient vector and Hessian tensor are evaluated at β_0. To eliminate the need to estimate the second derivatives required for the Hessian tensor, it is common to approximate the function used to define E by a linear approximation at β. For the special case where E is a least squares error estimator, the approximations are shown in the following section.

D.1.2 Hessian Matrix Approximation and Least Squares

To overcome numerical difficulties when estimating second derivatives required for the Hessian matrix, it is common to perform a linear approximation for the underlying function in the vicinity of β. Letting $E = \mathbf{f}(\beta) \bullet \mathbf{f}(\beta)$, where $\mathbf{f}(\beta) = \mathbf{x} - \mathbf{X}(\beta)$ in a least squares case with $\mathbf{x}$ as the measurement and $\mathbf{X}(\beta)$ representing the model prediction, we can write the following:

$$E(\beta) = \{\mathbf{x} - \mathbf{X}(\beta)\} \bullet \{\mathbf{x} - \mathbf{X}(\beta)\} = \mathbf{f} \bullet \mathbf{f}$$

$\mathbf{x}$ = vector of observations

$\mathbf{X}$ = vector of predictions

β = vector of N variables $\beta_1, \beta_2, \ldots, \beta_N$

$$\begin{aligned}
E(\beta) &= E(\beta_0 + \mathbf{d}\beta) \\
&= \mathbf{f}(\beta_0 + \mathbf{d}\beta) \bullet \mathbf{f}(\beta_0 + \mathbf{d}\beta) \\
&= [\mathbf{f}(\beta_0) + \nabla\mathbf{f}(\beta_0) \bullet \mathbf{d}\beta] \bullet [\mathbf{f}(\beta_0) + \nabla\mathbf{f}(\beta_0) \bullet \mathbf{d}\beta] \\
&= [\mathbf{f}(\beta_0) + \nabla\mathbf{f}(\beta_0) \bullet \mathbf{d}\beta]^T [\mathbf{f}(\beta_0) + \nabla\mathbf{f}(\beta_0) \bullet \mathbf{d}\beta] \text{ using notation for matrix operations} \\
&= \mathbf{f}(\beta_0)^T \bullet \mathbf{f}(\beta_0) + \left[\mathbf{f}(\beta_0)^T \bullet \nabla\mathbf{f}(\beta_0) \bullet \mathbf{d}\beta + (\nabla\mathbf{f}(\beta_0) \bullet \mathbf{d}\beta)^T \bullet \mathbf{f}(\beta_0)\right] \\
&\quad + (\nabla\mathbf{f}(\beta_0) \bullet \mathbf{d}\beta)^T \bullet (\nabla\mathbf{f}(\beta_0) \bullet \mathbf{d}\beta) \\
&= \mathbf{f}(\beta_0)^T \bullet \mathbf{f}(\beta_0) + \left[(\mathbf{f}(\beta_0)^T \bullet \nabla\mathbf{f}(\beta_0)) \bullet \mathbf{d}\beta + \mathbf{d}\beta^T \bullet (\nabla\mathbf{f}(\beta_0)^T \bullet \mathbf{f}(\beta_0))\right] \\
&\quad + \mathbf{d}\beta^T \bullet \nabla\mathbf{f}(\beta_0)^T \bullet (\nabla\mathbf{f}(\beta_0) \bullet \mathbf{d}\beta) \\
&= \mathbf{f}(\beta_0)^T \bullet \mathbf{f}(\beta_0) + 2\left(\mathbf{f}(\beta_0)^T \bullet \nabla\mathbf{f}(\beta_0)\right) \bullet \mathbf{d}\beta \\
&\quad + \mathbf{d}\beta^T \bullet \nabla\mathbf{f}(\beta_0)^T \bullet (\nabla\mathbf{f}(\beta_0) \bullet \mathbf{d}\beta)
\end{aligned} \tag{D.2}$$

Direct comparison of Eqs. (D.1) and (D.2), where $\mathbf{d}\beta = \beta - \beta_0$, leads to the following expressions:

$$E(\beta_0) = \mathbf{f}^T \bullet \mathbf{f}$$

$$\Rightarrow E(\beta_0) \cong \sum_{k=1}^{n} f_k(\beta_0) f_k(\beta_0)$$

$$\nabla_\beta \mathbf{E}(\beta_0) = 2\mathbf{f}^T \bullet \nabla \mathbf{f}(\beta_0)$$

$$\Rightarrow \nabla_\beta E(\beta_0)_i \cong \sum_{k=1}^{n} 2 f_k(\beta_0) \frac{\partial f_k}{\partial \beta_i}(\beta_0) \tag{D.3}$$

$$\nabla\nabla_\beta \mathbf{E}(\beta_0) \cong 2\nabla \mathbf{f}(\beta_0)^T \bullet \nabla \mathbf{f}(\beta_0)$$

$$\Rightarrow \nabla\nabla_\beta E(\beta_0)_{ij} \cong 2 \sum_{k=1}^{n} \frac{\partial f_k}{\partial \beta_i}(\beta_0) \frac{\partial f_k}{\partial \beta_j}(\beta_0)$$

$n =$ dimension of $\mathbf{f}$

D.1.3 Gradient Search Process

A typical straight-forward, **steepest descent algorithm** assumes that the new parameter value β^{i+1} is located down the gradient of the function, so that we can write:

$$\beta^{i+1} = \beta^i - \kappa \nabla_\beta \mathbf{E}(\beta^i) \tag{D.4}$$

$\kappa =$ relaxation factor, retarding or accelerating step size along the gradient of E

D.1.4 Local Quadratic Functional Form

Suppose that we can locally describe the behavior of E in the vicinity of β_0 using a quadratic function of β. Since $\nabla_\beta \mathbf{E}(\beta = \beta_{\min}) = 0$, then we can differentiate Eq. (D.1) to give the following:

$$\nabla_\beta \mathbf{E}(\beta) = \nabla_\beta \mathbf{E}(\beta_0) + \nabla\nabla_\beta \mathbf{E}(\beta_0) \bullet (\beta - \beta_0)$$

$$\nabla_\beta \mathbf{E}(\beta)_{\beta \min} = 0 \Rightarrow \nabla\nabla_\beta \mathbf{E}(\beta_0) \bullet (\beta - \beta_0) = -\nabla_\beta \mathbf{E}(\beta_0)$$

$$\text{If} \quad \nabla_\beta \mathbf{E}(\beta) = \mathbf{f} \bullet \mathbf{f}$$

$$\mathbf{f}(\beta_0 + \mathbf{d}\beta) = [\mathbf{f}(\beta_0) + \nabla \mathbf{f}(\beta_0) \bullet \mathbf{d}\beta]$$

$$\Rightarrow \nabla_\beta \mathbf{E}(\beta_0) = 2\mathbf{f}^T \bullet \nabla \mathbf{f}(\beta_0)$$

$$\nabla\nabla_{\beta}\mathbf{E}(\beta_0) \cong 2\nabla\mathbf{f}(\beta_0)^T \bullet \nabla\mathbf{f}(\beta_0)$$

$$\Rightarrow \left[\nabla\mathbf{f}(\beta_0)^T \bullet \nabla\mathbf{f}(\beta_0)\right] \bullet (\beta - \beta_0) = -\mathbf{f}^T \bullet \nabla\mathbf{f}(\beta_0) = -\nabla\mathbf{f}(\beta_0)^T \bullet \mathbf{f}$$

$$\Rightarrow \sum_{k=1}^{n} \left(\frac{\partial f_k}{\partial \beta_i}(\beta_0)\frac{\partial f_k}{\partial \beta_j}(\beta_0)\right) \Delta\beta_j = \sum_{k=1}^{n} \frac{\partial f_k}{\partial \beta_i}(\beta_0) f_k(\beta_0)$$

$$\Rightarrow (\beta - \beta_0) = -\left(\left[\nabla\mathbf{f}(\beta_0)^T \bullet \nabla\mathbf{f}(\beta_0)\right]^{-1}\right) \bullet \nabla\mathbf{f}(\beta_0)^T \bullet \mathbf{f}$$

$$\Rightarrow \Delta\beta_i = \left(\sum_{k=1}^{n} \left(\frac{\partial f_k}{\partial \beta_i}(\beta_0)\frac{\partial f_k}{\partial \beta_j}(\beta_0)\right)\right)^{-1} \sum_{k=1}^{n} \frac{\partial f_k}{\partial \beta_j}(\beta_0) f_k(\beta_0) \tag{D.5}$$

In this approach, which is a form of Newton–Raphson, we require information regarding the first derivatives of the basic function as an approximation for the Hessian matrix. As shown above, Eq. (D.5) provides an estimate for the location of the minimum. This approach will converge much more rapidly when close to the minimum than using steepest descent. By repeating the process at each estimated minimum location, we will approach the actual minimum relatively quickly once we are "sufficiently close".

D.1.5 Combined Steepest Descent and Newton–Raphson

The Levenberg–Marquardt approach mixes both steepest-descent (for rapid approach to minimum when removed from the correct location) and Newton–Raphson to maximize the speed of convergence as the location of the minimum is approached. A form of the Levenberg–Marquardt algorithm is written as follows:

$$\begin{aligned}
&\Delta\beta = -(\mathbf{H} + \kappa\mathbf{I})^{-1} \bullet \nabla_{\beta}\mathbf{E}(\beta_0)\\
&\kappa = \text{Relaxation factor}\\
&\mathbf{H} \cong \nabla\mathbf{f}(\beta_0)^T \bullet \nabla\mathbf{f}(\beta_0) = \text{Hessian approximation for } \mathbf{f}\bullet\mathbf{f} \text{ at } \beta_0\\
&\nabla_{\beta}\mathbf{E}(\beta_0) = \mathbf{f}^T \bullet \nabla\mathbf{f}(\beta_0) = \nabla\mathbf{f}(\beta_0)^T \bullet \mathbf{f}\\
&\mathbf{I} = \text{identity tensor}
\end{aligned} \tag{D.6}$$

As developed by Levenberg, and revisited and improved by Marquardt nearly 20 years later, the update rule in (D.6) attempts to move smoothly between steepest descent and the Hessian update.

Marquardt improved the method as it transitioned between the two extreme cases. Recognizing that, even in regions removed from the minimum, the Hessian matrix provides information that could "direct" the search process away from error valleys and more towards regions which have lower gradient (and hence closer to the correct minimum location), Marquardt suggested including $\kappa\mathbf{I}$ with the diagonal of the Hessian matrix. In this form, the modified method uses predominantly a steepest descent process away from the minimum location ($\kappa \rightarrow \infty$), while gradually increasing the importance of the Hessian matrix as the minimum is approached and the error decreases when $\kappa \rightarrow 0$.

Thus, the current form of the Levenberg–Marquardt algorithm is as follows:

$$\Delta\boldsymbol{\beta} = -(\mathbf{H} + \kappa\mathbf{I})^{-1} \bullet \nabla_{\boldsymbol{\beta}}\mathbf{E}(\boldsymbol{\beta}_0)$$

$$\mathbf{H} \cong \nabla\mathbf{f}(\boldsymbol{\beta}_0)^T \bullet \nabla\mathbf{f}(\boldsymbol{\beta}_0) \tag{D.7}$$

$$\mathbf{I} = \text{identity tensor}$$

The approach defined by (D.7) works extremely well in practice and is now considered a standard approach for non-linear optimization, even though it is not optimal for typical metrics on speed or error. For moderate sized problems with M parameters ($M \simeq 10^3$), the method is both efficient and effective. For larger problems ($M \simeq 10^4$), the requirement for matrix inversion which scales as M^3 can rapidly increase solution times while also causing potential problems with ill-conditioned matrices, resulting in a requirement to use more sophisticated inverse procedures (e.g. singular value decomposition) during the solution process.

D.1.6 Least Squares with 2D Image Positions

Let E be the sum of the squares of the difference between 2D measured values and 2D model predictions. For use in computer vision, the formulation is directly applicable for a single view of N target points. The function can be defined as follows:

$$E = \sum_{n=1}^{N} (\mathbf{f_n} \bullet \mathbf{f_n}) = \sum_{n=1}^{N} \left((x_n - X_n(\boldsymbol{\beta}))^2 + (y_n - Y_n(\boldsymbol{\beta}))^2 \right)$$

$$\mathbf{f} = \mathbf{x} - \mathbf{X}$$

$$\mathbf{x_n} = (x_n, y_n) = \text{image location in sensor plane } (x_n, y_n) \text{ for } n^{\text{th}} \text{ 3D object position}$$

$$\mathbf{X} = \text{predicted image position for a specified 3D object position}$$

$$\mathbf{X_n} = (X_n(\boldsymbol{\beta}), Y_n(\boldsymbol{\beta})) = n^{\text{th}} \text{ model prediction of image position}$$

$$\boldsymbol{\beta} = \text{vector of } NP \text{ model parameters}$$

$$\boldsymbol{\nabla}_{\boldsymbol{\beta}}\mathbf{E} = 2\mathbf{f}\bullet\frac{\partial \mathbf{f}}{\partial \boldsymbol{\beta}} = -2\mathbf{f}\bullet\frac{\partial \mathbf{X_n}}{\partial \boldsymbol{\beta}} = -2\left(x_n - X_n(\boldsymbol{\beta})\right)\frac{\partial X_n}{\partial \boldsymbol{\beta}} - 2\left(y_n - Y_n(\boldsymbol{\beta})\right)\frac{\partial Y_n}{\partial \boldsymbol{\beta}}$$

$$\nabla_{\beta}E_j = -2\left(x_n - X_n(\boldsymbol{\beta})\right)\frac{\partial X_n}{\partial \beta_j} - 2\left(y_n - Y_n(\boldsymbol{\beta})\right)\frac{\partial Y_n}{\partial \beta_j}$$

$$\begin{aligned}
\boldsymbol{\nabla\nabla}\mathbf{E} &= \mathbf{f}^T \bullet \boldsymbol{\nabla\nabla}\mathbf{f} + 2\boldsymbol{\nabla}\mathbf{f}^T \bullet \boldsymbol{\nabla}\mathbf{f} + \boldsymbol{\nabla\nabla}\mathbf{f}^T \bullet \mathbf{f} \\
&= -2(x_n - X_n)\frac{\partial^2 X_n}{\partial \boldsymbol{\beta}\partial \boldsymbol{\beta}} - 2(y_n - Y_n)\frac{\partial^2 Y_n}{\partial \boldsymbol{\beta}\partial \boldsymbol{\beta}} - 2\frac{\partial X_n}{\partial \boldsymbol{\beta}}\frac{\partial X_n}{\partial \boldsymbol{\beta}} - 2\frac{\partial Y_n}{\partial \boldsymbol{\beta}}\frac{\partial Y_n}{\partial \boldsymbol{\beta}} \\
&\cong 2\boldsymbol{\nabla}\mathbf{f}^T \bullet \boldsymbol{\nabla}\mathbf{f} \\
&= -2\frac{\partial X_n}{\partial \boldsymbol{\beta}}\frac{\partial X_n}{\partial \boldsymbol{\beta}} - 2\frac{\partial Y_n}{\partial \boldsymbol{\beta}}\frac{\partial Y_n}{\partial \boldsymbol{\beta}}
\end{aligned}$$

$$\begin{aligned}
\nabla\nabla E_{ij} &= -2\left(y_n - Y_n(\boldsymbol{\beta})\right)\frac{\partial^2 Y_n}{\partial \beta_i \partial \beta_i} - 2\left(x_n - X_n(\boldsymbol{\beta})\right)\frac{\partial^2 X_n}{\partial \beta_i \partial \beta_i} - 2\frac{\partial Y_n}{\partial \beta_i}\frac{\partial Y_n}{\partial \beta_j} \\
&\quad - 2\frac{\partial X_n}{\partial \beta_i}\frac{\partial X_n}{\partial \beta_j} \qquad i = 1,2,\ldots,NP; j = 1,2,\ldots,NP; i \neq j
\end{aligned}$$

$$\Rightarrow \nabla\nabla E_{ij} \cong \left\{-2\frac{\partial Y_n}{\partial \beta_i}\frac{\partial Y_n}{\partial \beta_j} - 2\frac{\partial X_n}{\partial \beta_i}\frac{\partial X_n}{\partial \beta_j}\right\} \tag{D.8}$$

$$\boldsymbol{\Delta\beta} = (\mathbf{H} + \kappa I)^{-1}\boldsymbol{\nabla}_{\boldsymbol{\beta}}\mathbf{E}(\boldsymbol{\beta_0})$$

where :

$$\mathbf{H}_{ij} = \overline{\mathbf{H}}_{ij} = \sum_{n=1}^{N}\left(\frac{\partial X_n}{\partial \beta_i}(\boldsymbol{\beta_0})\frac{\partial X_n}{\partial \beta_j}(\boldsymbol{\beta_0}) + \frac{\partial Y_n}{\partial \beta_i}(\boldsymbol{\beta_0})\frac{\partial Y_n}{\partial \beta_j}(\boldsymbol{\beta_0})\right)$$

$$[\kappa I]_{ii} \equiv \kappa\delta_{ij} \qquad i = 1,2,\cdots,\text{NPAR}$$

$$\delta_{ij} = \begin{Bmatrix} 1, & i = j \\ 0, & i \neq j \end{Bmatrix}$$

$$\nabla_{\beta}E(\beta_0)_i = -\sum_{n=1}^{N}\left[\left(x_n - X_n(\boldsymbol{\beta_0})\right)\frac{\partial X_n}{\partial \beta_i}(\boldsymbol{\beta_0}) + \left(y_n - Y_n(\boldsymbol{\beta_0})\right)\frac{\partial Y_n}{\partial \beta_i}(\boldsymbol{\beta_0})\right] \tag{D.9}$$

The optimization procedure using Eq. (D.9) proceeds as follows:

1. Define the error E function for the problem of interest.
2. Estimate the "initial" parameter set $\boldsymbol{\beta_0}$ to initiate the process and compute the initial value for $E(\boldsymbol{\beta_0})$.
3. Update the parameter set as indicated in (D.8) and (D.9) by defining $\boldsymbol{\beta} = \boldsymbol{\beta_0} + \boldsymbol{\Delta\beta}$ and compute the new error value, $E(\boldsymbol{\beta})$.
4. Compare the value of $E(\boldsymbol{\beta})$ after updating the parameter set to $E(\boldsymbol{\beta_0})$. If $E(\boldsymbol{\beta}) \geq E(\boldsymbol{\beta_0})$, initial step is further from the minimum location. In this case, reset the parameter set to its previous values and increase κ by a factor of 2–20 to emphasize steepest descent and repeat steps $1 \rightarrow 3$.
5. Repeat 1–4 until $E(\boldsymbol{\beta}) < E(\boldsymbol{\beta_0})$ so that the step is closer to the location of the minimum.

6. Retain the updated parameter set and replace the initial value by letting the new $\boldsymbol{\beta}_0 = \boldsymbol{\beta}$. Then, decrease κ by a factor of 2–20 to increase Newton–Raphson search. Repeat process until $E(\boldsymbol{\beta}) < e$, the pre-specified tolerance on the error and $|\beta_i - \beta_{0_i}|/|\beta_i| < \eta$, the pre-specified tolerance on the change in each parameter.

D.2 Least Squares for Optimal Parameter Estimation

Consider a cost function defined as the difference between the measurements and a model of the physical phenomenon, where the model is defined in terms of parameters $\boldsymbol{\beta}$.

$$E(\boldsymbol{\beta}) = \sum_{i=1}^{N} (F(\mathbf{x}_i;\boldsymbol{\beta}) - f(\mathbf{x}_i))^2 \tag{D.10}$$

$\mathbf{x}_i = i$th vector location with components (x_1, x_2, x_3)

$f(\mathbf{x}_i) =$ measurement at location $\mathbf{x}_i$

$F(\mathbf{x}_i;\boldsymbol{\beta}) =$ model of physical process as function of with unknown parameters, $\boldsymbol{\beta}$

Minimization of E with respect to $\boldsymbol{\beta}$ provides a least squares best estimate for the parameter set. In equation form, this is written:

$$\frac{\partial E}{\partial \boldsymbol{\beta}} = 0 = \sum_{i=1}^{N} \left(\frac{\partial F}{\partial \boldsymbol{\beta}}(\mathbf{x}_i;\boldsymbol{\beta}) F(\mathbf{x}_i;\boldsymbol{\beta}) - \frac{\partial F}{\partial \boldsymbol{\beta}}(\mathbf{x}_i;\boldsymbol{\beta}) f(\mathbf{x}_i) \right) \tag{D.11}$$

If $F(x_i, \boldsymbol{\beta}) = \beta_k g_k(\mathbf{x}_i)$, where $g_k(\mathbf{x}_i)$ are the basic functions used to represent the model function, then one can write:

$$\sum_{k=1}^{M} \beta_k \left[\sum_{i=1}^{N} (g_k(\mathbf{x}_i) g_L(\mathbf{x}_i)) \right] = \sum_{i=1}^{N} (g_L(\mathbf{x}_i) f(\mathbf{x}_i)); \qquad L = 1, 2, \cdots, M \tag{D.12}$$

For example, consider a component of displacement, u, on a planar surface. Then $u = u(x_1, x_2)$. Assuming that the displacement component has unknown spatial variations during loading, one approach for estimating the local deformations in the vicinity of a point is to assume a specific function form. If a quadratic function is assumed we have the following:

$$u(x_1, x_2) \cong F(\mathbf{x}_i, \boldsymbol{\beta}) = \beta_1 + \beta_2 x_1 + \beta_3 x_2 + \beta_4 x_1 x_2 + \beta_5 x_1^2 + \beta_6 x_2^2 \tag{D.13}$$

Using terminology provided in (D.1) $\rightarrow$ (D.3), then $L = 1 \rightarrow 6$ and the basic functions are as follows:

$$g_1 = 1; g_2 = x_1; g_3 = x_2; g_4 = x_1 x_2; g_5 = x_1^2; g_6 = x_2^2 \tag{D.14}$$

The resulting set of linear equations can be written in the following form:

$$\begin{bmatrix} M_{11} & M_{12} & M_{13} & M_{14} & M_{15} & M_{16} \\ & M_{22} & M_{23} & M_{24} & M_{25} & M_{26} \\ & & M_{33} & M_{34} & M_{35} & M_{36} \\ & & & M_{44} & M_{45} & M_{46} \\ & \text{SYMMETRIC} & & & M_{55} & M_{56} \\ & & & & & M_{66} \end{bmatrix} \begin{Bmatrix} \beta_1 \\ \beta_2 \\ \beta_3 \\ \beta_4 \\ \beta_5 \\ \beta_6 \end{Bmatrix} = \begin{Bmatrix} Q_1 \\ Q_2 \\ Q_3 \\ Q_4 \\ Q_5 \\ Q_6 \end{Bmatrix} \tag{D.15}$$

$$M_{KL} = \sum_{i=1}^{N} \left(g_K(x_1^i, x_2^i)\, g_L(x_1^i, x_2^i) \right); \qquad K = 1 \to 6; \quad L = 1 \to 6$$

$$Q_L = \sum_{i=1}^{N} \left(g_L(x_1^i, x_2^i)\, f(x_1^i, x_2^i) \right); \qquad L = 1 \to 6$$

Figure D.1 shows the process graphically, where the resulting surface fit to the measured experimental data is the basis for additional local estimations. For the local quadratic fit, at least six measured displacements are required to provide estimates for the fitting parameters. Once the local surface fit is obtained, estimates for local gradients can be readily computed. As shown in Fig. D.1, gradients in $u(x,y)$ may be computed at the center-point of the region. For a quadratic fit:

$$\begin{aligned} \frac{\partial u}{\partial x_1}\left(x_1^{CP}, y_1^{CP}\right) &\cong \frac{\partial F}{\partial x_1}\left(x_1^{CP}, y_1^{CP}\right) = \beta_2 + \beta_4 y_1^{CP} + 2\beta_5 x_1^{CP} \\ \frac{\partial u}{\partial x_2}\left(x_1^{CP}, y_1^{CP}\right) &\cong \frac{\partial F}{\partial x_2}\left(x_1^{CP}, y_1^{CP}\right) = \beta_3 + \beta_4 x_1^{CP} + 2\beta_6 y_1^{CP} \end{aligned} \tag{D.16}$$

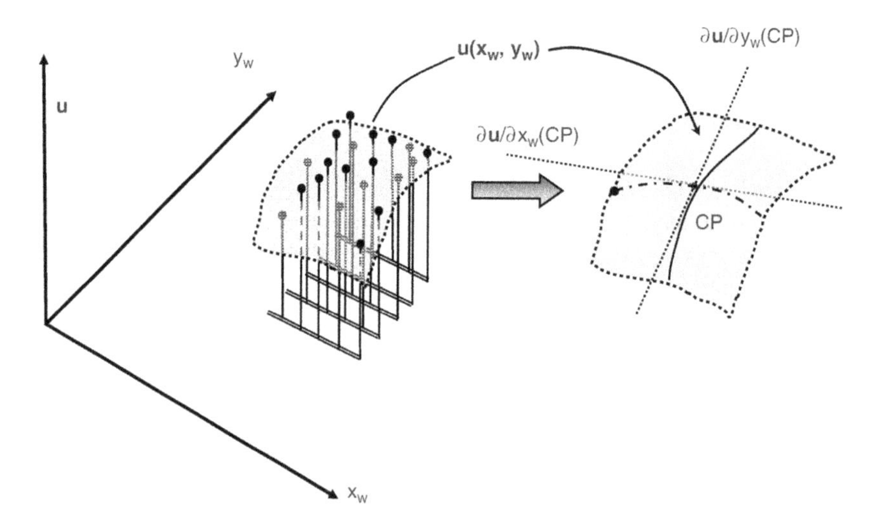

Fig. D.1 Schematic of local least squares fitting process. Derivatives obtained at center-point of region, CP, using fitted functional form

Appendix E
Terminology in Statistics and Probability

For many variables used to describe experiments, it is oftentimes difficult to obtain the functional form for the probability distribution function. As a result, estimates for the expectation and variance (standard deviation) are usually employed to describe the probabilistic characteristics of the quantities of interest.

E.1 Expectation

Let X be a discrete random variable with probability density function $\rho_X(X)$. The expectation $E(X)$ is written:

$$E(X) = \sum_{i=1}^{N} X_i \rho_X(X_i) \tag{E.1}$$

E.2 Mean Value

If the measurement results have a probability density function that is given by the following uniform distribution:

$$\rho_X(X_1) = \rho_X(X_2) = \cdots = \rho_X(X_N) = \frac{1}{N} \tag{E.2}$$

then the expectation is oftentimes designated the *mean value* of a set of N statistically independent observations and is written:

$$E(X) = \overline{X} = \frac{1}{N} \sum_{i=1}^{N} X_i \tag{E.3}$$

For a continuous random variable X with probability density function $f_X(X)$, the mean value is written:

$$E(X) = \int_{-\infty}^{\infty} X f_X(X)\, dX \tag{E.4}$$

Let $[\mathbf{M}]$ be an $N \times M$ matrix. Each of the M columns in the matrix can be considered to be a vector X_j with N components. The ith component of the jth column is X_{ij}; Assuming that each component has statistical variation with probability mass function $\rho_{X_{ij}}(X_{ij})$, then one can define the mean value of each term as follows:

$$\mathbf{E}(\mathbf{X}_{ij}) = \overline{\mathbf{X}}_{ij} = \left\{ \sum_{k=1}^{K} X_{ij}^{k} p_{ij}(X_{ij}^{k}) \right\} \tag{E.5}$$

If $\mathbf{X}$ is any column of the matrix, then the mean value of the column vector can be written:

$$\mathbf{E}(\mathbf{X}) = \overline{\mathbf{X}} = \left\{ \begin{matrix} \sum_{k=1}^{K} X_{1k} p_1(X_{1k}) \\ \sum_{k=1}^{K} X_{2k} p_2(X_{2k}) \\ \vdots \\ \vdots \\ \sum_{k=1}^{K} X_{Nk} p_N(X_{Nk}) \end{matrix} \right\} \tag{E.6}$$

If each column vector, $\mathbf{X}$, has continuously varying components with probability density functions for the ith component, $f_{X_i}(X_i)$, then one can write:

$$\mathbf{E}(\mathbf{X}) = \left\{ \begin{matrix} E(X_1) \\ E(X_2) \\ \vdots \\ \vdots \\ E(X_N) \end{matrix} \right\} = \left\{ \begin{matrix} \int_{-\infty}^{\infty} X_1 f_{X_1}(X_1)\, dX_1 \\ \int_{-\infty}^{\infty} X_2 f_{X_2}(X_2)\, dX_2 \\ \vdots \\ \int_{-\infty}^{\infty} X_N f_{X_N}(X_N)\, dX_N \end{matrix} \right\} \tag{E.7}$$

E.3 Variance

The variance of a continuous random variable X is its second central moment and is generally denoted $Var(X)$.

$$Var(X) = E\left[X - E(X)\right]^2 = E(X^2) - \left[E(X)\right]^2 \tag{E.8}$$

The positive square root of $Var()$ is the standard deviation of X and is generally written:

$$\sigma_X = \sqrt{Var(X)} \tag{E.9}$$

The covariance of X and Y is defined by:

$$Cov(X,Y) = E\left[(X - E(X))(Y - E(Y))\right] = E(XY) - E(X)E(Y) \tag{E.10}$$

Assuming the random variables X and Y are statistically independent, $Cov(X,Y) = 0$.

When $\mathbf{X}$ is an N-dimensional vector, the covariance matrix (also called the variance–covariance matrix) is usually noted $\mathbf{Var}(\mathbf{X})$ or $\Sigma_{\mathbf{X}}$ and has the expression:

$$\begin{aligned}
\mathbf{Var}(\mathbf{X}) &= \mathbf{E}\left[(\mathbf{X} - \mathbf{E}(\mathbf{X}))(\mathbf{X} - \mathbf{E}(\mathbf{X}))^T\right] \\
&= \mathbf{E}\left(\mathbf{X} \bullet \mathbf{X}^{\mathbf{T}}\right) - \mathbf{E}(\mathbf{X}) \bullet \mathbf{E}(\mathbf{X})^{\mathbf{T}} \\
&= \begin{bmatrix}
\sigma_{X_1}^2 & Cov(X_1,X_2) & \cdot & Cov(X_1,X_N) \\
Cov(X_2,X_1) & \sigma_{X_2}^2 & \cdot & \cdot \\
\cdot & \cdot & \cdot & \cdot \\
\cdot & \cdot & \cdot & \cdot \\
Cov(X_N,X_1) & \cdot & \cdot & \sigma_{X_N}^2
\end{bmatrix}
\end{aligned} \tag{E.11}$$

An experimental covariance matrix can be computed from a set of P observations. A general term in the matrix can be written as follows:

$$S_{X_i - X_j} = \frac{1}{P-1}\left[\sum_{k=1}^{P}\left[(X_{ik} - \overline{X}_i)\left(X_{jk} - \overline{X}_j\right)^T\right]\right] \tag{E.12}$$

where $\mathbf{X}$ is an M-dimensional column vector of measurements.

E.4 Approximate Expectation and Variance Expressions

Consider a vector function, $\mathbf{Y} = \mathbf{g}(\mathbf{X})$, where $\mathbf{X} = (X_1, X_2, \cdots, X_N)$. Let $\mathbf{E}(\mathbf{X})$ be defined by Eq. (E.7) with $\mathbf{E}(\mathbf{X}) = \mu$. Let $\mathbf{Var}(\mathbf{X})$ be given by the matrix in Eq. (E.11).

Expanding the function $\mathbf{Y} = \mathbf{g}(\mathbf{X})$ about the point $\mathbf{X_0}$, one can write:

$$\mathbf{Y} \cong \mathbf{g}(\mathbf{X_0}) + \begin{bmatrix} \left.\frac{\partial g_1}{\partial X_1}\right|_{\mathbf{X_0}} & \left.\frac{\partial g_1}{\partial X_2}\right|_{\mathbf{X_0}} & \cdots & \left.\frac{\partial g_1}{\partial X_N}\right|_{\mathbf{X_0}} \\ \left.\frac{\partial g_2}{\partial X_1}\right|_{\mathbf{X_0}} & \left.\frac{\partial g_2}{\partial X_2}\right|_{\mathbf{X_0}} & \cdots & \left.\frac{\partial g_2}{\partial X_N}\right|_{\mathbf{X_0}} \\ \cdots & \cdots & \cdots & \cdots \\ \left.\frac{\partial g_m}{\partial X_1}\right|_{\mathbf{X_0}} & \left.\frac{\partial g_m}{\partial X_2}\right|_{\mathbf{X_0}} & \cdots & \left.\frac{\partial g_m}{\partial X_N}\right|_{\mathbf{X_0}} \end{bmatrix} \bullet \begin{Bmatrix} X_1 - X_0 \\ X_2 - X_0 \\ \vdots \\ X_N - X_0 \end{Bmatrix} \tag{E.13}$$

For the purposes of determining the expectation, $\mathbf{E}(\mathbf{Y})$, and the variance matrix, $\mathbf{Var}(\mathbf{Y})$, we perform the expansion about the expectation of $\mathbf{X}$, so that $\mathbf{X_0} = \mu$. Applying Eqs. (E.4) and (E.10), we obtain the following expressions for $\mathbf{E}(\mathbf{Y})$ and $\mathbf{Var}(\mathbf{Y})$:

$$\mathbf{E}(\mathbf{Y}) \cong \mathbf{g}(\mu)$$

$$\mathbf{Var}(\mathbf{Y}) \cong [\mathbf{J}] \bullet \mathbf{Var}(\mathbf{X}) \bullet [\mathbf{J}]^{\mathbf{T}}$$

$$[\mathbf{J}] = \begin{bmatrix} \left.\frac{\partial g_1}{\partial X_1}\right|_{\mu} & \left.\frac{\partial g_1}{\partial X_2}\right|_{\mu} & \cdots & \left.\frac{\partial g_1}{\partial X_N}\right|_{\mu} \\ \left.\frac{\partial g_2}{\partial X_1}\right|_{\mu} & \left.\frac{\partial g_2}{\partial X_2}\right|_{\mu} & \cdots & \left.\frac{\partial g_2}{\partial X_N}\right|_{\mu} \\ \cdots & \cdots & \cdots & \cdots \\ \left.\frac{\partial g_m}{\partial X_1}\right|_{\mu} & \left.\frac{\partial g_m}{\partial X_2}\right|_{\mu} & \cdots & \left.\frac{\partial g_m}{\partial X_N}\right|_{\mu} \end{bmatrix} \tag{E.14}$$

where $[\mathbf{J}]$ is the Jacobian of the transformation, $\mathbf{Y} = \mathbf{g}(\mathbf{X})$.

Appendix F
Basics of Projective Geometry

For more details on Projective Geometry see Chapters 2 and 3 of [69, 70, 105].

F.1 Homogeneous Coordinates

In inhomogeneous coordinates a point $\mathbf{p}$ in 2D is represented as a coordinate pair $\mathbf{p} = (x, y)$. This is a "common" representation of a plane as a subset of $\mathbb{R}^2$.

Let a 2D point $\mathbf{p}$ be represented by the *homogeneous* form (x, y, z). Then, if two points are the same (equivalent) if they differ only by scale, we can write $\mathbf{p} = (x, y, z) = k(x, y, z)$, where k is the scale factor. With this structure, every vector represents an equivalence class of vectors, differing only in scale. The vectors defined by this structure form the homogeneous class, and the set of all homogeneous vectors in $\mathbb{R}^3$ (except the origin at $(0,0,0)$ form the 2D projective space $\mathcal{P}^2$.

F.2 Why are Homogeneous Coordinates Interesting in Computer Vision: An Example

Let's consider the pinhole camera model given on Fig. F.1.

If the world coordinates of a point are (X, Y, Z) and the image coordinates are (x, y), then:

$$\begin{aligned} x &= f\tfrac{X}{Z} \\ y &= f\tfrac{Y}{Z} \end{aligned} \tag{F.1}$$

The pinhole camera model given by Eq. (F.1) is **non-linear**.

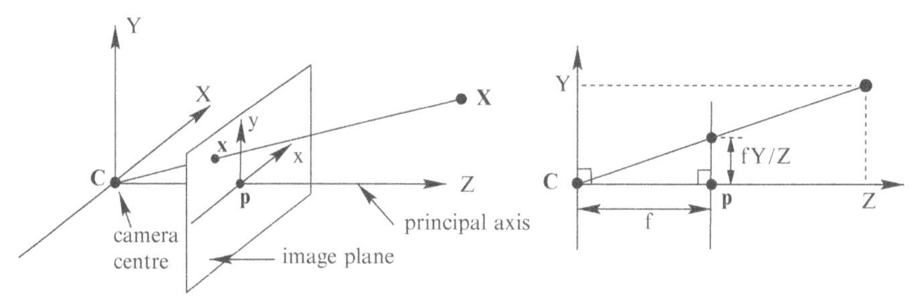

Fig. F.1 The pinhole camera model

If we consider now the homogeneous coordinates:

$$\begin{bmatrix} x \\ y \\ 1 \end{bmatrix} \in \mathcal{P}^2 \qquad \text{and} \qquad \begin{bmatrix} X \\ Y \\ Z \\ 1 \end{bmatrix} \in \mathcal{P}^3$$

the pinhole camera model can be written:

$$\lambda \begin{bmatrix} x \\ y \\ 1 \end{bmatrix} = \begin{bmatrix} f\,0\,0\,0 \\ 0\,f\,0\,0 \\ 0\,0\,1\,0 \end{bmatrix} \begin{bmatrix} X \\ Y \\ Z \\ 1 \end{bmatrix} \tag{F.2}$$

In terms of homogeneous coordinates, the model is now **linear**.

F.3 Lines in $\mathcal{P}^2$

A line equation in $\mathbb{R}^2$ is:

$$a_1x_1 + a_2x_2 + a_3 = 0 \tag{F.3}$$

If we consider $\mathbf{X} \in \mathcal{P}^2$ and by substituting $x_i = X_i/X_3$ by homogeneous coordinates, we get a homogeneous linear equation:

$$(a_1, a_2, a_3) \cdot (X_1, X_2, X_3) = \sum_{i=1}^{3} a_i X_i = 0 \tag{F.4}$$

A line $\mathbf{L}$ in $\mathcal{P}^2$ is represented by a homogeneous three-vector (a_1, a_2, a_3).
A point $\mathbf{X} \in \mathcal{P}^2$ lies on line $\mathbf{L} \in \mathcal{P}^2$ if

$$\mathbf{a} \cdot \mathbf{X} = 0 \quad \text{or} \quad \mathbf{a}^T \cdot \mathbf{X} = 0 \quad \text{or} \quad \mathbf{X}^T \cdot \mathbf{a} = 0 \tag{F.5}$$

Appendix G
Rotation Tensor Formulations

The rotation matrix $[\mathbf{R}]$ can be written in several ways. One approach is use nine terms and six constraints, an approach that generally introduces unwanted difficulties in the solution process. Another approach uses a minimal parameterization of $[\mathbf{R}]$ so that constraints are not required. There are two commonly used minimal parameterizations. One approach is to define $[\mathbf{R}]$ in terms of Euler angles. Here, three successive rotations about specific axes are used to describe the orientation of an object. Assuming the first rotation, Θ_x, is about X_w and the second rotation angle, Θ_y, is about the rotated Y_w axis and the last rotation angle, Θ_z, is about the twice-rotated Z_w axis, then the rotation matrix can be written as the multiplication of three successive rotations to give:

$$[\boldsymbol{R}] = \begin{bmatrix} 1 & 0 & 0 \\ 0 & \cos\Theta_x & \sin\Theta_x \\ 0 & -\sin\Theta_x & cos\Theta_x \end{bmatrix} \begin{bmatrix} \cos\Theta_y & 0 & -\sin\Theta_y \\ 0 & 1 & 0 \\ \sin\Theta_y & 0 & cos\Theta_y \end{bmatrix} \begin{bmatrix} \cos\Theta_z & sin\Theta_z & 0 \\ -\sin\Theta_z & cos\Theta_z & 0 \\ 0 & 0 & 1 \end{bmatrix} \tag{G.1}$$

A slight disadvantage for this formulation is that the order of rotation must be maintained when defining the physical position of an object using the three angles. Another potential problem with this approach has been identified when defining angular rotation rates in gyroscope systems. Known as "gymbal lock", situations exist where one of the Euler angles $\rightarrow \pi/2$, resulting in a singular situation and an inability to integrate a set of first order differential equations for the current angular position of a rotating body. Fortunately, such situations are unlikely to occur when optimizing for the Euler angles and this approach has been used successfully in many applications.

To eliminate the potential pitfall associated with the use of Euler angles, a second minimal parameterization approach that has found widespread application in the computer vision field is the use of the finite rotation vector, $\boldsymbol{\Theta}\{\tilde{\mathbf{n}}\} = [\mathbf{R}] \cdot \{\mathbf{n}\}$, where $\{\tilde{\mathbf{n}}\}$ is an unknown unit vector along a fixed axis of rotation and Θ is a finite rotation angle. In this form, the three unknown parameters are

$\{\Theta\tilde{n}_x, \Theta\tilde{n}_y, \Theta\tilde{n}_z\} = (n_x, n_y, n_z)$. If one uses these parameters to define the extrinsic orientation, then a matrix form in terms of $(\tilde{n}_x, \tilde{n}_y, \tilde{n}_z)$ and Θ can be written as follows:

$$\begin{bmatrix} \tilde{n}_x^2(1-\cos\Theta)+\cos\Theta & \tilde{n}_x\tilde{n}_y(1-\cos\Theta)-\tilde{n}_z\sin\Theta & \tilde{n}_x\tilde{n}_z(1-\cos\Theta)+\tilde{n}_y\sin\Theta \\ \tilde{n}_x\tilde{n}_y(1-\cos\Theta)+\tilde{n}_z\sin\Theta & \tilde{n}_y^2(1-\cos\Theta)+\cos\Theta & \tilde{n}_y\tilde{n}_z(1-\cos\Theta)-\tilde{n}_x\sin\Theta \\ \tilde{n}_x\tilde{n}_z(1-\cos\Theta)-\tilde{n}_y\sin\Theta & \tilde{n}_y\tilde{n}_z(1-\cos\Theta)+\tilde{n}_x\sin\Theta & \tilde{n}_z^2(1-\cos\Theta)+\cos\Theta \end{bmatrix} \tag{G.2}$$

Note that, when $\Theta = 0$, the above formulation is the identity matrix and hence it does not provide any information regarding the fixed axis orientation (i.e., it is arbitrary). When $\Theta \neq 0$, there are three unknown parameters since $n_x^2 + n_y^2 + n_z^2 = \Theta^2$ can be used to eliminate one of the components of the matrix.

Appendix H
Spline Functions

A spline representation is defined by a piecewise polynomial function. Due to the simplicity of their representation, splines are widely used in computer graphics for curve and surface fitting (interpolation or approximation)[1] and particularly for modeling complex functions. In this book, spline surfaces are used to correct complex image distortions in images (see Chapter 3).

Note that this appendix is only an introduction to spline functions providing basic information in order to understand the use of spline functions in this book. The reader may refer to [67,203] for a more complete overview about splines.

H.1 Two-Dimensional Case: Spline Curve

H.1.1 Spline Definition

Given $m+1$ points $t_0, \ldots, t_m$ (called knots) in an interval $[a,b]$ such that $a = t_0 \leq t_1 \leq \ldots \leq t_{m-1} \leq t_m = b$, the function S is a spline of degree p if:

- S and its first $p-1$ derivatives are continuous in the interval $[a,b]$:

$$\mathrm{S} \in \mathbf{C}^{p-1}(a,b)$$

- S is a polynomial of degree at most p in the m subintervals (called knot spans) $[t_i, t_{i+1}]$ with $i = 0 \ldots m-1$:

$$\mathrm{S}_{[t_i,t_{i+1}]} \in \Pi^p, i = 0 \ldots m-1$$

[1] The difference between interpolation and approximation is that interpolation provides a model which passes exactly through all the data points whereas approximation minimizes a function in order to provide a model which passes the closest possible to all the data points in some sense. For example, minimizing the l_2 norm leads to an approximation in the least-squares sense, minimizing the l_∞ norm leads to an approximation in the min-max sense, etc.

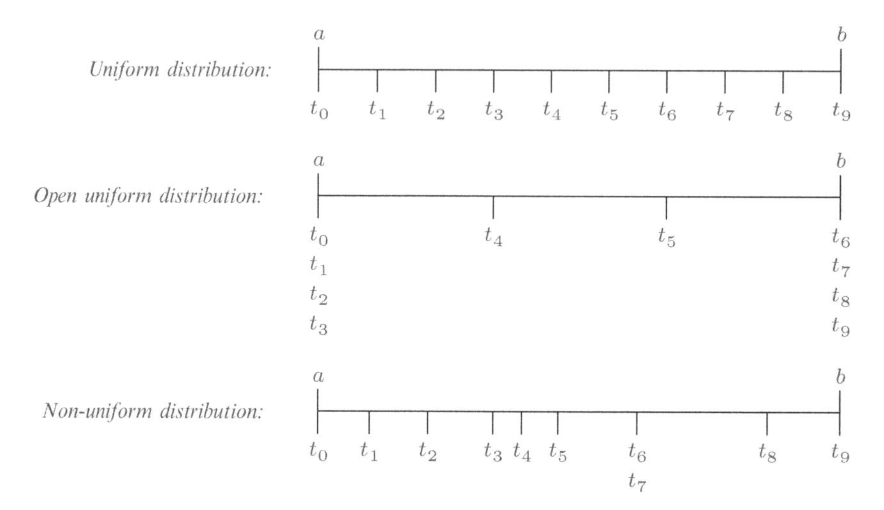

Fig. H.1 Example of the three possible knot distribution for a cubic B-spline (degree $p = 3$) with $m+1 = 10$ knots. From *top* to *bottom*: uniform, open-uniform and non-uniform distribution

Depending on the knot distribution within $[a,b]$ (see Fig. H.1), the spline is:

- **Uniform:** The knots are equidistantly distributed $t_1 - t_0 = t_2 - t_1 = \cdots = t_m - t_{m-1}$
- **Open uniform:** The $p+1$ first knots are all equal $t_0 = t_1 = \cdots = t_p$ and the $p+1$ last knots as well $t_{m-p} = t_{m-p+1} = \cdots = t_m$. The other knots (from t_{p+1} to t_{m-p-1}) are called interior knots and are equidistantly distributed.
- **Non-uniform:** The knots are only subject to the general constraint $t_i \leq t_{i+1}$.

H.1.2 B-Spline Definition

Splines used in our work are actually uniform B-splines (short for basis-splines), a special case of spline. B-spline functions are spline functions defined such that:

$$S(t) = \sum_{i=0}^{n} \alpha_i N_{i,p}(t)$$

where:

- $n = m - p - 1$ with $m + 1$, the number of knots and p, the degree of the B-spline
- α_i are the $(n+1)$ spline coefficients (also called control points or de Boor points)
- $N_{i,p}$ are the basis functions (also called kernels) of the B-spline, defined recursively as follows:

$$N_{i,0}(t) = \begin{cases} 1 & \text{if } t \in [t_i, t_{i+1}[\\ 0 & \text{otherwise} \end{cases}$$

$$N_{i,p\neq 0}(t) = \frac{t - t_i}{t_{i+p} - t_i} N_{i,p-1}(t) + \frac{t_{i+p+1} - t}{t_{i+p+1} - t_{i+1}} N_{i+1,p-1}(t) \qquad \text{(H.1)}$$

Equation H.1 shows that the shape of the basis functions are determined by the relative spacing between the knots. In our work, the B-splines are uniform and Fig. H.2 illustrates the shapes of the basis functions in this case.

Each point of the B-spline curve is influenced by $p+1$ basis functions $N_{i,p}$ weighted by their associated coefficients α_i. The B-spline curve is then not available for the p first and p last knots (see Fig. H.2) and is composed of $n-p+1$ curves joined $\mathbf{C}^{p-1}$ continuously at the interior knots.

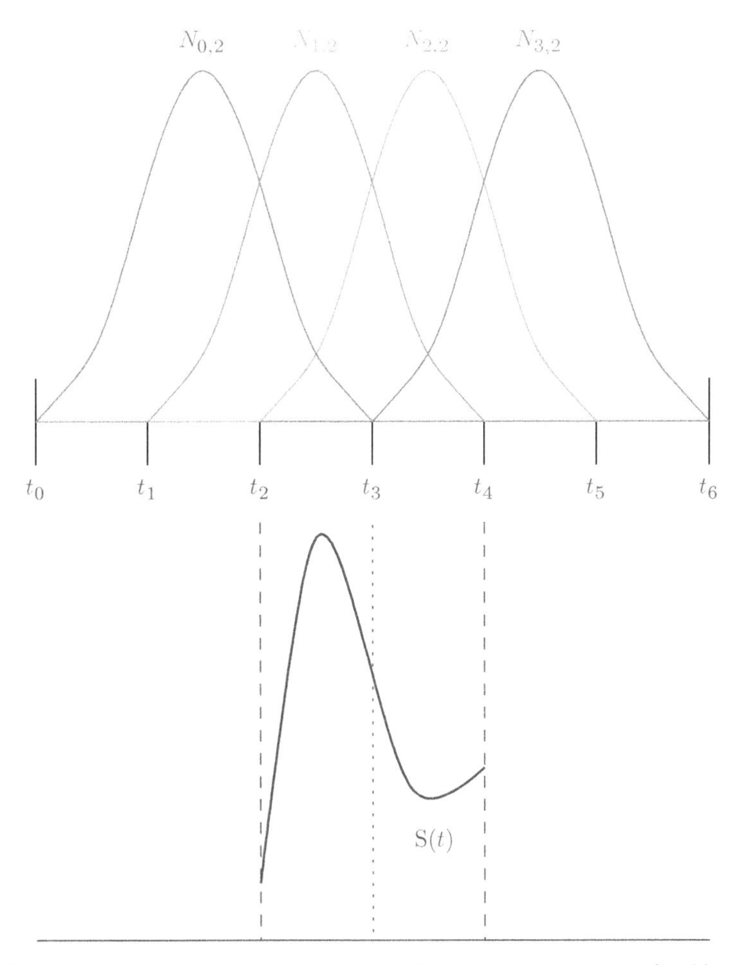

Fig. H.2 Example of basis functions of a uniform quadratic B-spline (degree $p = 2$) with $m+1 = 7$ knots. The B-spline is defined by $n+1 = m-p = 4$ control points and is composed of $n-p+1 = 2$ curves

H.2 Three-Dimensional Case: Spline Surface

A B-spline surface is defined as the tensor product of two B-splines curves:

$$S(x_s, y_s) = \sum_{i=0}^{n_i} \sum_{j=0}^{n_j} \alpha_{i,j} N_{i,p}(x_s) N_{j,p}(y_s)$$

where:

- $n_i = m_i - p - 1$ and $n_j = m_j - p - 1$ with $m_i + 1$ (respectively $m_j + 1$), the number of horizontal (respectively vertical) knots and p, the degree of the B-spline
- $\alpha_{i,j}$ are the $n_i + 1$ (respectively $(n_j + 1)$) spline coefficients
- $N_{i,p}$ (respectively $N_{j,p}$) are the basis functions of the B-spline along x-axis (respectively along y-axis)

Similarly to the B-spline curve, the points of a B-spline surface are affected by $(p+1) \times (p+1)$ basis functions and their coefficients. The surface is only available for the horizontal and vertical interior knots (see Fig. H.3). The rectangular area defined by an interior horizontal and vertical knot span is called a patch.

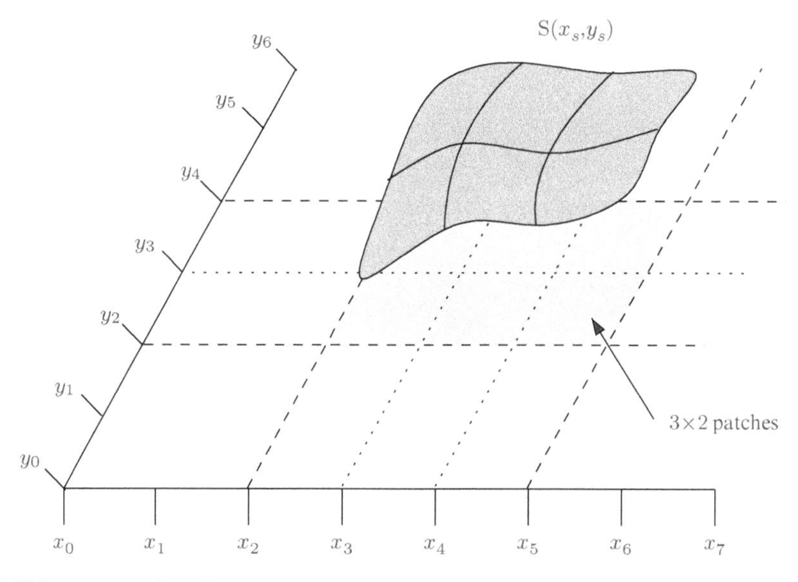

Fig. H.3 Example of a uniform quadratic B-spline surface (degree $p = 2$) with $m_i + 1 = 7$ horizontal knots and $m_j + 1 = 6$ vertical knots. The B-spline surface is available on 3×2 patches

H.3 Spline Derivatives

Evaluating the derivative of a spline along the x-axis (respectively y-axis) is equivalent to evaluating the spline with the kernel (basis function) for x-axis (respectively y-axis) replaced by its derivative.[2] Indeed:

$$\mathrm{S}(x,y,\alpha) = \sum_{i=0}^{n_i}\sum_{j=0}^{n_j} \alpha_{i,j} M_i(x) M_j(y) \Rightarrow \begin{cases} \dfrac{\partial \mathrm{S}(x,y,\alpha)}{\partial x} = \displaystyle\sum_{i=0}^{n_i}\sum_{j=0}^{n_j} \alpha_{i,j} \frac{\partial M_i}{\partial x} M_j(y) \\ \dfrac{\partial \mathrm{S}(x,y,\alpha)}{\partial y} = \displaystyle\sum_{i=0}^{n_i}\sum_{j=0}^{n_j} \alpha_{i,j} M_i(x) \frac{\partial M_j}{\partial y} \end{cases}$$

Actually, to simplify the expressions of the kernels, the spacing between knots is considered constant (uniform spline) and equal to 1. The variables x and y are then replaced by $f_x = \frac{x}{k_x}$ and $f_y = \frac{y}{k_y}$ where k_x is the width of the patches and k_y the height. The equations become:

$$\mathrm{S}(x,y,\alpha) = \sum_{i=0}^{n_i}\sum_{j=0}^{n_j} \alpha_{i,j} N_i(f_x) N_j(f_y) \Rightarrow \begin{cases} \dfrac{\partial \mathrm{S}(x,y,\alpha)}{\partial x} = \dfrac{1}{k_x} \displaystyle\sum_{i=0}^{n_i}\sum_{j=0}^{n_j} \alpha_{i,j} \frac{\partial N_i}{\partial f_x} N_j(f_y) \\ \dfrac{\partial \mathrm{S}(x,y,\alpha)}{\partial y} = \dfrac{1}{k_y} \displaystyle\sum_{i=0}^{n_i}\sum_{j=0}^{n_j} \alpha_{i,j} N_i(f_x) \frac{\partial N_j}{\partial f_y} \end{cases}$$

For uniform splines, expressions of the basis functions and their derivatives depending on the degree of spline are given in Table H.1.

[2] To simplify the expressions, $N_{i,p}$ has been written N_i.

Table H.1 Expressions of the kernels and their respective derivative depending on the degree of spline

Degree of spline	Kernel: $N_*(f)$	Derivative: $\frac{\partial N_*(f)}{\partial f}$
1	$N_i(f) = 1 - f$	$\frac{\partial N_i}{\partial f} = -1$
	$N_{i+1}(f) = f$	$\frac{\partial N_{i+1}}{\partial f} = 1$
2	$N_i(f) = \frac{1-2f+f^2}{2} = \frac{(1-f)^2}{2}$	$\frac{\partial N_i}{\partial f} = f - 1$
	$N_{i+1}(f) = \frac{1+2f-2f^2}{2}$	$\frac{\partial N_{i+1}}{\partial f} = 1 - 2f$
	$N_{i+2}(f) = \frac{f^2}{2}$	$\frac{\partial N_{i+2}}{\partial f} = f$
3	$N_i(f) = \frac{1-3f+3f^2-f^3}{6} = \frac{(1-f)^3}{6}$	$\frac{\partial N_i}{\partial f} = \frac{-1+2f-f^2}{2} = -\frac{(1-f)^2}{2}$
	$N_{i+1}(f) = \frac{4-6f^2+3f^3}{6}$	$\frac{\partial N_{i+1}}{\partial f} = \frac{-4f+3f^2}{2}$
	$N_{i+2}(f) = \frac{1+3f+3f^2-3f^3}{6}$	$\frac{\partial N_{i+2}}{\partial f} = \frac{1+2f-3f^2}{2}$
	$N_{i+3}(f) = \frac{f^3}{6}$	$\frac{\partial N_{i+3}}{\partial f} = \frac{f^2}{2}$
4	$N_i(f) = \frac{1-4f+6f^2-4f^3+f^4}{24} = \frac{(1-f)^4}{24}$	$\frac{\partial N_i}{\partial f} = \frac{-1+3f-3f^2+f^3}{6} = -\frac{(1-f)^3}{6}$
	$N_{i+1}(f) = \frac{11-12f-6f^2+12f^3-4f^4}{24}$	$\frac{\partial N_{i+1}}{\partial f} = \frac{-3-3f+9f^2-4f^3}{6}$
	$N_{i+2}(f) = \frac{11+12f-6f^2-12f^3+6f^4}{24}$	$\frac{\partial N_{i+2}}{\partial f} = \frac{1-f-3f^2+2f^3}{2}$
	$N_{i+3}(f) = \frac{1+4f+6f^2+4f^3-4f^4}{24}$	$\frac{\partial N_{i+3}}{\partial f} = \frac{1+3f+3f^2-4f^3}{6}$
	$N_{i+4}(f) = \frac{f^4}{24}$	$\frac{\partial N_{i+4}}{\partial f} = \frac{f^3}{6}$
5	$N_i(f) = \frac{1-5f+10f^2-10f^3+5f^4-f^5}{120} = \frac{(1-f)^5}{120}$	$\frac{\partial N_i}{\partial f} = \frac{-1+4f-6f^2+4f^3-f^4}{24} = -\frac{(1-f)^4}{24}$
	$N_{i+1}(f) = \frac{26-50f+20f^2+20f^3-20f^4+5f^5}{120}$	$\frac{\partial N_{i+1}}{\partial f} = \frac{-10+8f+12f^2-16f^3+5f^4}{24}$
	$N_{i+2}(f) = \frac{66-60f^2+30f^4-10f^5}{120}$	$\frac{\partial N_{i+2}}{\partial f} = \frac{-12f+12f^3-5f^4}{12}$
	$N_{i+3}(f) = \frac{26+50f+20f^2-20f^3-20f^4+10f^5}{120}$	$\frac{\partial N_{i+3}}{\partial f} = \frac{5+4f-6f^2-8f^3+5f^4}{12}$
	$N_{i+4}(f) = \frac{1+5f+10f^2+10f^3+5f^4-5f^5}{120}$	$\frac{\partial N_{i+4}}{\partial f} = \frac{1+4f+6f^2+4f^3-5f^4}{24}$
	$N_{i+5}(f) = \frac{f^5}{120}$	$\frac{\partial N_{i+5}}{\partial f} = \frac{f^4}{24}$

Appendix I
Triangulation – Location of 3D Position with Skew Rays

Figure I.1 shows a schematic of the basic concept needed to extract a best estimate of the three dimensional position of a point Q. Here, we assume that the subset shown in the camera 1 sensor plane, with center-point at (x_s^1, y_s^1), is an optimal match to the subset shown in camera 2. Thus, these two subsets are the projections into the camera 1 and camera 2 sensor planes of the points in the sub-region surrounding point Q.

Assuming that camera 1 is the reference image, image correlation is used to determine the matching center-point location, (x_s^2, y_s^2), in camera 2. Since the image matching process will introduce slight errors, it is likely that the two rays projected from the pinholes of camera 1 and camera 2 and passing through the centers of the individual subsets shown in Fig. I.1, will be skew rays; they will pass close to each other but generally will not intersect.

First, we assume that (a) both cameras have been calibrated independently using procedures described in Section 3.2 or (b) the combined stereovision system has been calibrated using procedures defined in Section 4.2.3. If we assume that camera 1 is the "reference" camera and the sensor plane is not skewed ($\theta = 0$), then the relationship between the three-dimensional point Q in camera 1 coordinates and the corresponding sensor position in camera 1 can be written in the following form:

$$\left\{ \begin{array}{l} (f^1 S_x^1) X_{c_1} + 0\, Y_{c_1} + (x_s^1 - c_x^1) Z_{c_1} = 0 \\ 0\, X_{c_1} + (f^1 S_y^1) Y_{c_1} + (y_s^1 - c_y^1) Z_{c_1} = 0 \end{array} \right\} \tag{I.1}$$

Assuming that the pinhole translation between camera 1 and camera 2 is given by $(t_{xc_1}, t_{yc_1}, t_{zc_1})$ and that the orientation of camera 1 relative to camera 1 is given by the rigid body rotation matrix, $[\overline{R}]$, then the three-dimensional position for Q in camera 2 and the corresponding sensor position in camera 2 can be written in the following form:

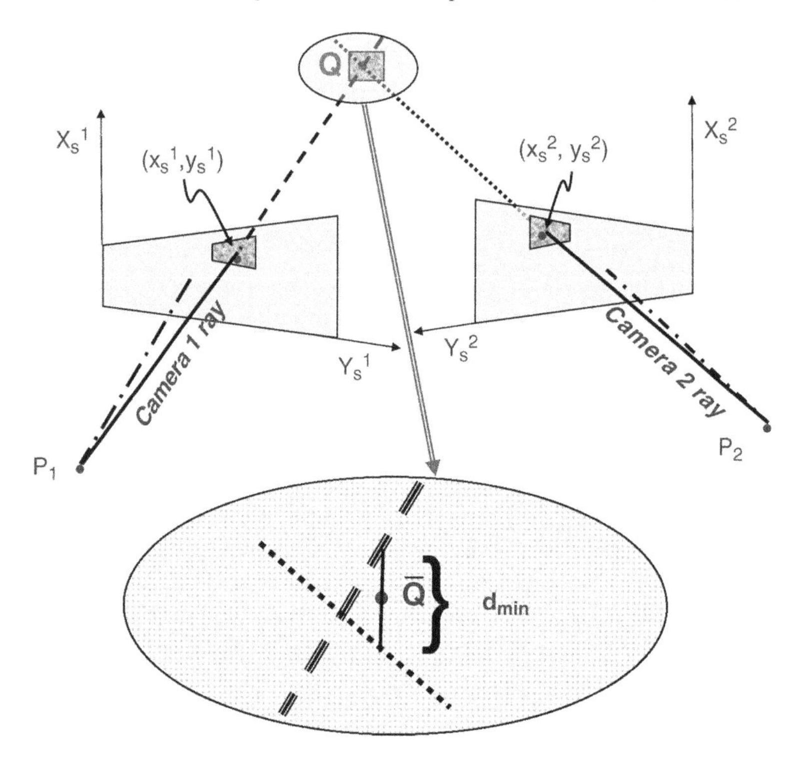

Fig. I.1 Minimum distance estimate for optimal 3D location with two skew rays passing through matched subset centers in a stereo-rig system

$$\left\{\begin{array}{c} \{(x_s^2-c_x^2)\overline{R}_{31}-(f^2S_x^2)\overline{R}_{11}\}X_{c_1}+\{(x_s^2-c_x^2)\overline{R}_{32}-(f^2S_x^2)\overline{R}_{12}\}Y_{c_1} \\ +\{(x_s^2-c_x^2)\overline{R}_{33}-(f^2S_x^2)\overline{R}_{13}\}Z_{c_1}=(f^2S_x^2)t_{xc_1}+(c_x^2-x_s^2)t_{zc_1} \\ \\ \{(y_s^2-c_y^2)\overline{R}_{31}-(f^2S_y^2)\overline{R}_{21}\}X_{c_1}+\{(y_s^2-c_y^2)\overline{R}_{32}-(f^2S_y^2)\overline{R}_{22}\}Y_{c_1} \\ +\{(y_s^2-c_y^2)\overline{R}_{33}-(f^2S_y^2)\overline{R}_{23}\}Z_{c_1}=(f^2S_y^2)t_{yc_1}+(c_y^2-y_s^2)t_{zc_1} \end{array}\right\} \tag{I.2}$$

It is noted that the "measured" positions (x_s^1, y_s^1) and (x_s^2, y_s^2) used in (I.1) and (I.2) should be corrected for image distortion whenever this is necessary; the first image in camera 1 (corresponding to an undeformed state) does not require correction.

Equations (I.1) and (I.2) include the camera parameters for camera 1, camera 2 and the relative position of camera 2 relative to camera 1. The relative orientation and position of the two cameras can be obtained during calibration of a combined stereo-vision (**stereo-rig**) system. The same transformation can be readily determined for independently calibrated cameras by appropriate combinations of the camera 1-to-world transformation and camera 2-to-world transformation.

In a general sense, Eqs. (I.1) and (I.2) can be combined to obtain an over-determined set of equations in the following form:

$$\begin{pmatrix} M_{11} & \cdots & M_{13} \\ \vdots & \ddots & \vdots \\ M_{41} & \cdots & M_{43} \end{pmatrix} \begin{Bmatrix} X_{c_1} \\ Y_{c_1} \\ Z_{c_1} \end{Bmatrix} = \begin{Bmatrix} 0 \\ 0 \\ r_3 \\ r_4 \end{Bmatrix} \tag{I.3}$$

For a common 3D position $(X_{c_1}, Y_{c_1}, Z_{c_1})$, (I.3) represents the difference between the distortion corrected sensor location measurements, (x_s^1, y_s^1) and (x_s^2, y_s^2), and the predicted sensor locations. Thus, a metric that can be used to optimally locate the 3D position is a least squares norm. One form of the least squares norm is in terms of the vector difference[1] is given by:

$$E(X_{c_1}, Y_{c_1}, Z_{c_1}) = \left\| \begin{pmatrix} M_{11} & \cdots & M_{13} \\ \vdots & \ddots & \vdots \\ M_{41} & \cdots & M_{43} \end{pmatrix} \begin{Bmatrix} X_{c_1} \\ Y_{c_1} \\ Z_{c_1} \end{Bmatrix} - \begin{Bmatrix} 0 \\ 0 \\ r_3 \\ r_4 \end{Bmatrix} \right\|^2$$

$$\left.\begin{matrix} r_1 = x^1_{Sm} - x^1_{Smodel} \cong 0 \\ r_2 = x^2_{Sm} - x^2_{Smodel} \cong 0 \end{matrix}\right\} \text{position of image point in } \textbf{master camera} \text{ is known}$$

$$\left.\begin{matrix} r_3 = x^3_{Sm} - x^3_{Smodel} \\ r_4 = x^4_{Sm} - x^4_{Smodel} \end{matrix}\right\} \text{position of image point in } \textbf{camera 2} \text{ is estimated} \tag{I.4}$$

Equation (I.4) can be solved using procedures outlined in Appendix D by setting the partial derivatives $\partial E/\partial X_{c_1} = \partial E/\partial Y_{c_1} = \partial E/\partial Z_{c_1} = 0$. An alternative form for the solution can be written in terms of the generalized inverse (see Appendix D for a brief summary) as:

$$\begin{Bmatrix} X_{c_1} \\ Y_{c_1} \\ Z_{c_1} \end{Bmatrix}_{\text{optimum}} = \left[\begin{pmatrix} M_{11} & \cdots & M_{13} \\ \vdots & \ddots & \vdots \\ M_{41} & \cdots & M_{43} \end{pmatrix}^T \begin{pmatrix} M_{11} & \cdots & M_{13} \\ \vdots & \ddots & \vdots \\ M_{41} & \cdots & M_{43} \end{pmatrix} \right]^{-1} \begin{pmatrix} M_{11} & \cdots & M_{13} \\ \vdots & \ddots & \vdots \\ M_{41} & \cdots & M_{43} \end{pmatrix}^T \begin{Bmatrix} 0 \\ 0 \\ r_3 \\ r_4 \end{Bmatrix} \tag{I.5}$$

[1] It is not required to set $r_1 = r_2 = 0$. However, since the sensor position in the master image is specified, then the "measured initial location" is oftentimes assumed to be known. In the same way, it is common to assume that the sensor location in the model matches the specified value.

References

1. J. Abanto-Bueno and J. Lambros. Investigation of crack growth in functionally graded materials using digital image correlation. *Engineering Fracture Mechanics*, 69(14):1695–1711, September 2002.
2. J. Abanto-Bueno and J. Lambros. Mechanical and fracture behavior of an artificially ultraviolet-irradiated poly(ethylene-carbon monoxide) copolymer. *Journal of Applied Polymer Science*, 92(1):139–148, April 2004.
3. J. Abanto-Bueno and J. Lambros. An experimental study of mixed mode crack initiation and growth in functionally graded materials. *Experimental Mechanics*, 46(2):179–196, April 2006.
4. J. Abanto-Bueno and J. Lambros. Parameters controlling fracture resistance in functionally graded materials under mode I loading. *International Journal of Solids and Structures*, 43(13):3920–3939, June 2006.
5. F. Ackermann. Digital image correlation: Performance and potential application in photogrammetry. *The Photogrammetric Record*, 11(64):429–439, October 1984.
6. R. J. Adrian. Particle imaging techniques for experimental fluid mechanics. *Annual Reviews in Fluid Mechanics*, 23:261–304, 1991.
7. R. Albertani, B. Stanford, J. P. Hubner, and P. G. Ifju. Aerodynamic coefficients and deformation measurements on flexible micro air vehicle wings. *Experimental Mechanics*, 47(5): 625–635, October 2007.
8. M. Ali-Ahmad, K. Subramaniam, and M. Ghosn. Experimental investigation and fracture analysis of debonding between concrete and FRP sheets. *Journal of Engineering Mechanics*, 132(9):914–923, 2006.
9. R. Ambu, F. Aymerich, and F. Bertolino. Investigation of the effect of damage on the strength of notched composite laminates by digital image correlation. *The Journal of Strain Analysis for Engineering Design*, 40(5):451–462, July 2005.
10. D. R. Ambur, N. Jaunky, and M. W. Hilburger. Progressive failure studies of stiffened panels subjected to shear loading. *Composite Structures*, 65(2):129–142, August 2004.
11. B. E. Amstutz, M. A. Sutton, and D. S. Dawicke. Experimental study of mixed mode I/II stable crack growth in thin 2024-T3 aluminum. *ASTM STP 1256 on Fatigue and Fracture*, 26:256–273, 1995.
12. B. E. Amstutz, M. A. Sutton, D. S. Dawicke, and M. L. Boone. Effects of mixed mode I/II loading and grain orientation on crack initiation and stable tearing in 2024-T3 aluminum. *ASTM STP 1296 on Fatigue and Fracture*, 27:105–125, 1997.
13. J. Anderson, W. H. Peters, M. A. Sutton, W. F. Ranson, and T. C. Chu. Application of digital correlation methods to rigid body mechanics. *Optical Engineering*, 22(6):238–243, 1984.
14. M. Andersson and D. Betsis. Point reconstruction from noisy images. *Journal of Mathematical Imaging and Vision*, 5(1):77–90, 1995. doi: 10.1007/BF01250254.

15. K. Andresen. Strain tensor for large three-dimensional surface deformation of sheet metal from an object grating. *Experimental Mechanics*, 39(1):30–35, March 1999.
16. N. P. Andrianopoulos. Full-field displacement measurement of a speckle grid by using a mesh-free deformation function. *Strain*, 42(4):265–271, November 2006.
17. Anon. Gilbert Louis Hobrough. *The Photogrammetric Record*, 18(104):337–340, 2003.
18. E. Archbold, J. M. Burch, and A. E. Ennos. Recording of in-plane surface displacements by double exposure speckle photography. *Optica Acta*, 17:883–898, 1970.
19. E. Archbold and A. E. Ennos. Laser photography to measure deformation of weld cracks under load. *Non-Destructive Testing*, 8:181–184, 1975.
20. X. Armangué and J. Salvi. Overall view regarding fundamental matrix estimation. *Image and Vision Computing*, 21(2):205–220, 2003.
21. S. Avril, M. Grédiac, and F. Pierron. Sensitivity of the virtual fields method to noisy data. *Computational Mechanics*, 34(6):439–452, 2004.
22. S. Avril and F. Pierron. General framework for the identification of elastic constitutive parameters from full-field measurements. *International Journal of Solids and Structures*, 44:4978–5002, 2007.
23. N. Bahlouli, S. Mguil-Touchal, S. Ahzi, and M. Laberge. Stress-strain response of biomaterials by a digital image correlation method: application to tecoflex. *Journal of Materials Science & Technology*, 20:114–116, 2004.
24. S. Baker and I. Matthews. Equivalence and efficiency of image alignment algorithms. In *Proceedings of the IEEE Conference on Computer Vision and Pattern Recognition*, 1:1090–1097, 2001.
25. S. Baker and I. Matthews. Lucas-Kanade 20 years on: A unifying framework. *International Journal of Computer Vision*, 56(3):221–255, March 2004.
26. B. K. Bay, T. S. Smith, D. P. Fyhrie, and M. Saad. Digital volume correlation: Three-dimensional strain mapping using X-ray tomography. *Experimental Mechanics*, 39(3):217–226, September 1999.
27. B. K. Bay. Texture correlation: A method for the measurement of detailed strain distributions within trabecular bone. *Journal of Orthopaedic Research*, 13(2):258–267, March 1995.
28. T. A. Berfield, J. K. Patel, R. G. Shimmin, P. V. Braun, J. Lambros, and N. R. Sottos. Micro- and nanoscale deformation measurement of surface and internal planes via digital image correlation. *Experimental Mechanics*, 47(1):51–62, February 2007.
29. S. Bergonnier, F. Hild, and S. Roux. Digital image correlation used for mechanical tests on crimped glass wool samples. *The Journal of Strain Analysis for Engineering Design*, 40(2):185–197, February 2005.
30. P. Bésuelle, G. Viggiani, N. Lenoir, J. Desrues, and M. Bornert. X-ray micro CT for studying strain localization in clay rocks under triaxial compression. In J. Desrues, G. Viggiani, and P. Bésuelle, editors, *Proceedings of 2nd International Workshop on X-Ray CT for Geomaterials (GeoX 2006)*, pages 35–52, Grenoble & Aussois, France, October 2006.
31. A. R. Bhandari and J. Indue. Experimental study of strain rates effects on strain localization characteristics of soft rocks. *Soils and Foundations*, 45(1):125–140, 2005.
32. B. Birgisson, A. Montepara, E. Romeo, R. Roncella, J. A. L. Napier, and G. Tebaldi. Determination and prediction of crack patterns in hot mix asphalt (HMA) mixtures. *Engineering Fracture Mechanics*, 75(3–4):664–673, February–March 2008.
33. C. Bontus, T. Kohler, and R. Proksa. En:Pi: Filtered back-projection algorithm for helical CT using an n-rm Pi acquisition. *Medical Imaging*, 24(8):977–986, 2005.
34. F. Boochs. Off-line compilation of photogrammetric stereo models using digital image correlation. *Photogrammetria*, 41(3):183–199, June 1987.
35. M. Born and E. Wolf. *Principles of Optics*. Pergamon Press, London, 3rd edition, 1965.
36. P. Brand, R. Mohr, and P. Bobet. Distorsions optiques: Correction dans un modèle projectif. In *Actes du 9ème Congrès AFCET (RFIA'94)*, pages 87–98, Paris, France, January 1994.
37. H. A. Bruck, S. R. McNeill, M. A. Sutton, and W. H. Peters. Digital image correlation using Newton-Raphson method of partial differential correction. *Experimental Mechanics*, 29(3):261–267, September 1989.

38. H. A. Bruck, C. L. Moore, and T. M. Valentine. Bending actuation in polyurethanes with a symmetrically graded distribution of one-way shape memory alloy wires. *Experimental Mechanics*, 44(1):62–70, February 2004.
39. J. F. Cárdenas-Garcia, H. G. Yao, and S. Zheng. 3D reconstruction of objects using stereo imaging. *Optics and Lasers in Engineering*, 22(3):193–213, 1995.
40. Y. J. Chao and M. A. Sutton. Measurement of strains in a paper tensile specimen using computer vision and digital image correlation – Part 2: Tensile specimen test. *Tappi Journal*, 70(4):153–156, 1988.
41. I. Chasiotis. Mechanics of thin films and microdevices. *IEEE Transactions on Device and Materials Reliability*, 4(2):176–188, June 2004.
42. I. Chasiotis and W. G. Knauss. A new microtensile tester for the study of MEMS materials with the aid of atomic force microscopy. *Experimental Mechanics*, 42(1):51–57, March 2002.
43. D. J. Cheng, F. P. Chang, Y. S. Tan, and H. S. Don. Digital speckle-displacement measurement using a complex spectrum method. *Applied Optics*, 32:1839, 1993.
44. P. Cheng, M. A. Sutton, H. W. Schreier, and S. R. McNeill. Full-field speckle pattern image correlation with B-Spline deformation function. *Experimental Mechanics*, 42(3):344–352, September 2002.
45. T. Cheng, C. Dai, and R. Gan. Viscoelastic properties of human tympanic membrane. *Annals of Biomedical Engineering*, 35(2):305–314, February 2007.
46. S. Choi and S. P. Shah. Measurement of deformations on concrete subjected to compression using image correlation. *Experimental Mechanics*, 37(3):307–313, 1997.
47. S. Choi and S. P. Shah. Fracture mechanism in cement-based materials subjected to compression. *Journal of Engineering Mechanics*, 124(1):94–102, January 1998.
48. T. C. Chu, W. F. Ranson, M. A. Sutton, and W. H. Peters. Applications of digital-image-correlation techniques to experimental mechanics. *Experimental Mechanics*, 25(3):232–244, 1985.
49. D. Coburn and J. A. Slevin. Development of a digital speckle correlation system for use in the non-destructive testing of advanced engineering ceramics. *Key Engineering Materials*, 86-87:237–244, 1993.
50. S. A. Collette, M. A. Sutton, P. Miney, A. P. Reynolds, X. Li, P. E. Colavita, W. A. Scrivens, Y. Luo, T. Sudarshan, P. Muzykov, and M. L. Myrick. Development of patterns for nanoscale strain measurements: I. Fabrication of imprinted Au webs for polymeric materials. *Nanotechnology*, 15(12):1812–1817, December 2004.
51. P. Compston, M. Styles, and S. Kalyanasundaram. Low energy impact damage modes in aluminum foam and polymer foam sandwich structures. *Journal of Sandwich Structures and Materials*, 8(5):365–379, September 2006.
52. N. Cornille. *Accurate 3D Shape and Displacement Measurement Using a Scanning Electron Microscope*. Ph.D. thesis, INSA/University of South Carolina, France, Columbia, USA, June 2005.
53. N. Cornille, D. Garcia, M. A. Sutton, S. R. McNeill, and J.-J. Orteu. Automated 3-D reconstruction using a scanning electron microscope. In *Proceedings of SEM 2003 Annual Conference on Experimental and Applied Mechanics*, Charlotte, N.C., USA, June 2003.
54. N. Cornille, D. Garcia, M. A. Sutton, S. R. McNeill, and J.-J. Orteu. 3D shape and strain measurements using a scanning electron microscope. In *Proceedings of International Conference on Experimental Mechanics (ICEM12)*, Bari, Italy, September 2004.
55. J. Cugnoni, J. Botsis, V. Sivasubramaniam, and J. Janczak-Rusch. Experimental and numerical studies on size and constraining effects in lead-free solder joints. *Fatigue & Fracture of Engineering Materials & Structures*, 30(5):387–399, May 2007.
56. Y. Z. Dai, C. J. Tay, and F. P. Chiang. Determination of the plastic zone by laser-speckle correlation. *Experimental Mechanics*, 31(4):348–352, December 1991.
57. J. C. Dainty, editor. *Laser Speckle and Related Phenomena*. Springer, Berlin, 1975.
58. S. Daly, A. Miller, G. Ravichandran, and K. Bhattacharya. An experimental investigation of crack initiation in thin sheets of nitinol. *Acta Materialia*, 55(18):6322–6330, October 2007.

59. S. Das and N. Ahuja. Performance analysis of stereo, vergence, and focus as depth cues for active vision. *IEEE Transactions on Pattern Analysis and Machine Intelligence*, 17(12):1213–1219, 1995.
60. D. L. Davidson and J. Lankford. Fatigue crack tip strains in 7075-T6 aluminum alloy using stereo-imaging and their use in fatigue crack growth Models. In J. Lankford, D. L. Davidson, and R. P. Morris, W. L. Wei, editors, *Fatigue Mechanisms: Advances in Quantitative Measurement of Physical Damage*, pages 371–399. ASTM STP 811, Philadelphia, PA, 1983.
61. D. L. Davidson, D. R. Williams, and J. E. Buckingham. Crack tip stresses as determined from strains determined by stereoimaging. *Experimental Mechanics*, 23(2):242–248, 1983.
62. D. S. Dawicke and M. A. Sutton. CTOA and crack-tunneling measurements in thin sheet 2024-T3 aluminum alloy. *Experimental Mechanics*, 34(4):357–368, December 1994.
63. M. Devy, V. Garric, and J.-J. Orteu. Camera calibration from multiple views of a 2D object using a global non linear minimization method. In *International Conference on Intelligent Robots and Systems (IROS'97)*, Grenoble, France, September 1997.
64. P. Doumalin, M. Bornert, and D. Caldemaison. Microextensometry by image correlation applied to micromechanical studies using the scanning electron microscopy. In The Japan Society of Mechanical Engineering, editor, *Proceedings of International Conference on Advanced Technology in Experimental Mechanics (ATEM 99)*, pages 81–86, Ube City, Japan, 1999.
65. F. J. Doyle. The historical development of analytical photogrammetry. *Photogrammetric Engineering*, pages 259–265, 1964.
66. I. Dutta, D. Pan, S. Ma, B. S. Majumdar, and S. Harris. Role of shape-memory alloy reinforcements on strain evolution in lead-free solder joints. *Journal of Electronic Materials*, 35(10):1902–1913, October 2006.
67. G. Farin. *Curves and Surfaces for Computer Aided Geometric Design – A Practical Guide*. Academic Press, San Diego, CA, 1993.
68. O. Faugeras and F. Devernay. Computing Differential Properties of 3D shapes from Stereoscopic Images without 3D Models. Technical Report 2304, INRIA (France), 1994.
69. O. Faugeras, Q. T. Luong, and T. Papadopoulos. *The Geometry of Multiple Images*. MIT Press, Cambridge, 2001. ISBN 0-262-06158-9.
70. O. Faugeras. *Three-Dimensional Computer Vision: A Geometric Viewpoint*. MIT Press, Cambridge, 1993. ISBN 0-262-06220-8.
71. C. Feichter, Z. Major, and R. W. Lang. Deformation analysis of notched rubber specimens. *Strain*, 42(4):299–304, November 2006.
72. R. J. Finno and A. L. Rechenmacher. Effects of consolidation history on critical state of sand. *Journal of Geotechnical and Geoenvironmental Engineering*, 129(4):350–360, April 2003.
73. J. N. Florando, M. M. LeBlanc, and D. H. Lassila. Multiple slip in copper single crystals deformed in compression under uniaxial stress. *Scripta Materialia*, 57(6):537–540, September 2007.
74. J. N. Florando, M. Rhee, A. Arsenlis, M. M. LeBlanc, and D. H. Lassila. Calculation of the slip system activity in deformed zinc single crystals using digital 3-D image correlation data. *Philosophical Magazine Letters*, 86(12):795–805, December 2006.
75. G. Florou and R. Mohr. What accuracy for 3D measurements with cameras? *Proceedings of International Conference on Pattern Recognition*, Vienna, Austria, pages 354–358, 1996.
76. C. Forno. White-light speckle photography for measuring deformation, strain, and shape. *Optics & Laser Technology*, 7(5):217–221, October 1975.
77. P. Forquin, L. Rota, Y. Charles, and F. Hild. A method to determine the macroscopic toughness scatter of brittle materials. *International Journal of Fracture*, 125(1):171–187, January 2004.
78. M. Fouinneteau and A. K. Pickett. Failure characterisation of heavy tow braided composites using digital image correlation (DIC). *Applied Mechanics and Materials*, 5-6:399–406, 2006.
79. M. R. C. Fouinneteau and A. K. Pickett. Shear mechanism modelling of heavy tow braided composites using a meso-mechanical damage model. *Composites Part A: Applied Science and Manufacturing*, 38(11):2294–2306, November 2007.

80. C. Franck, S. Hong, S. A. Maskarinec, D. A. Tirrell, and G. Ravichandran. Three-dimensional full-field measurements of large deformations in soft materials using confocal microscopy and digital volume correlation. *Experimental Mechanics*, 47(3):427–438, June 2007.
81. A. Fusiello, M. Farenzena, A. Busti, and A. Benedetti. Computing rigorous bounds to the accuracy of calibrated stereo reconstruction. *IEEE Proceedings of Vision, Image and Signal Processing*, 152(6):695–701, December 2005. doi:10.1049/ip-vis:20041054.
82. D. Gabor. Microscopy by reconstructed wavefronts. In *Proceedings of the Royal Society*, volume A197, pages 454–487, 1949.
83. D. Garcia. *Mesure de formes et de champs de déplacements tridimentionnels par stéréo-corrélation d'images*. Ph.D. thesis, Institut National Polytechnique de Toulouse, France, December 2001.
84. D. Garcia, J.-J. Orteu, and M. Devy. Accurate calibration of a stereovision sensor: comparison of different approaches. In *Proceedings of 5th Workshop on Vision, Modeling, and Visualization (VMV'2000)*, pages 25–32, Saarbrücken, Germany, November 2000.
85. M. G. D. Geers, R. de Borst, and T. Peijs. Mixed numerical-experimental identification of non-local characteristics of random-fibre-reinforced composites. *Composites Science and Technology*, 59(10):1569–1578, August 1999.
86. A. Germaneau, P. Doumalin, and J.-C. Dupré. Improvement of accuracy of strain measurement by digital volume correlation for transparent materials. In *Proceedings of Photomechanics 2006*, Clermont-Ferrand, France, July 2006.
87. A. Germaneau, P. Doumalin, and J.-C. Dupré. 3D strain field measurement by correlation of volume images using scattered light: Recording of images and choice of marks. *Strain*, 43(3):207–218, August 2007.
88. A. Germaneau, P. Doumalin, and J.-C. Dupré. Full 3D measurement of strain field by scattered light for analysis of structures. *Experimental Mechanics*, 47(4):523–532, August 2007.
89. A. Germaneau. *Développement de techniques de mesure dans le volume : Photoélasticimétrie 3D par découpage optique et corrélation volumique par tomographie optique et rayons X. Application à l'étude des effets mécaniques 3D dans les structures et les biomatériaux*. Ph.D. thesis, Université de Poitiers, November 2007.
90. G. Geymonat, F. Hild, and S. Pagano. Identification of elastic parameters by displacement field measurement. *Comptes Rendus Mécanique*, 330(6):403–408, 2002.
91. G. Geymonat and S. Pagano. Identification of mechanical properties by displacement field measurement: A variational approach. *Meccanica*, 38(5):535–545, October 2003.
92. M. Giton, A.-S. Caro-Bretelle, and P. Ienny. Hyperelastic behaviour identification by a forward problem resolution: Application to a tear test of a silicone-rubber. *Strain*, 42(4):291–297, November 2006.
93. J. Gonzalez and W. G. Knauss. Strain inhomogeneity and discontinuous crack growth in a particulate composite. *Journal of the Mechanics and Physics of Solids*, 46(10):1981–1995, October 1998.
94. M. Grédiac. Principe des travaux virtuels et identification/principle of virtual work and identification. *Comptes Rendus de l'Académie des Sciences*, pages 1–5, 1989. In French with abridged English version.
95. M. Grédiac, F. Pierron, S. Avril, and E. Toussaint. The virtual fields method for extracting constitutive parameters from full-field measurements: a review. *Strain*, 42(4):233–253, 2006.
96. H. Gruner. Reinhard Hugershoff. *Photogrammetric Engineering*, 37(9):939–947, 1971.
97. R. Guastavino and P. Göoransson. A 3D displacement measurement methodology for anisotropic porous cellular foam materials. *Polymer Testing*, 26(6):711–719, September 2007.
98. M. Gutierrez and I. Vardoulakis. Energy dissipation and post-bifurcation behaviour of granular soils. *International Journal for Numerical and Analytical Methods in Geomechanics*, 31(3):435–455, March 2007.
99. K. Ha and R. A. Schapery. A three-dimensional viscoelastic constitutive model for particulate composites with growing damage and its experimental validation. *International Journal of Solids and Structures*, 35(26):3497–3517, 1998.

100. K. Haines and B. P. Hildebrand. Contour generation by wavefront construction. *Physics Letters*, 21:422–423, 1965.
101. G. Han, M. A. Sutton, and Y. J. Chao. A study of stationary crack tip deformation fields in thin sheets by computer vision. *Experimental Mechanics*, 34(2):751–761, June 1994.
102. G. Han, M. A. Sutton, and Y. J. Chao. A study of stable crack growth in thin SEC specimens of 304 stainless steel. *Engineering Fracture Mechanics*, 52(3):525–555, 1995.
103. J. Han and T. Siegmund. A combined experimental-numerical investigation of crack growth in a carbon/carbon composite. *Fatigue & Fracture of Engineering Materials & Structures*, 29(8):632–645, August 2006.
104. R. M. Haralick. Propagating covariance in computer vision. *Workshop on Performance Characteristics of Vision Algorithms*, Cambridge, UK, edited by H. Christensen, W. Forstner and C. Madsen, pages 1–12, 1996.
105. R. I. Hartley and A. Zisserman. *Multiple View Geometry in Computer Vision*. Cambridge University Press, Cambridge, 2nd edition, 2004, ISBN 0521540518.
106. R. S. He, C. T. Horn, H. J. Wang, and S. F. Hwang. Deformation measurement by a digital image correlation method combined with a hybrid genetic algorithm. *Key Engineering Materials*, 326-328(1):139–142, 2006.
107. E. Hecht and A. Zajac. *Optics*. Addison-Wesley, Reading, MA, 1st edition, 1974.
108. U. V. Helava. Digital correlation in photogrammetric instruments. *Photogrammetria*, 34(1): 18–41, January 1978.
109. J. D. Helm, S. R. McNeill, and M. A. Sutton. Improved three-dimensional image correlation for surface displacement measurement. *Optical Engineering*, 35(7):1911–1920, 1996.
110. J. D. Helm, M. A. Sutton, D. S. Dawicke, and G. Hanna. Three-dimensional computer vision applications for aircraft fuselage materials and structures. In *Proceedings of 1st Joint DoD/FAA/NASA Conference on Aging Aircraft*, 1:1327–1341, Ogden, UT, USA, 1997.
111. J. D. Helm, M. A. Sutton, and S. R. McNeill. Deformations in wide, center-notched, thin panels, part I: Three-dimensional shape and deformation measurements by computer vision. *Optical Engineering*, 42(5):1293–1305, 2003.
112. J. D. Helm, M. A. Sutton, and S. R. McNeill. Deformations in wide, center-notched, thin panels, part II: finite element analysis and comparison to experimental measurements. *Optical Engineering*, 42(5):1306–1320, 2003.
113. E. Héripré, M. Dexet, J. Crépin, L. Gélébart, A. Roos, M. Bornert, and D. Caldemaison. Coupling between experimental measurements and polycrystal finite element calculations for micromechanical study of metallic materials. *International Journal of Plasticity*, 23(9): 1512–1539, September 2007.
114. F. Hild. Correli: A Software for Displacement Field Measurements by Digital Image Correlation. Technical Report 254, LMT-Cachan, 2002.
115. F. Hild, B. Raka, M. Baudequin, S. Roux, and F. Cantelaube. Multiscale displacement field measurements of compressed mineral-wool samples by digital image correlation. *Applied Optics*, 41(32):6815–6828, November 2002.
116. F. Hild and S. Roux. Digital image correlation: From displacement measurement to identification of elastic properties – a review. *Strain*, 42(2):69–80, May 2006.
117. B. P. Hildebrand and K. Haines. Multiple wavelength and multiple source holography applied to contour generation. *Journal of the Optical Society of America*, 57:155–162, 1967.
118. K. D. Hinsch. Three-dimensional particle image velocimetry. *Measurement Science and Technology*, 6:742–753, 1995.
119. K. D. Hinsch and N. J. Lawson. Three-dimensional particle image velocimetry: Experimental error analysis of a digital angular stereoscopic system. *Measurement Science and Technology*, 8:1455–1464, 1995.
120. T. Hoc, L. Henry, M. Verdier, D. Aubry, L. Sedel, and A. Meunier. Effect of microstructure on the mechanical properties of Haversian cortical bone. *Bone*, 38(4):466–474, April 2006.
121. P. Hof. Stress determination on glass by digital image correlation (DIC). *IABSE Reports*, 92:38–39, 2006.

122. D. Holstein, L. Salbut, M. Kujawinska, and W. Jüptner. Hybrid experimental-numerical concept of residual stress analysis in laser weldments. *Experimental Mechanics*, 41(4), December 2001.
123. Z. Hou and Y. Qin. The study of fractal correlation method in the displacement measurement and its application. *Optics and Lasers in Engineering*, 39(4):465–472, April 2003.
124. T. S. Huang and S. D. Blostein. Error analysis in stereo determination of 3-D point positions. *IEEE Transactions on Pattern Analysis and Machine Intelligence*, 9(6):752–765, 1987.
125. Y. Y. Hung and C. E. Taylor. Speckle shearing interferometric camera-a tool for measurement of derivatives of surface displacement. In *SPIE Proceedings*, volume 41, pages 169–175, 1973.
126. V. Huon, B. Wattrisse, M. S. El Youssoufi, and A. Chrysochoos. Elastic behavior of anisotropic terra cotta ceramics determined by kinematic full-field measurements. *Journal of the European Ceramic Society*, 27(5):2303–2310, 2007.
127. C. Husson, N. Bahlouli, S. Mguil-Touchal, and S. Ahzi. Mechanical characterization of thin metal sheets by a digital image correlation method. *Journal de Physique IV*, 12(11):393–400, 2002.
128. S.-F. Hwang, J.-T. Horn, and H.-J. Wang. Strain measurement of nickel thin film by a digital image correlation method. *Strain*, 44(3):215–222, June 2008.
129. M. Ikeda, M. Masuda, K. Murata, and S. Ukyo. Analysis of in-plane shear behavior of wood based panels by digital image correlation. *Journal of the Society of Materials Science, Japan*, 55(6):569–575, 2006.
130. Correlated Solutions Incorporated. Calibrated sensor and method for calibrating same, 2006. Patent Filed 2/6/2003: Patent Issued 11/7/2006: US Patent No. 7,133,570.
131. E. Jiménez-Piqué, L. J. M. G. Dortmans, and G. De With. Fracture and crack profile fictitious crack modeling of porous poly(methyl methacrylate). *Journal of Polymer Science Part B: Polymer Physics*, 41(10):1112–1122, May 2003.
132. H. Jin and H. A. Bruck. A new method for characterizing nonlinearity in scanning probe microscopes using digital image correlation. *Nanotechnology*, 16(9):1849–1855, September 2005.
133. H. Jin, W.-Y. Lu, S. Scheffel, T. D. Hinnerichs, and M. K. Neilsen. Full-field characterization of mechanical behavior of polyurethane foams. *International Journal of Solids and Structures*, 44(21):6930–6944, October 2007.
134. H. Jin and H. A. Bruck. Pointwise digital image correlation using genetic algorithms. *Experimental Techniques*, 29(1):36–39, January 2005.
135. H. Jin and H. A. Bruck. Theoretical development for pointwise digital image correlation. *Optical Engineering*, 44(6):067003, June 2005.
136. A. Jones, J. Shaw, and A. Wineman. An experimental facility to measure the chemorheological response of inflated elastomer membranes at high temperature. *Experimental Mechanics*, 46(5):579–587, 2006.
137. Z. L. Kahn-Jetter and T. C. Chu. Three-dimensional displacement measurements using digital image correlation and photogrammic analysis. *Experimental Mechanics*, 30(1):10–16, March 1990.
138. J. Kang, M. Jain, D. S. Wilkinson, and J. D. Embury. Microscopic strain mapping using scanning electron microscopy topography image correlation at large strain. *The Journal of Strain Analysis for Engineering Design*, 40(6):559–570, 2005.
139. N. R. Kay, S. Ghosh, I. Guven, and E. Madenci. A combined experimental and analytical approach for interface fracture parameters of dissimilar materials in electronic packages. *Materials Science and Engineering: A*, 421(1–2):57–67, April 2006.
140. M. V. Klein. *Optics*. Wiley, New York, 1970.
141. W. G. Knauss and W. Zhu. Nonlinearly viscoelastic behavior of polycarbonate. I. Response under pure shear. *Mechanics of Time-Dependent Materials*, 6(3):231–269, 2002.
142. T. Kodaka, Y. Higo, S. Kimoto, and F. Oka. Effects of sample shape on the strain localization of water-saturated clay. *International Journal for Numerical and Analytical Methods in Geomechanics*, 31(3):483–521, March 2007.

143. O. Kolednik and H. P. Stüwe. The stereophotogrammetric determination of the critical crack tip opening displacement. *Engineering Fracture Mechanics*, 21(1):145–155, 1985.
144. G. Konecny. The international society for photogrammetry and remote sensing – 75 years old, or 75 years young. *Photogrammetric Engineering and Remote Sensing*, 51(7):919–933, 1985.
145. A. J. Lacey, N. A. Thacker, S. Crossley, and R. B Yates. A multi-stage approach to the dense estimation of disparity from stereo SEM images. *Image and Vision Computing*, 16(1):373–383, 1998.
146. F. Lagattu, F. Bridier, P. Villechaise, and J. Brillaud. In-plane strain measurements on a microscopic scale by coupling digital image correlation and an in situ SEM technique. *Materials Characterization*, 56(1):10–18, January 2006.
147. L. Larsson, M. Sjödahl, and F. Thuvander. Microscopic 3-D displacement field measurements using digital speckle photography. *Optics and Lasers in Engineering*, 41(5):767–777, May 2004.
148. J. M. Lavest, M. Viala, and M. Dhome. Do we really need an accurate calibration pattern to achieve a reliable camera calibration? In *Proceedings of European Conference on Computer Vision*, pages 158–174, Frieburg, Germany, 1998.
149. D. Lecompte, J. Vantomme, and H. Sol. Crack detection in a concrete beam using two different camera techniques. *Structural Health Monitoring*, 5(1):59–68, 2006.
150. D. Lecompte, A. Smits, H. Sol, J. Vantomme, and D. Van Hemelrijck. Mixed numerical-experimental technique for orthotropic parameter identification using biaxial tensile tests on cruciform specimens. *International Journal of Solids and Structures*, 44(5):1643–1656, March 2007.
151. J.-H. Lee and K.-J. Kang. In situ measurement of lateral side-necking of a fracture specimen using a stereo vision and digital image correlation. *Journal of the Korean Society of Precision Engineering*, 21(10):154–161, 2004.
152. S. Lee, F. Barthelat, N. Moldovan, H. D. Espinosa, and H. N. G. Wadley. Deformation rate effects on failure modes of open-cell Al foams and textile cellular materials. *International Journal of Solids and Structures*, 43(1):53–73, January 2006.
153. J. A. Leendertz and J. N. Butters. An image shearing speckle pattern interferometer for measuring bending moments. *Journal of Physics E: Scientific Instruments*, 7:1107–1110, 1973.
154. E. N. Leith and J. Upatnieks. Reconstructed wavefronts and communication theory. *Journal of the Optical Society of America*, 25:1123–1130, 1962.
155. N. Lenoir, M. Bornert, J. Desrues, P. Bésuelle, and G. Viggiani. 3D digital correlation applied to X-ray microtomography images of in-situ triaxial tests on argiliceous rock. In *Proceedings of Photomechanics 2006*, Clermont-Ferrand, France, 2006.
156. N. Lenoir, M. Bornert, J. Desrues, P. Bésuelle, and G. Viggiani. Volumetric digital image correlation applied to X-ray microtomography images from triaxial compression tests on argillaceaous rocks. *Strain*, 43(3):193–205, August 2007.
157. J. Li and R. Shi. Stretch-induced nerve conduction deficits in guinea pig ex vivo nerve. *Journal of Biomechanics*, 40(3):569–578, 2007.
158. N. Li, M. A. Sutton, X. Li, and H. W. Schreier. Full-field thermal deformation measurements in a scanning electron microscope by 2D digital image correlation. *Experimental Mechanics*, 48(5):635–646, October 2008.
159. X. Li, W. Xu, M. A. Sutton, and M. Mello. In situ nanoscale in-plane deformation studies of ultrathin polymeric films during tensile deformation using atomic force microscopy and digital image correlation techniques. *IEEE Transactions on Nanotechnology*, 6(1):4–12, January 2007.
160. X. Li, W. Xu, M. A. Sutton, and M. Mello. Nanoscale deformation and cracking studies of advanced metal evaporated magnetic tapes using atomic force microscopy and digital image correlation techniques. *Materials Science and Technology*, 22(7):835–844, July 2006.
161. J. Liu, M. A. Sutton, and J. S. Lyons. Experimental characterization of crack tip deformations in alloy 718 at high temperatures. *ASME Journal of Engineering Materials and Technology*, 20(1):71–78, 1998.

162. J. Liu and M. Iskander. Adaptive cross correlation for imaging displacements in soils. *Journal of Computing in Civil Engineering*, 18(1):46–57, 2004.
163. W. D. Lockwood and A. P. Reynolds. Simulation of the global response of a friction stir weld using local constitutive behavior. *Materials Science and Engineering: A*, 339(1):35–42, January 2003.
164. W. D. Lockwood, B. Tomaz, and A. P. Reynolds. Mechanical response of friction stir welded AA2024: experiment and modeling. *Materials Science and Engineering: A*, 323(1):348–353, January 2002.
165. W. D. Lockwood and A. P. Reynolds. Use and verification of digital image correlation for automated 3-D surface characterization in the scanning electron microscope. *Materials Characterization*, 42(2):123–134, February 1999.
166. L. Louis, T.-F. Wong, and P. Baud. Imaging strain localization by X-ray radiography and digital image correlation: Deformation bands in Rothbach sandstone. *Journal of Structural Geology*, 29(1):129–140, January 2007.
167. H. Lu and D. Cary. Deformation measurements by digital image correlation: Implementation of a second-order displacement gradient. *Experimental Mechanics*, 40(4):393–400, December 2000.
168. H. Lu, X. Zhang, and W. G. Knauss. Uniaxial, shear, and poisson relaxation and their conversion to bulk relaxation: Studies on poly(methyl methacrylate). *Polymer Composites*, 18(2):211–222, April 1997.
169. H. Lu, X. Zhang, and W. G. Knauss. Uniaxial, shear, and poisson relaxation and their conversion to bulk relaxation: Studies on poly(methyl methacrylate). *Polymer Engineering & Science*, 37(6):1053–1064, June 1997.
170. B. D. Lucas and T. Kanade. An iterative image registration technique with an application to stereo vision. In *Proceedings of the 1981 DARPA Image Understanding Workshop*, pages 121–130, April 1981.
171. O. Ludwig, M. Dimichiel, L. Salvo, M. Suéry, and P. Falus. In-situ three-dimensional microstructural investigation of solidification of an Al-Cu alloy by ultrafast X-ray microtomography. *Metallurgical and Materials Transactions A*, 36(6):1515–1523, June 2005.
172. P. F. Luo, Y. J. Chao, and M. A. Sutton. Application of stereo vision to three-dimensional deformation analyses in fracture experiments. *Optical Engineering*, 33(3):981–990, 1994.
173. P. F. Luo, Y. J. Chao, M. A. Sutton, and W. H. Peters. Accurate measurement of three-dimensional deformations in deformable and rigid bodies using computer vision. *Experimental Mechanics*, 33(2):123–132, June 1993.
174. P. F. Luo and F. C. Huang. Application of stereo vision to the study of mixed-mode crack-tip deformations. *Optics and Lasers in Engineering*, 33(5):349–368, May 2000.
175. A. R. Luxmoore, F. A. A. Amin, and W. T. Evans. In-plane strain measurement by speckle photography: A practical assessment of the use of Young's fringes. *Journal of Strain Analysis*, 9:26–34, 1974.
176. J. S. Lyons, J. Liu, and M. A. Sutton. High-temperature deformation measurements using digital-image correlation. *Experimental Mechanics*, 36(1):64–70, March 1996.
177. S. P. Ma, L. G. Wang, and G. C. Jin. Damage evolution inspection of rock using digital speckle correlation method (DSCM). *Key Engineering Materials*, 326-328:1117–1120, 2006.
178. Y. Ma, S. Soatto, J. Kosecka, and S. S. Sastry. *An Invitation to 3-D Vision: From Images to Geometric Models*. Springer, Berlin, 2003.
179. K. Machida, M. Sato, and S. Ogihara. Application of stress analysis by digital image correlation and intelligent hybrid method to unidirectional CFRP laminate. *Key Engineering Materials*, pages 1121–1124, 2007.
180. S. Mallik and M. L. Roblin. Speckle pattern interferometry applied to the study of phase objects. *Optics Communications*, 6:45–49, 1972.
181. L. E. Malvern. *Introduction to the Mechanics of a Continuous Medium*. Prentice-Hall, Englewood Cliffs, NJ, 1969.
182. Y. Marco, L. Chevalier, and M. Chaouche. WAXD study of induced crystallization and orientation in poly(ethylene terephthalate) during biaxial elongation. *Polymer*, 43(24):6569–6574, November 2002.

183. R. Marik, J. Kittler, and M. Petrou. Error sensitivity assessment of vision algorithm based on direct error-propagation. *Workshop on Performance Characteristics of Vision Algorithms*, Cambridge, UK, edited by H. Christensen, W. Forstner and C. Madsen, pages 45–58, 1996.
184. M. Marya, L. G. Hector, R. Verma, and W. Tong. Microstructural effects of AZ31 magnesium alloy on its tensile deformation and failure behaviors. *Materials Science and Engineering: A*, 418(1–2):341–356, February 2006.
185. M. A. Matin, J. G. A. Theeven, W. P. Vellinga, and M. G. D. Geers. Correlation between localized strain and damage in shear-loaded Pb-free solders. *Microelectronics Reliability*, 47(8):1262–1272, August 2007.
186. M. J. McGinnis, S. Pessiki, and H. Turker. Application of three-dimensional digital image correlation to the core-drilling method. *Experimental Mechanics*, 45(4):359–367, August 2005.
187. D. M. McGowan, D. R. Ambur, T. Glen Hanna, and S. R. McNeill. Evaluating the compressive response of notched Composite panels using full-field displacements. *Journal of Aircraft*, 38(1):122–129, 2001.
188. T. O. McKinley and B. K. Bay. Trabecular bone strain changes associated with subchondral stiffening of the proximal tibia. *Journal of Biomechanics*, 36(2):155–163, February 2003.
189. S. R. McNeill, W. H. Peters, and M. A. Sutton. Estimation of stress intensity factor by digital image correlation. *Engineering Fracture Mechanics*, 28(1):101–112, 1987.
190. W. Mekky and P. S. Nicholson. The fracture toughness of Ni/Al_2O_3 laminates by digital image correlation II: Bridging-stresses and R-curve models. *Engineering Fracture Mechanics*, 73(5):583–592, March 2006.
191. W. Mekky and P. S. Nicholson. Modeling constrained metal behavior in ceramic/metal laminates. *Composites Part B: Engineering*, 39(3):497–504, April 2008.
192. A. Montepara, E. Romeo, R. Roncella, and G. Tebaldi. Local strain inhomogeneity in hot mix asphalt by digital image correlation. In *Proceedings of Photomechanics 2006*, Clermont-Ferrand, France, 2006.
193. C. L. Moore and H. A. Bruck. A fundamental investigation into large strain recovery of one-way shape memory alloy wires embedded in flexible polyurethanes. *Smart Materials and Structures*, 11(1):130–139, February 2002.
194. Y. Morimoto and M. Fujigaki. Automated analysis of 3-D shape and surface strain distributions of a moving object using stereo vision. *Optics and Lasers in Engineering*, 18(3):195–212, 1993.
195. R. Moser and J. G. Lightner III. Using three-dimensional digital imaging correlation techniques to validate tire finite-element model. *Experimental Techniques*, 31(4):29–36, July/August 2007.
196. G. Murasawa and S. Yoneyama. Local strain distribution arising in shape memory alloy composite subjected to thermal loading. *Materials Transactions*, 47(3):766–771, 2006.
197. G. Murasawa, S. Yoneyama, and T. Sakuma. Nucleation, bifurcation and propagation of local deformation arising in NiTi shape memory alloy. *Smart Materials and Structures*, 16(1):160–167, February 2007.
198. K. Murata and M. Masuda. Microscopic observation of transverse swelling of latewood tracheid: Effect of macroscopic/mesoscopic structure. *Journal of Wood Science*, 52(4):283–289, 2006.
199. L. Muszyński. Empirical data for modeling: Methodological aspects in experimentation involving hygromechanical characteristics of wood. *Drying Technology*, 24(9):1115–1120, 2006.
200. D. V. Nelson, A. Makino, and T. Schmidt. Residual Stress Determination Using Hole Drilling and 3D Image Correlation. *Experimental Mechanics*, 46(1):31–38, February 2006.
201. D. P. Nicolella, A. E. Nicholls, J. Lankford, and D. T. Davy. Machine vision photogrammetry: a technique for measurement of microstructural strain in cortical bone. *Journal of Biomechanics*, 34(1):135–139, January 2001.
202. G. Nicoletto, G. Anzelotti, E. Riva, and R. Roncella. Mesomechanic strain fields in woven composites: Experiments vs. FEM modeling. In *Proceedings of 3rd Workshop on Optical*

Measurement Techniques for Structures and Systems (OPTIMESS'2007), Leuven, Belgium, May 2007.
203. H. B. Nielsen. Cubic splines. Lecture notes, Department of Mathematical Modeling, Technical University of Denmark, 1998.
204. J.-J. Orteu, T. Cutard, D. Garcia, E. Cailleux, and L. Robert. Application of Stereovision to the Mechanical Characterisation of Ceramic Refractories Reinforced with Metallic Fibres. *Strain*, 43(2):96–108, May 2007.
205. J.-J. Orteu, V. Garric, and M. Devy. Camera calibration for 3D reconstruction: Application to the measure of 3D deformations on sheet metal parts. In *SPIE Proceedings of European Symposium on Lasers, Optics and Vision in Manufacturing (EUROPTO)*, volume 3101, Munich, Germany, 1997.
206. J.-J. Orteu, Y. Rotrou, T. Sentenac, and L. Robert. An Innovative Method for 3-D Shape, Strain and Temperature Full-Field Measurement using a Single Type of Camera: Principle and Preliminary Results. *Experimental Mechanics*, 48(2):163–179, April 2008.
207. G. M. Owolabi and M. N. K. Singh. A comparison between two analytical models that approximate notch-root elastic-plastic strain-stress components in two-phase, particle-reinforced, metal matrix composites under multiaxial cyclic loading: experiments. *International Journal of Fatigue*, 28(8):918–925, August 2006.
208. G. M. Owolabi and M. N. K. Singh. Erratum to "A comparison between two analytical models that approximate notch-root elasticplastic strainstress components in two-phase, particle-reinforced, metal matrix composites under multiaxial cyclic loading: Experiments [Int. J. Fatigue 28 (2009) 918-925]". *International Journal of Fatigue*, 29(4):792, April 2007.
209. J. D. Pack, F. Noo, and R. Clackdoyle. Cone-beam reconstruction using the backprojection of locally filtered projections. *Medical Imaging*, 24(1):70–85, 2005.
210. B. Pan and H.-M. Xie. Digital image correlation method with differential evolution. *Journal of Optoelectronics Laser*, 18(1):100–103, 2007.
211. A. V. Panin, A. A. Panina, and Y. F. Ivanov. Deformation macrolocalisation and fracture in ultrafine-grained armco iron. *Materials Science and Engineering: A*, 486(1–2):267–272, July 2008.
212. S. Park, R. Dhakal, L. Lehman, and E. J. Cotts. Grain deformation and strain in board level snAgCu Solder Interconnects Under Deep Thermal Cycling. *IEEE Transactions on Components and Packaging Technologies*, 30(1):178–185, March 2007.
213. E. Parsons, M. C. Boyce, and D. M. Parks. An experimental investigation of the large-strain tensile behavior of neat and rubber-toughened polycarbonate. *Polymer*, 45:2665–2684, 2004.
214. J.-N. Périé, S. Calloch, C. Cluzel, and F. Hild. Analysis of a multiaxial test on a C/C composite by using digital image correlation and a damage model. *Experimental Mechanics*, 42(3):318–328, September 2002.
215. W. H. Peters and W. F. Ranson. Digital imaging techniques in experimental stress analysis. *Optical Engineering*, 21(3):427–431, 1982.
216. W. H. Peters, Z.-H. He, M. A. Sutton, and W. F. Ranson. Two-dimensional fluid velocity measurements by use of digital speckle correlation techniques. *Experimental Mechanics*, 24(2):171–121, 1984.
217. B. Peuchot. Camera virtual equivalent model – 0.01 pixel detector. In *Proceedings of Computerized Medical Imaging and Graphics*, 17:289–294, July 1993.
218. B. Peuchot. Utilisation de détecteurs subpixels dans la modélisation d'une caméra. In *Actes du 9ème Congrès AFCET (RFIA'94)*, pages 691–695, Paris, France, January 1994.
219. N. S. Phatak, S. E. Kim, Q. Sun, D. L. Parker, R. K. Sanders, A. I. Veress, B. J. Ellis, and J. A. Weiss. Noninvasive determination of ligament strain with deformable image registration. *Annals of Biomedical Engineering*, 35(7):1175–1187, 2007.
220. G. P. Pires, M. H. Robert, and R. Arrieux. Studies on drawing of the aluminium A5052 alloy in the thixocast condition. *Journal of Materials Processing Technology*, 157–158:596–603, December 2004.
221. M. Pitter, C. W. See, and M. Somekh. Subpixel microscopic deformation analysis using correlation and artificial neural networks. *Optics Express*, 8(6):322–327, 2001.

222. O. Ple, E. Astudillo de la Vega, G. Bernier, and O. Bayard. Biaxial tensile behavior of the reactive powder concrete. *ACI Special Publication*, 209:369–388, September 2002.
223. A. Poitou, A. Ammar, Y. Marco, L. Chevalier, and M. Chaouche. Crystallization of polymers under strain: From molecular properties to macroscopic models. *Computer Methods in Applied Mechanics and Engineering*, 192(28-30):3245–3264, July 2003.
224. D. Post. White light moiré interferometry. *Applied Optics*, 24:4163–4167, 1979.
225. D. Post, P. Ifju, and B. T. Han. *High Sensitivity Moiré*. Springer, New York, 1994.
226. P. J. Rae, S. J. P. Palmer, H. T. Goldrein, A. L. Lewis, and J. E. Field. White-light digital image cross-correlation (DICC) analysis of the deformation of composite materials with random microstructure. *Optics and Lasers in Engineering*, 41(4):635–648, April 2004.
227. J. F. Rakow and A. M. Waas. Size effects in metal foam cores for sandwich structures. *AIAA Journal*, 42(7):1331–1337, July 2004.
228. R. S. Ramakrishna and B. Vaidvanathan. Error analysis in stereo vision. *Proceedings ACCV'98, Book chapter with Lecture Notes in Computer Science*, 1351:296–304, 1997. doi: 10.1007/3-540-63930-6.
229. O. Ravn, N. A. Andersen, and A. T. Sorensen. Auto-calibration in Automation Systems using Vision. In *3rd International Symposium on Experimental Robotics (ISER'93)*, pages 206–218, Japan, 1993.
230. D. T. Read, Y.-W. Cheng, R. R. Keller, and J. David McColskey. Tensile properties of free-standing aluminum thin films. *Scripta Materialia*, 45(5):583–589, September 2001.
231. A. L. Rechenmacher and R. J. Finno. Digital image correlation to evaluate shear banding in dilative sands. *ASTM Geotechnical Testing Journal*, 27(1):13–22, January 2004.
232. A. L. Rechenmacher. Grain-scale processes governing shear band initiation and evolution in sands. *Journal of the Mechanics and Physics of Solids*, 54(1):22–45, January 2006.
233. J. N. Reddy. *An Introduction to Continuum Mechanics*. Cambridge University Press, Cambridge, 2007.
234. J. Réthoré, A. Gravouil, F. Morestin, and A. Combescure. Estimation of mixed-mode stress intensity factors using digital image correlation and an interaction integral. *International Journal of Fracture*, 132(1):65–79, March 2005.
235. J. Réthoré, F. Hild, and S. Roux. Shear-band capturing using a multiscale extended digital image correlation technique. *Computer Methods in Applied Mechanics and Engineering*, 196(49–52):5016–5030, November 2007.
236. J. Réthoré, F. Hild, and S. Roux. Extended digital image correlation with crack shape optimization. *International Journal for Numerical Methods in Engineering*, 73(2):258–272, 2008.
237. J. Réthoré, S. Roux, and F. Hild. An extended and integrated digital image correlation technique applied to the analysis fractured samples. *European Journal of Computational Mechanics*, in press, 2008.
238. J. Réthoré, S. Roux, and F. Hild. From pictures to extended finite elements: extended digital image correlation (X-DIC). *Comptes Rendus Mécanique*, 335(3):131–137, March 2007.
239. J. Réthoré, S. Roux, and F. Hild. Noise-robust stress intensity factor determination from kinematic field measurements. *Engineering Fracture Mechanics*, 75(13):3763–3781, September 2008.
240. M. R. Riahi, H. Latifi, and M. Sajjadi. Speckle correlation photography for the study of water content and sap flow in plant leaves. *Applied Optics*, 45(29):7674–7678, October 2006.
241. W. T. Riddell, R. S. Piascik, M. A. Sutton, W. Zhao, S. R. McNeill, and J. D. Helm. Local Crack Closure Measurements: Determination of Fatigue Crack Opening Loads from Near-Crack-Tip Measurements. In R. C. McClung and J. C. Newman, Jr., editors, *Advances in Fatigue Crack Closure Measurements and Analysis*, pages 157–174. ASTM STP 1343, West Conshohocken, PA, 1999.
242. L. Robert, F. Nazaret, T. Cutard, and J.-J. Orteu. Use of 3-D digital image correlation to characterize the mechanical behavior of a fiber reinforced refractory castable. *Experimental Mechanics*, 47(6):761–773, December 2007.
243. A. Rosenfeld. From image analysis to computer vision: An annotated bibliography, 1955-1979. *Computer Vision and Image Understanding*, 84:298–324, 2001.

244. K. L. Ruggy and B. N. Cox. Deformation mechanisms of dry textile preforms under mixed compressive and shear loading. *Journal of Reinforced Plastics and Composites*, 23(13): 1425–1442, September 2004.
245. S. S. Russell, M. A. Sutton, and H. S. Chen. Image correlation quantitative NDE of impact and fabrication damage in a glass fiber reinforced composite system. *Journal for Materials Evaluation*, 47(5):550–558, 1989.
246. N. Sabaté, D. Vogel, A. Gollhardt, J. Marcos, I. Gràcia, C. Cané, and B. Michel. Digital image correlation of nanoscale deformation fields for local stress measurement in thin films. *Nanotechnology*, 17(20):5264–5270, October 2006.
247. L. Salvo, P. Belestin, E. Maire, M. Jacquesson, C. Vecchionaci, E. Boller, M. Bornert, and P. Doumalin. Structure and mechanical properties of AFS sandwiches studied by in situ compression tests in X-ray microtomography. *Advanced Engineering Materials*, 6(6):411–415, 2004.
248. F. M. Sánchez and G. Pulos. Micro and macromechanical study of stress-induced martensitic transformation in a Cu-Al-Be polycrystalline shape memory alloy. *Materials Science Forum*, 509:87–92, 2006.
249. A. Savitzky and M. J. E. Golay. Smoothing and differentiation of data by simplified least squares procedures. *Analytical Chemistry*, 36:1627–1639, 1964.
250. T. E. Schmidt, J. Tyson, and K. Galanulis. Full-field dynamic displacement and strain measurement – specific examples using advanced 3D image correlation photogrammetry: Part II. *Experimental Techniques*, 27(4):22–26, July 2003.
251. T. E. Schmidt, J. Tyson, and K. Galanulis. Full-field dynamic displacement and strain measurement using advanced 3D image correlation photogrammetry: Part I. *Experimental Techniques*, 27(3):47–50, May 2003.
252. H. W. Schreier, J. Braasch, and M. A. Sutton. Systematic errors in digital image correlation caused by intensity interpolation. *Optical Engineering*, 39(11):2915–2921, 2000.
253. H. W. Schreier and M. A. Sutton. Systematic errors in digital image correlation due to undermatched subset shape functions. *Experimental Mechanics*, 42(3):303–310, September 2002.
254. H. W. Schreier, D. Garcia, and M. A. Sutton. Advances in light microscope stereo vision. *Experimental Mechanics*, 44(3):278–288, June 2004.
255. W. A. Scrivens, Y. Luo, M. A. Sutton, S. A. Collette, M. L. Myrick, P. Miney, P. E. Colavita, A. P. Reynolds, and X. Li. Development of patterns for digital image correlation measurements at reduced length scales. *Experimental Mechanics*, 47(1):63–77, February 2007.
256. Y. Seo, Y. R. Kim, M. W. Witczak, and R. Bonaquist. Application of digital image correlation method to mechanical testing of asphalt-aggregate mixtures. *Transportation Research Record*, 1789:162–172, 2002.
257. William N. Sharpe, Jr., editor, *Springer Handbook of Experimental Solid Mechanics*. Springer, Berlin, 2008. ISBN: 0-387-34362-8.
258. X. Q. Shi, Z. P. Wang, and J. P. Pickering. A new methodology for the characterization of fracture toughness of filled epoxy films involved in microelectronics packages. *Microelectronics Reliability*, 43(7):1105–1115, July 2003.
259. X. Q. Shi, Y. L. Zhang, and W. Zhou. Determination of fracture toughness of underfill/chip interface with digital image speckle correlation technique. *IEEE Transactions on Components and Packaging Technologies*, 30(1):101–109, March 2007.
260. T. Siebert, T. Becker, K. Spiltthof, I. Neumann, and R. Krupka. High-speed digital image correlation: Error estimations and applications. *Optical Engineering*, 46(5):051004 (7 pages), May 2007.
261. G. Sierra, B. Wattrisse, and C. Bordreuil. Structural analysis of steel to aluminum welded overlap joint by digital image correlation. *Experimental Mechanics*, 48(2):213–223, April 2008.
262. M. Sjödahl. Electronic speckle photography: Increased accuracy by nonintegral pixel shifting. *Applied Optics*, 33:6667–6673, October 1994.
263. M. Sjödahl. Accuracy in electronic speckle photography. *Applied Optics*, 36(13):2875–2885, May 1997.

264. M. Sjödahl. Some recent advances in electronic speckle photography. *Optics and Lasers in Engineering*, 29(2–3):125–144, March 1998.
265. T. S. Smith, B. K. Bay, and M. M. Rashid. Digital volume correlation including rotational degrees of freedom during minimization. *Experimental Mechanics*, 42(3):272–278, September 2002.
266. B. Stanford, R. Albertani, and P. Ifju. Static Finite Element Validation of a Flexible Micro Air Vehicle. *Experimental Mechanics*, 47(2):283–294, April 2007.
267. K. A. Stetson. New design for laser image speckle interferometer. *Optics and Laser Technology*, 2:179–181, 1970.
268. G. Strang. *Linear Algebra and Its Applications*. Academic Press, 2nd edition, New York, 1980.
269. K. V. Subramaniam, C. Carloni, and L. Nobile. Width effect in the interface fracture during shear debonding of FRP sheets from concrete. *Engineering Fracture Mechanics*, 74(4):578–594, March 2007.
270. Y. Sun and J. H. L. Pang. AFM image reconstruction for deformation measurements by digital image correlation. *Nanotechnology*, 17(4):933–939, February 2006.
271. Y. Sun and J. H. L. Pang. Study of optimal subset size in digital image correlation of speckle pattern images. *Optics and Lasers in Engineering*, 45(9):967–974, September 2007.
272. Y. Sun and J. H. L. Pang. Digital image correlation for solder joint fatigue reliability in microelectronics packages. *Microelectronics Reliability*, 48(2):310–318, 2008.
273. Y. Sun, J. H. L. Pang, C. K. Wong, and F. Su. Finite element formulation for a digital image correlation method. *Applied Optics*, 44(34):7357–7363, December 2005.
274. Z. Sun, J. S. Lyons, and S. R. McNeill. Measuring microscopic deformations with digital image correlation. *Optics and Lasers in Engineering*, 27(4):409–428, July 1997.
275. H. J. Sung, S.-H. Park, and M. S. Kim. Accuracy of correlation-based image registration for pressure-sensitive paint. *Experiments in Fluids*, 39(3):630–635, September 2005.
276. H. Jin Sung, S.-H. Park, and M. S. Kim. Erratum to “accuracy of correlation-based image registration for pressure-sensitive paint” [Exp. in Fluids 39 (2005) 630–635]. *Experiments in Fluids*, 40(4):664–664, April 2006.
277. M. A. Sutton. Digital image correlation. In W. N. Sharpe, Jr., editor, *Springer Handbook of Experimental Solid Mechanics*, Springer, Berlin, 2008. ISBN: 978-0-387-26883-5.
278. M. A. Sutton, T. L. Chae, J. L. Turner, and H. A. Bruck. Development of a Computer Vision Methodology for the Analysis of Surface Deformations in Magnified Images. In G. F. Vander Voort, editor, *MiCon 90: Advances in Video Technology for Microstructural Control*, pages 109–132. ASTM STP 1094, Philadelphia, PA, 1991.
279. M. A. Sutton and Y. J. Chao. Measurement of strains in a paper tensile specimen using computer vision and digital image correlation – part 1: Data acquisition and image analysis system. *Tappi Journal*, 70(3):173–175, 1988.
280. M. A. Sutton, M. Q. Cheng, W. H. Peters, Y. J. Chao, and S. R. McNeill. Application of an optimized digital correlation method to planar deformation analysis. *Image and Vision Computing*, 4(3):143–150, August 1986.
281. M. A. Sutton, X. Ke, S. M. Lessner, M. Goldbach, M. Yost, F. Zhao, and H. W. Schreier. Strain field measurements on mouse carotid arteries using microscopic three dimensional digital image correlation. *Journal of Biomedical Materials Research Part A*, 84(1):178–190, July 2008.
282. M. A. Sutton, N. Li, D. Garcia, N. Cornille, J.-J. Orteu, S. R. McNeill, H. W. Schreier, and X. Li. Metrology in a scanning electron microscope: Theoretical developments and experimental validation. *Measurement Science and Technology*, 17(10):2613–2622, August 2006. ISSN 0957-0233.
283. M. A. Sutton, N. Li, D. Garcia, N. Cornille, J.-J. Orteu, S. R. McNeill, H. W. Schreier, X. Li, and A. P. Reynolds. Scanning electron microscopy for quantitative small and large deformation measurements part II: Experimental validation for magnifications from 200 to 10,000. *Experimental Mechanics*, 47(6):789–804, December 2007.

284. M. A. Sutton, N. Li, D. C. Joy, A. P. Reynolds, and X. Li. Scanning Electron microscopy for quantitative small and large deformation measurements part I: SEM imaging at magnifications from 200 to 10,000. *Experimental Mechanics*, 47(6):775–787, December 2007.
285. M. A. Sutton, S. R. McNeill, J. D. Helm, and Y. J. Chao. Advances in two-dimensional and three-dimensional computer vision. In P. K. Rastogi, editor, *Photomechanics, Topics in Applied Physics*. Springer, Berlin, 2000.
286. M. A. Sutton, S. R. McNeill, J. D. Helm, and H. W. Schreier. Computer vision applied to shape and deformation measurement. In P. K. Rastogi and D. Inaudi, editors, *Trends in Optical Nondestructive Testing and Inspection*. Elsevier Science, Amsterdam, 2000.
287. M. A. Sutton, S. R. McNeill, J. Jang, and M. Babai. Effects of sub-pixel image restoration on digital correlation error. *Optical Engineering*, 27(10):870–877, 1988.
288. M. A. Sutton, W. J. Wolters, W. H. Peters, W. F. Ranson, and S. R. McNeill. Determination of displacements using an improved digital correlation method. *Image and Vision Computing*, 1(3):133–139, August 1983.
289. M. A. Sutton, J. H. Yan, V. Tiwari, H. W. Schreier, and J.-J. Orteu. The effect of out of plane motion on 2D and 3D digital image correlation measurements. *Optics and Lasers in Engineering*, 46(10):746–757, October 2008.
290. M. A. Sutton, W. Zhao, S. R. McNeill, J. D. Helm, R. S. Piascik, and W. T. Riddell. Local crack closure measurements: Development of a measurement system using computer vision and a far-field microscopy. In R. C. McClung and J. C. Newman, Jr., editors. *Advances in Fatigue Crack Closure Measurement and Analysis*, pages 145–156, ASTM STP 1343, West Conshohocken, PA, USA, November, 1997.
291. M. A. Sutton and Y. J. Chao. Computer vision in fracture mechanics. *Experimental Techniques in Fracture*, pages 59–94, 1993.
292. M. A. Sutton, Y. J. Chao, and J. S. Lyons. Computer vision methods for surface deformation measurements in fracture mechanics. In J. S. Epstein, editor, *ASME-AMD Novel Experimental Methods in Fracture*, 176:123–133, 1993. ISBN: 978-0-471-18865-0.
293. M. A. Sutton, S. R. McNeill, J. D. Helm, and Y. J. Chao. Advances in two-dimensional and three-dimensional computer vision. In P. K. Rastogi, editor, *Photomechanics, Topics in Applied Physics*. Springer, Berlin, 2000.
294. M. A. Sutton, J. L. Turner, T. L. Chae, and H. A. Bruck. Full field representation of discretely sampled surface deformation for displacement and strain analysis. *Experimental Mechanics*, 31(2):168–177, 1991.
295. M. A. Sutton, J. L. Turner, Y. J. Chao, H. A. Bruck, and T. L. Chae. Experimental investigations of three-dimensional effects near a crack tip using computer vision. *International Journal of Fracture*, 53(3):201–228, February 1992.
296. M. A. Sutton, S. R. McNeill, J. D. Helm, and M. L. Boone. Measurement of crack tip opening displacement and full-field deformations during fracture of aerospace materials using 2D and 3D image correlation methods. In A. Lagarde, editor, *IUTAM Symposium on Advanced Optical Methods and Applications in Solid Mechanics*, 82:571–580, Springer, Netherlands, 1998.
297. M. A. Sutton, B. Wan, M. F. Petrou, and K. Harries. Two-dimensional computer vision to investigate FRP-concrete bond toughness. *Advanced Measurement Methods*, 1, 2002.
298. M. A. Sutton, J. Yan, X. Deng, C.-S. Cheng, and P. Zavattieri. Three-dimensional digital image correlation to quantify deformation and crack-opening displacement in ductile aluminum under mixed-mode I/III loading. *Optical Engineering*, 46(5):051003 (17 pages), May 2007.
299. M. A. Sutton, B. Yang, A. P. Reynolds, and J. Yan. Banded microstructure in 2024-T351 and 2524-T351 aluminum friction stir welds: Part II. Mechanical characterization. *Materials Science and Engineering: A*, 364(1–2):66–74, January 2004.
300. P. Synnergren and M. Sjödahl. A stereoscopic digital speckle photography system for 3-D displacement field measurements. *Optics and Lasers in Engineering*, 31(6):425–443, June 1999.
301. P. Synnergren. Measurement of three-dimensional displacement fields and shape using electronic speckle photography. *Optical Engineering*, 36(8):2302–2310, August 1997.

302. P. Synnergren and H. Timothy Goldrein. Dynamic measurements of internal three-dimensional displacement fields with digital speckle photography and flash X rays. *Applied Optics*, 38(28):5956–5961, October 1999.
303. H. Tan, C. Liu, Y. Huang, and P. H. Geubelle. The cohesive law for the particle/matrix interfaces in high explosives. *Journal of the Mechanics and Physics of Solids*, 53(8):1892–1917, August 2005.
304. C. Tang. A new synthetical algorithm for digital image correlation method and its application in the fracture. *Optical Technique*, 29(1):69–72, 2003.
305. Q. Tian and M. N. Huhns. Algorithms for subpixel registration. *Computer Vision, Graphics, and Image Processing*, 35(2):220–233, August 1986.
306. V. Tiwari, M. A. Sutton, and S. R. McNeill. Assessment of high speed imaging systems for 2D and 3D deformation measurements: Methodology development and validation. *Experimental Mechanics*, 47(4):561–579, August 2007.
307. W. Tong. Strain characterization of propagative deformation bands. *Journal of the Mechanics and Physics of Solids*, 46(10):2087–2102, October 1998.
308. W. Tong. Detection of plastic deformation patterns in a binary aluminum alloy. *Experimental Mechanics*, 37(4):452–459, December 1997.
309. W. Tong, H. Tao, X. Jiang, N. Zhang, M. P. Marya, L. G. Hector, and X. Q. Gayden. Deformation and fracture of miniature tensile bars with resistance-spot-weld microstructures. *Metallurgical and Materials Transactions A*, 36(10):2651–2669, October 2005.
310. V. Tvergaard. Cohesive zone representations of failure between elastic or rigid solids and ductile solids. *Engineering Fracture Mechanics*, 70(14):1859–1868, 2003.
311. J. Tyson, T. Schmidt, and K. Galanulis. Biomechanics deformation and strain measurements with 3D image correlation photogrammetry. *Experimental Techniques*, 26(5):39–42, September 2002.
312. S. Ukyo and M. Masuda. Fracture analysis of wood in moment-resisting joints with four drift-pins using digital image correlation method (DIC). *Journal of the Society of Materials Science, Japan*, 51(4):367–372, 2002.
313. S. Ukyo and M. Masuda. Investigation of the true stress-strain relation in shear using the digital image correlation method. *Journal of the Japan Wood Research Society*, 50(3):146–150, 2004.
314. E. Umezaki and K. Futase. Deformation behaviour of the cross-section of a heat-sealed area in plastic film used for liquid package bags. *The Journal of Strain Analysis for Engineering Design*, 32(1):29–35, 1997.
315. E. Umezaki and T. Ichikawa. Measurement of deformation of epoxy resin plates with an embedded SMA wire using digital image correlation. *International Journal of Modern Physics B*, 17(8–9):1750–1755, 2003.
316. M. Unser, A. Aldroubi, and M. Eden. B–spline processing part 2 — efficient design and applications. *IEEE Trans. Signal Processing*, 41(2):834–847, 1993.
317. M. Unser, A. Aldroubi, and M. Eden. B–spline processing part I — theory. *IEEE Transactions on Signal Processing*, 41(2):821–833, 1993.
318. A. Vagnon, O. Lame, D. Bouvard, M. Di Michiel, D. Bellet, and G. Kapelski. Deformation of steel powder compacts during sintering: Correlation between macroscopic measurement and in situ microtomography analysis. *Acta Materialia*, 54(2):513–522, January 2006.
319. M. F. Vallat, P. Martz, J. Fontaine, and J. Schultz. The application of coherent optics to the study of adhesive joints. I. Speckle photography. *Journal of Applied Polymer Science*, 31(2):309–321, February 1986.
320. D. VanArsdale. Homogeneous transformation matrices for computer graphics. *Computers & Graphics*, 18(2):177–191, 1994.
321. G. Vendroux and W. G. Knauss. Submicron deformation field measurements: Part 1. Developing a digital scanning tunneling microscope. *Experimental Mechanics*, 38(1):18–23, March 1998.
322. G. Vendroux and W. G. Knauss. Submicron deformation field: Part 2. Improved digital image correlation. *Experimental Mechanics*, 38(2):86–92, June 1998.

323. G. Vendroux, N. Schmidt, and W. G. Knauss. Submicron deformation field measurements: Part 3. Demonstration of deformation determinations. *Experimental Mechanics*, 38(3):154–160, September 1998.
324. A. I. Veress, N. S. Phatak, and J. A. Weiss. Deformable image registration with hyperelastic warping. In Evangelia Micheli-Tzanakou, Jasjit S. Suri, David L. Wilson, and Swamy Laxminarayan, editors, *Handbook of Medical Image Analysis: Advanced Segmentation and Registration Models*, 2003.
325. A. I. Veress, J. A. Weiss, G. T. Gullberg, D. G. Vince, and R. D. Rabbitt. Strain measurement in coronary arteries using intravascular ultrasound and deformable images. *ASME Journal of Biomech Engineering*, 124:734–741, 2003.
326. E. Verhulp, B. van Rietbergen, and R. Huiskes. A three-dimensional digital image correlation technique for strain measurements in microstructures. *Journal of Biomechanics*, 37(9):1313–1320, September 2004.
327. C. M. Vest. *Holographic Interferometry*. Wiley, New York, 1979.
328. P. Vialettes, J.-M. Siguier, P. Guigue, M. Karama, S. Mistou, O. Dalverny, S. Granier, and F. Petitjean. Experimental and numerical simulation of super-pressure balloon apex section: Mechanical behavior in realistic flight conditions. *Advances in Space Research*, 37(11):2082–2086, 2006.
329. J. H. Vogel and D. Lee. An automated two-view method for determining strain distributions on deformed surfaces. *Journal of Material Shaping Technology*, 6(4):205–216, 1989.
330. B. Wan, M. A. Sutton, M. F. Petrou, K. A. Harries, and N. Li. Investigation of bond between fiber reinforced polymer and concrete undergoing global mixed mode I/II loading. *Journal of Engineering Mechanics*, 130(12):1467–1475, 2004.
331. Christopher C.-B. Wang, N. O. Chahine, C. T. Hung, and G. A. Ateshian. Optical determination of anisotropic material properties of bovine articular cartilage in compression. *Journal of Biomechanics*, 36(3):339–353, March 2003.
332. Christopher C.-B. Wang, Jian-Ming Deng, Gerard A. Ateshian, and Clark T. Hung. An automated approach for direct measurement of two-dimensional strain distributions within articular cartilage under unconfined compression. *Journal of Biomechanical Engineering*, 124(5):557–567, 2002.
333. Y. Wang and A. M. Cuitiño. Full-field measurements of heterogeneous deformation patterns on polymeric foams using digital image correlation. *International Journal of Solids and Structures*, 39(13):3777–3796, June 2002.
334. Y. Wang, G. Gioia, and A. M. Cuitiño. The deformation habits of compressed open-cell solid foams. *Journal of Engineering Materials and Technology*, 122(4):376–378, October 2000.
335. Y. Q. Wang, M. A. Sutton, and H. W. Schreier, Quantitative error assessment in 3D stereovision, in review.
336. Y. Q. Wang, M. A. Sutton, and H. W. Schreier. Quantitative error assessment in pattern matching: Effect of interpolation, intensity pattern noise and contrast, subset size on motion accuracy. *International Journal of Strain*, in press 2008.
337. A. Wasil and D.C. Merchant. Plate-deflection measurement by photogrammetric methods. *Experimental Mechanics*, 4(3):77–83, March 1964.
338. G. O. Williams, V. Randle, J. R. Cowan, and P. Spellward. The role of misorientation and phosphorus content on grain growth and intergranular fracture in iron-carbon-phosphorus alloys. *Journal of Microscopy*, 213(3):321–327, March 2004.
339. H. Wolf, D. König, and Th. Triantafyllidis. The influence of the stress-strain behavior of non-cohesive soils on the geometry of shear band systems under extensional strain. *Engineering Structures*, 28(13):1760–1773, November 2006.
340. L. Xu and J. H. L. Pang. In-situ electromigration study on Sn-Ag-Cu solder joint by digital image speckle analysis. *Journal of Electronic Materials*, 35(11):1993–1999, November 2006.
341. Z.-H. Xu, M. A. Sutton, and X. D. Li. Drift and spatial distortion elimination in atomic force microscopy images by digital image correlation techniques. *The Journal of Strain Analysis for Engineering Design*, 43(8):729–743, 2008.

342. Z.-H. Xu, M. A. Sutton, and X. D. Li. Mapping nanoscale wear field by combined atomic force microscopy and digital image correlation techniques. *Acta Materialia*, 56(20):6304–6309, December 2008.
343. I. Yamaguchi. Automatic measurement of in-plane translation by speckle correlation using a linear image sensor. *Journal of Physics E - Scientific Instruments*, 19(11):944–948, November 1986.
344. S. Yoneyama, Y. Morimoto, and M. Takashi. Automatic evaluation of mixed-mode stress intensity factors utilizing digital image correlation and nonlinear least-squares. *Journal of Japanese Society for Non-Destructive Inspection*, 51(10):667–673, 2002.
345. S. Yoneyama, Y. Morimoto, and M. Takashi. Automatic evaluation of mixed-mode stress intensity factors utilizing digital image correlation. *Strain*, 42(1):21–29, February 2006.
346. S. Yoneyama, T. Ogawa, and Y. Kobayashi. Evaluating mixed-mode stress intensity factors from full-field displacement fields obtained by optical methods. *Engineering Fracture Mechanics*, 74(9):1399–1412, June 2007.
347. D. Zhang, D. D. Arola, R. K. Reprogel, W. Zheng, U. Tasch, and R. M. Dyer. A method for characterizing the mechanical behaviour of hoof horn. *Journal of Materials Science*, 42(4):1108–1115, January 2007.
348. D. Zhang, C. D. Eggleton, and D. D. Arola. Evaluating the mechanical behavior of arterial tissue using digital image correlation. *Experimental Mechanics*, 42(4):409–416, December 2002.
349. J. Zhang, M. Li, C. Y. Xiong, J. Fang, and S. Yi. Thermal deformation analysis of BGA package by digital image correlation technique. *Microelectronics International*, 22(1):34–42, January 2005.
350. J. Zhang, G. Jin, S. Ma, and L. Meng. Application of an improved subpixel registration algorithm on digital speckle correlation measurement. *Optics & Laser Technology*, 35(7):533–542, October 2003.
351. W. Zhang and X. Deng. Formulation of a cohesive zone model for a mode III crack. *Engineering Fracture Mechanics*, 72(12):1818–1829, 2005.
352. Z.-F. Zhang, Y.-L. Kang, H.-W. Wang, Q.-H. Qin, Y. Qiu, and X.-Q. Li. A novel coarse-fine search scheme for digital image correlation method. *Measurement*, 39(8):710–718, October 2006.
353. J. Zhou, Z. Gao, S. Allameh, E. Akpan, A. Cuitiño, and W. Soboyejo. Multiscale deformation of open cell aluminum foams. *Mechanics of Advanced Materials and Structures*, 12(3):201–216, May–June 2005.
354. J. Zhou, Z. Gao, A. M. Cuitiño, and W. O. Soboyejo. Effects of heat treatment on the compressive deformation behavior of open cell aluminum foams. *Materials Science and Engineering: A*, 386(1–2):118–128, November 2004.
355. A. G. Zink, R. W. Davidson, and R. B. Hanna. Strain measurement in wood using a digital image correlation technique. *Wood and Fiber Science*, 27(4):346–359, 1995.

Index

2D-DIC, *see* two-dimensional digital image correlation
3D position, 226
3D-DIC, *see* three-dimensional digital image correlation

a priori, *see* distortion, non-parametric
aberrations, *see* distortion
aliasing, 95, 233, 234, 236
amplitude attenuation, *see* interpolation
angle of view, 19
aperture
  diaphragm, 17
  practical considerations, 230, 247–249
  problem, 81, 86
  relative, 18
astigmatism, *see* distortion
autocorrelation, 233

B-spline, 290
bi-normal, 270
bias, 114, 226
  interpolation, 100, 103–110, 113
  noise-induced, 110–113
  strain, 109
box filter, 90
brightness change constraint equation, 85
bundle adjustment, 48, 67, 76, 180, 250

calibration, 44–50, 228
  constraints, 8, 50
  single camera, 44–60
  stereo, 73–80
  target, 44
camera
  calibration, *see* calibration
  coordinates, 28
  high-speed, 191
  pinhole, 27
  selection, 247
camera parameters
  extrinsic, 46, 67, 228
  intrinsic, 46, 67, 228
cannula, 200
cantilever beam, 151
catadioptric, 148
CCD resolution, 177
center
  lens, 22
  sensor coordinates, 29, 30
circle of confusion, 18
COD, *see* crack opening displacement
compositional, *see* update rules
computed tomography, 209
  image slices, 209–211
confidence margin, 113, 116
contrast, 246
convolution, 100, 104, 105
coordinate system, 28–33
  camera, 28
  common, 182
  sensor, 29
  world, 28
coordinate transformation
  camera and image system, 29
  image and sensor system, 29
  world and camera systems, 28
correlation criterion, *see* optimization criterion
correspondence problem, 82, 117
covariance, 228
covariance matrix, 116
crack
  closure load, 148
  extension, 177

crack (*continued*)
growth, 176, 186
opening displacement, 145
cross-correlation
normalized, 96
CT scanner, *see* computed tomography
curvature of field, *see* distortion

dead pixels, 231
decorrelation, 88, 92
defocus zone, 16
deformation gradient tensor, 255
depth of field, 17, 148, 229, 245
far limit, 19
near limit, 19
derivative filter, 100
DIC, *see* dimensional digital image correlation
differential line element, 255
differential motion estimation, 84
digital image correlation, 1–3
three-dimensional digital image correlation, 7, 9, 69, 96, 150, 229, 247
two-dimensional digital image correlation, 3, 65, 81, 96, 99, 140, 150, 246
volumetric digital image correlation, 10, 209
dilatation, 265
disparity, *see* gray value difference
distortion, 33, 150
astigmatism, 38
calibration, 67
coma, 36
curvature of field, 39
de-centering, 40
drift-induced, 168
linear, 39
non-parametric, 42–63
parametric, 35–41
radial, 40
spatial, 168
spherical, 36
volumetric, 213
DOF, *see* depth of field
drop tower, 193
dust, 231
dwell time, 161

E-beam rastering, 161
effective object distance, 128
efficient image matching methods, 99
general shape function, 101
update rules, 99
entrance pupil, 18
diameter, 18

epipolar
constraint, 72
line, 72, 94
error estimation
covariance for 3D position, 228
optimal 3D position, 225
Euler angles, 287
Eulerian strain tensor, 257
expectation, 114, 227
exposure time, 243
extensometer, 120

f-number, 18
far-field microscope, *see* microscope
field of view, 19, 229
fill-factor, 236
finite-element based smoothing, 156
fluorescent beads, 200
focal length, 14
front, 22
rear, 22
focal point, 14, 22
front, 22
rear, 22
focus plane, 15
Fourier filter, 108
FOV, *see* field of view
fracture, 195
front image plane model, 25

Gauss
approximation, 13
Gaussian distribution, *see* weighting function
gradient search, 275
gray card, 231
gray value difference, 87
grid, *see* calibration target, aperture problem

Hessian
approximation, 274
calibration matrix structure, 56–60
histogram, 232
homogeneous coordinates, 21, 285
homography, 93
hyperfocal distance, 19

illumination, *see* lighting
image artifacts, 231
image blur, 16
image distance, 15, 22
image integration, 167
image matching errors
interpolation bias, 100, 103–110, 113
noise with interpolation, 110–113

practical considerations, 245, 246
statistical covariance for error estimation, 116–117
statistical error analysis, 113–116
image matching methods
deformable subset with shape function, 88
differential, 84
template, 87
image matching metrices, *see* optimization criterion
image plane, 25
front, 27
image plane coordinates, 27, 29
image rectification, *see* rectification
image registration, *see* image matching methods
image shift, 165
intensity pattern matching, *see* image matching methods and optimization criterion
interpolation, 245
amplitude attenuation, 106
B-spline, 106
cubic polynomial, 105, 113
linear, 104
optimized, 107–109
phase error, 105, 236
polynomial, 104
inverse compositional
see update rules, 317
inverse methods, 150
isotropic, 83

Lagrangian strain tensor, 122, 257
least squares, 226, 274
lens
selection, 248
thick, 22
thin, 13
lensmaker's equation, 15
Levenberg–Marquardt, 273
lighting, 95
linear algebra, 261
fundamental theorem, 262
lithography, 140
local strain estimation, 269
low-pass filter, 90, 91
Lucas–Kanade, 87, 99

magnetic resonance imaging, 209
matching, 84
material property estimation
experiments, 150–159
virtual fields, 150–154
micro-damage, 218
microscale, 138
microscope
confocal, 209
far-field, 138–150
scanning electron, 159
mixed-mode loading, 191
mouse carotid, 200, 205
MRI
see magnetic resonance imaging, 317

NCC, *see* optimization criterion
neighborhood, *see* subset
Newton–Raphson, 276
noise, 110, 113
thermal, 245
non-linear optimization, 273
NSSD, *see* optimization criterion

optical flow, 85
optimal 3D estimator, 226
optimization criterion
NCC, 98
NSSD, 97
offset in lighting, 96
SAD, 98
scale in lighting, 97
SSD, 96
volumetric sum of squared differences, 214
ZNSSD, 98
ZSSD, 97
optimization search methods
gradient, 275
Least squares, 277–279
Levenberg–Marquardt, 273
Newton–Raphson, 276
steepest descent, 276
orthonormal basis, 263
out-of-plane motion
experiments, 127–138
practicle considerations, 244, 245

pan angle, 132, 248
parameter estimation, 273
paraxial approximation, *see* Gauss approximation
particle image velocimetry (PIV), 5
pattern application, *see* speckle pattern
pattern matching, *see* image matching methods and optimization criterion
perspective
model, 21
projection, 22, 27, 29, 70
phase error, *see* interpolation

photogrammetry, 48
photometric transformation, 95–98
pinhole
  model, 24
  system, 14
planarity constraint, 119
plane-to-plane imaging, *see* two-dimensional computer vision
Poisson's ratio, 125, 157
power spectrum, 111
pre-load, 244
pressurization, 205
principal plane, 22
principal point, 14, 22
projection
  affine, 65
  perspective, 22
  pinhole, 24
  prospective, 24
projective transformation, 22

quantization, 138
quasi-static, 196

radiodensity, 210
ray tracing, 13
rectification, 94
reflection, 232
reflectivity, 95
refraction, 208
  index of, 246
retinal plane, *see* image plane
rotation tensor
  definition, 28, 287
  partial derivatives, 52, 53

SAD, *see* optimization criterion
saturation, 232, 246
Savitzky–Golay, 91
scale factors, 30
scanline, 94
SEM, *see* microscope
sensor coordinates, 27, 225
shape function
  affine, 89
  general, 100
  homographic, 92–95
  polynomial, 90
  stereo matching, *see* shape function, homographic
  two-dimensional, 88–92
  volumetric, 214–217
shift theorem, 105, 108
singular value decomposition, 266
skew
  scale factor, 30, 31
skew rays, 295
spatial resolution, 206
speckle, *see* speckle pattern
speckle pattern
  isotropy, 83, 247
  practical considerations, 233–243
speckle size, 233, 236
spectral decomposition theorem, 267
spline, 42, 289
  basis function, 290
  curve, 289
  kernel, 290
  surface, 292
spray paint, 177, 239
SSD, *see* optimization criterion
statistical error analysis, *see* image matching errors
statistical parameters
  covariance, 283
  expectation, 281
  mean, 281
  variance, 282
steepest descent algorithm, 275
steepest-descent, 276
stereo angle, 94, *see* pan angle
stereo-rig, *see* three dimensional computer vision
stereomicroscope, 199
stereovision system, *see* three dimensional computer vision
strain rate tensors
  Eulerian, 260
  Langrangian, 260
strain tensor, 257
strobe light, 199
sub-optimal 3D estimator, 225
subset, 83, 88
SVD, *see* singular value decomposition
swing angle, 132
synchronization, 122

tangent vector, 269
telecentric lens, 128
template, 87
tension-torsion, 191
thermal effects, 246
thermal strain, 173
thin lens, 13
three-dimensional computer vision
  camera and lens selection, 247
  coordinate transformations, 71
  epipolar constraint, 72

independent camera calibration, 73
stereo system calibration, *see* stereo-rig calibration
stereo system configuration, 249, 250
stereo-rig calibration, 75
tilt angle, 132
time dependent distortion, *see* drift distortion
toner powder, 201
translation vector, 28
triangulation, 295
two-dimensional computer vision, 65

uniaxial tension, 120
update rule, 99–102
forwards, 100
inverse, 100
inverse compositional, 102, 115

vacuum, 246
variance, 114, 228
VDIC, *see* volumetric digital image correlation
virtual fields method, 150
volumetric image reconstruction, 218
voxel, 210
VSSD, *see* optimization criterion

water spots, 231
weighting function, 92
window, *see* subset

Young's modulus, 157

ZNSSD, *see* optimization criterion
ZSSD, *see* optimization criterion

CPSIA information can be obtained
at www.ICGtesting.com
Printed in the USA
BVHW03*0722100518
515877BV00007B/10/P